EMPIRICAL KNOWLEDGE

EMPIRICAL KNOWLEDGE

Readings in Contemporary Epistemology

SECOND EDITION

Paul K. Moser

ROWMAN & LITTLEFIELD PUBLISHERS, INC.

ROWMAN & LITTLEFIELD PUBLISHERS, INC.

Published in the United States of America
by Rowman & Littlefield Publishers, Inc.
4720 Boston Way, Lanham, Maryland 20706

3 Henrietta Street
London WC2E 8LU, England

British Cataloging in Publication Information Available

Library of Congress Cataloging-in-Publication Data

Empirical knowledge : readings in contemporary epistemology / [edited
by] Paul K. Moser. — 2nd ed.
p. cm.
Includes bibliographical references and index.
1. Knowledge, Theory of. 2. Empiricism. 3. Justification (Theory
of knowledge) I. Moser, Paul K., 1957– .
BD161.E57 1996 96-7696 CIP

ISBN 0–8476–8203–X (cloth : alk. paper)
ISBN 0–8476–8204–8 (pbk. : alk. paper)

Printed in the United States of America

∞ ™ The paper used in this publication meets the minimum requirements of
American National Standard for Information Sciences—Permanence of
Paper for Printed Library Materials, ANSI Z39.48–1984.

Contents

Preface

This book makes available some of the most important contemporary essays on the nature of empirical knowledge and provides a comprehensive and nontechnical overview of the central issues of contemporary epistemology. Part I contains eight essays focusing on the nature of empirical justification and knowledge. Part II contains four essays on the Gettier Problem concerning the analysis of propositional knowledge. Part III consists of three essays on skepticism about propositional knowledge. Part IV consists of three essays on naturalized epistemology.

The book's selections are accessible to middle- and upper-level undergraduate philosophy students. Each selection is, moreover, sufficiently substantial and original to be important reading for graduate-level philosophy students and even professional philosophers. The book's introduction draws some important distinctions for the theory of empirical knowledge and summarizes the selections. It can serve as a beginning orientation for students who are not specialists in epistemology. The concluding bibliography lists in topical arrangement many important books and articles on empirical knowledge.

I thank Jason Silverman, my research assistant at Loyola University of Chicago, for his fine assistance in the preparation of the book's manuscript. I thank Dwayne Mulder for excellent help with the permissions and the summaries of the selections. I also thank Jennifer Ruark of Rowman & Littlefield for encouraging this new edition.

Paul K. Moser
Chicago, Illinois

Introduction

Empirical Knowledge

Paul K. Moser

Contemporary philosophers investigating empirical, or perceptual, knowledge have focused on three main topics concerning the nature, origin, and scope of such knowledge: (a) what empirical knowledge *consists in* (e.g., justified true perceptual belief of a special nature), (b) what empirical knowledge is *based on* (e.g., nonbelief sensory experience), and (c) what the *extent* of human empirical knowledge is (e.g., objective, conceiver-independent facts as well as subjective, conceiver-dependent facts). Topics (a)–(c) have engaged epistemologists since the time of Plato. Contemporary epistemologists have not, by any means, put such topics to rest, but they have made some distinctive contributions.

A typical philosophical answer to the question, "What is empirical knowledge?", sets forth logically necessary and sufficient conditions for empirical knowledge: i.e., conditions that are individually required by and jointly sufficient for empirical knowledge. According to the traditional analysis of propositional knowledge, suggested in Plato's *Theaetetus,* you know that *P* if and only if you have a *justified* true belief that *P*. You might believe a true groundless conjecture, but would not thereby *know* that this conjecture is true. Standardly construed, knowledge requires not only that a belief condition and a truth condition be satisfied, but also that the satisfaction of the belief condition be *appropriately related* to the satisfaction of the truth condition. The latter requirement leads to a justification condition for knowledge, a condition that excludes such coincidental phenomena as lucky guesswork. Regarding empirical knowledge, the justification condition must be empirical in a manner needing specification. Accordingly, the traditional analysis of empirical knowledge states that you know that *P* if and only if you have empirically justified true belief that *P*.

1

A central question occupying contemporary epistemologists concerns the nature of the justification condition for empirical knowledge. One way to put this question is: What are the individually necessary and jointly sufficient conditions for empirical justification, the kind of justification crucial to empirical knowledge? Associated questions include: What is empirical justification based on (e.g., nonbelief sensory experience), and what is the extent of empirical justification? Selections 1–12 bear on questions about the conditions for empirical justification and knowledge, and selections 13–15 bear on questions about the extent of empirical knowledge. Selections 16–18 examine how a theory of empirical knowledge is related to a naturalistic theory, such as a naturalistic psychological theory.

I. Epistemic Justification

Contemporary epistemologists typically allow for empirically justified false beliefs, and this allowance is called *fallibilism* about empirical justification. Fallibilism allows, for example, that the Ptolemaic astronomers before Copernicus were justified in holding their geocentric model of the universe even though it was a false model. Justification for a proposition, according to most contemporary epistemologists, need not logically entail the proposition justified. That is, it is not the case that necessarily if the justifying proposition is true, then the justified proposition is true too. When justification does logically entail what it justifies, we have *deductive* justification. *Inductive* justification, in contrast, does not logically entail what it justifies. It obtains when if the justifying proposition is true, then the justified proposition is, to some extent, *probably* true. Contemporary epistemologists do not share a single account of the sort of probability appropriate to inductive justification.

Most contemporary epistemologists hold that epistemic justification is *defeasible*. That is, they hold that a justifying proposition can cease to be justifying for a person when that person acquires additional justification. For instance, your justification for thinking that there is water on the road ahead can be overridden by new evidence acquired by approaching the relevant spot on the road. Justification is thus subject to change with the acquisition of new evidence.

Contemporary epistemologists have examined the kind of justification we have for our beliefs about the external world, including the belief that household physical objects exist. Many epistemologists maintain that such beliefs are justified inductively, in terms of justification that does not logically entail the beliefs justified. Some skeptics have demanded deductive support for the belief that external objects exist. Other skeptics have questioned whether we can have even inductive, probabilistic justification for this belief.

Some skeptics have used a *regress argument* to contend that we are not justified in believing anything about the external, conceiver-independent world (see Oakley 1976). This argument highlights the question whether, and if so how, we are justified in holding any belief about the external world on the basis of other beliefs, i.e., by means of *inferential justification*. A skeptic's use of the regress argument aims to show that each of the available accounts of inferential justification fails, and that such justification is not to be had, or at least that we cannot reasonably presume to have such justification. The initial skeptical worry is: If one's belief that external objects exist is supposedly justified on the basis of another belief, how is the latter, allegedly justifying belief itself justified? Is it supposedly justified by some other belief? If so, how is the latter belief itself justified?

We seem threatened by an endless regress of required justifying beliefs—a regress too complex to employ in our actual everyday reasoning. Our problem, according to many epistemologists, is: Either (i) explain why an endless regress of required justifying beliefs is not actually troublesome, (ii) show how we can terminate the threatening regress, or (iii) accept the skeptical conclusion that inferential justification is impossible, or at least not actual.

An example will illustrate the problem of inferential justification. While walking along the shore of Lake Michigan, we decide that swimming today would be nice, but that swimming outdoors is dangerous now. Our belief that swimming outdoors is dangerous today gets support from other beliefs we have. We believe, for example, that (a) meteorologists have predicted lightning storms today in our area, (b) there are cumulonimbus clouds overhead, and (c) the meteorologists' reports and the presence of the cumulonimbus clouds are reliable indicators of impending lightning. Our belief that swimming outdoors is dangerous today receives support from our belief that (a)–(c) are true. What, however, supports (a)–(c) for us? Other beliefs we have will contribute support. So, the chain of inferential justification will continue. Support for (a) might include our belief that (d) we heard radio reports today from some meteorologists. Support for (b) might include our belief that (e) we see dark thunderclouds overhead. Our support for (d) and (e) might likewise be inferential, thus extending the chain of inferential justification even further.

Skepticism aside, epistemologists have offered four noteworthy replies to the regress problem. The first reply, called *epistemic infinitism,* proposes that regresses of inferential justification are indeed infinite, but that this does not preclude genuine justification. Our belief that swimming is dangerous today, given this reply, would be justified by belief (a) above, belief (a) would be justified by belief (d) above, belief (d) would be justified by a further belief, and so on without end. While attracting very few proponents, such infinitism was supported by Charles Peirce (1868, pp. 36–8), the founder of American pragmatism. Infinitism apparently requires that one must have an infinity of justifying beliefs to have any inferentially justified belief.

Skeptics will argue that infinite chains of supposed inferential justification cannot yield genuine justification. They will contend that no matter how far back we go in an infinite regress of inferential justification, we find beliefs that are only *conditionally* justified: that is, justified *if,* and *only if,* their supporting beliefs are justified. The problem is that the supporting beliefs themselves are at most conditionally justified too. They are justified if, and only if, *their* supporting beliefs are justified. At every point in the never-ending chain we find a belief that is merely conditionally justified, and not actually justified.

Skeptics will note, in addition, that one's having an infinity of supporting beliefs apparently requires an infinite amount of time, given that belief-formation for each of the supporting beliefs takes a certain amount of time. We humans, of course, do not have an infinite amount of time. So, it is doubtful that our actual justification includes infinite regresses of justifying beliefs. Proponents of infinitism thus have some difficult explaining to do. As a result, infinitism has attracted very few supporters throughout the history of epistemology.

A second, highly influential reply to the regress problem is *epistemic coherentism:* the view that all justification is systematic in virtue of "coherence relations" among beliefs. Justification for any belief, according to epistemic coherentism, ends in a system of beliefs with which the justified belief coheres. Coherentists thus deny that justification is linear in the way suggested by infinitism.

A coherence theory of *justification*—so-called epistemic coherentism—differs from a coherence theory of *truth.* A coherence theory of truth, of the sort endorsed by Brand Blanshard (1939, p. 268; 1980, p. 590), aims to specify the meaning of 'truth', or the essential nature of truth. A coherence theory of justification, in contrast, aims to explain the nature not of truth, but of the kind of justification crucial to knowledge. Recent proponents of epistemic coherentism, of one version or another, include: Wilfrid Sellars (1956, 1975, 1979), Nicholas Rescher (1973, 1979), Gilbert Harman (1973, 1986), Keith Lehrer (1974, 1990, selection 4 below), and Laurence BonJour (1985). In the present volume, the selection by Keith Lehrer defends a version of epistemic coherentism, and the selection by Robert Audi critically assesses epistemic coherentism.

Epistemic coherentists have pursued two important questions: First, what kind of coherence relation is crucial to justified belief? Second, what kind of belief-system must a justified belief cohere with? Regarding the first question, many epistemic coherentists acknowledge logical-entailment relations and explanation relations as coherence relations among beliefs. Explanatory coherence relations obtain when some of one's beliefs effectively explain why some other of one's beliefs are true. For example, my belief that it is snowing outside might effectively explain the truth of my belief that my office-windows are white. Regarding the second question, not just any belief-system will serve the purpose of epistemic coherentism. Some belief-systems, including those consisting of science-fiction propositions, seem obviously false and thus unable to offer a basis for epistemically justified belief.

Epistemic coherentism implies that the justification of any belief depends on that belief's coherence relations to other beliefs. Such coherentism is thus systematic, stressing the role of interconnectedness of beliefs in epistemic justification. Skeptics will ask why we should regard coherence among one's beliefs as a reliable indication of empirical truth, of how things actually are in the empirical world. (We shall return to the matter of skepticism below.) Consider also the *isolation objection* to epistemic coherentism: Epistemic coherentism entails that one can be epistemically justified in accepting a contingent empirical proposition that is incompatible with, or at least improbable given, one's total empirical evidence. (For elaboration, see Moser 1985, pp. 84–103; 1989a; 1989b, pp. 176–82). A proponent of this objection does not restrict empirical evidence to empirical propositions believed or accepted.

The isolation objection is universally applicable to coherence theories of justification so long as empirical evidence goes beyond the propositions (or, judgments) believed or accepted by a person. Suppose that one's empirical evidence includes the subjective nonpropositional contents (e.g., visual images) of one's *non*belief perceptual and sensory awareness-states, such as one's seeming to perceive something or one's feeling a pain. Such contents, being nonpropositional, are not among what one believes or accepts. One can, of course, accept that one is having a particular visual image, but this does not mean that the image itself is a proposition one accepts. If the nonprositional contents of nonbelief perceptual and sensory states are among one's empirical evidence, the isolation objection will bear directly on all coherence theories of justification.

Coherence theories, by definition, make epistemic justification depend just on coherence relations among propositions one believes or accepts. They thus neglect, as a matter of principle, the evidential significance of the nonpropositional contents of nonbelief perceptual and sensory states. Epistemic coherentists have not yet offered a uniform response to the isolation objection.

A third reply to the regress problem is *epistemic foundationalism.* Foundationalism about epistemic justification states that such justification is two-tiered: Some instances of justification are noninferential, or foundational; all other instances of justification are inferential, or nonfoundational, in that they derive ultimately from foundational justification. This structural view was proposed by Aristotle's *Posterior Analytics* (as a view about knowledge), received an extreme formulation in Descartes's *Meditations,* and is represented, in various forms, by the following: Bertrand Russell (1940), C. I. Lewis (1929, 1946), Roderick Chisholm (1964, 1977, 1989), William Alston (1989), John Pollock (1974, 1986), James Cornman (1980), Marshall Swain (1981), Robert Audi (1993), Richard Foley (1987), and Paul Moser (1985, 1989b, 1993), among many others. In this volume, the essays by Chisholm and Audi represent foundationalism.

Foundationalists differ on two matters: the explanation of noninferential, foundational justification, and the explanation of how justification can be transmitted from foundational to nonfoundational beliefs. Some epistemologists, following Descartes, have assumed that foundational beliefs must be *certain* (e.g., indubitable or infallible). This assumption underlies *radical* foundationalism, which requires not only that foundational beliefs be certain, but also that such beliefs guarantee the certainty of the nonfoundational beliefs they support. Two considerations explain why radical foundationalism attracts few contemporary epistemologists. First, very few, if any, of our perceptual beliefs are certain. (For a survey and assessment of some prominent views about the certainty of subjective beliefs about sensations, see Meyers (1988, chap. 3) and Alan Goldman (1988, chap. 7); cf. Alston (1989, chaps. 10, 11).) Second, the beliefs that are the best candidates for certainty (e.g., the belief that I am thinking) are not informative enough to guarantee the certainty of our specific inferential beliefs about the external world (e.g., our standard beliefs about physics, chemistry, and biology).

Contemporary foundationalists typically endorse *modest* foundationalism, the view that foundational beliefs need not possess or yield certainty, and need not deductively support justified nonfoundational beliefs. Foundationalists typically characterize a *noninferentially justified, foundational* belief as a belief whose epistemic justification does not derive from other beliefs. They leave open whether the *causal* basis of (the existence of) foundational beliefs includes other beliefs. They typically hold, in addition, that foundationalism is an account of a belief's (or a proposition's) *having* justification for a person, not of one's *showing* that a belief (or a proposition) has justification or is true.

Modest foundationalists have offered three notable approaches to noninferential, foundational justification: (i) self-justification, (ii) justification by nonbelief, nonpropositional experiences, and (iii) justification by a reliable nonbelief origin of a belief. Recent proponents of self-justification have included Roderick Chisholm (see his essay below) and C. J. Ducasse (1968). They contend that a foundational belief can justify itself, apart from any evidential support from something else. In contrast, proponents of foundational justification by nonbelief experiences reject literal self-justification. They hold, following C. I. Lewis (1929, 1946), that foundational perceptual beliefs can be justified by nonbelief sensory or perceptual experiences (e.g., my nonbelief experience involving seeming to see a book) that either make true, are best explained by, or otherwise support those foundational beliefs (e.g., the belief that there is, or at least appears to be, a book here). (For an exposition of Lewis's views on the given element in experience, see Firth (1969) and Moser (1988).)

Proponents of foundational justification by reliable origins hold that noninferential justification depends on nonbelief belief-forming processes (e.g., perception, memory, introspection) that are truth-conducive to some extent, in virtue of tending to produce true rather than false beliefs. The view that reliable belief-forming processes confer epistemic justification has come to be known as *epistemic reliabilism*. (Epistemic reliabilism of one sort or another has been defended by Swain (1981), Alvin Goldman (1986), Alston (1989), and Sosa (1991, selection 6 below).) Reliabilism about foundational justification invokes the reliability of a belief's nonbelief origin, whereas the previous nonreliabilist view invokes the particular sensory or perceptual experiences that underlie a foundational belief. Despite the aforementioned disagreements, proponents of modest foundationalism typically agree that noninferential justification, at least in many cases, can be defeated upon expansion of one's justified beliefs. The justification for your belief that there is an orange football in the corner, for example, might be overridden by the introduction of new evidence that there is an orange light shining on the ball.

Wilfrid Sellars (1975) and Laurence BonJour (1985, selection 3 below) have offered an influential argument against foundationalist claims about noninferential justification. They contend that one cannot be noninferentially epistemically justified in holding any belief, since one is epistemically justified in holding a belief only if one has good reason to think that the belief is true. This, they claim, entails that the justification of an alleged foundational belief will actually depend on an argument of the following form:

(a) My foundational belief that P has feature F.
(b) Beliefs having feature F are likely to be true.
(c) Hence, my foundational belief that P is likely to be true.

If the justification of one's foundational beliefs depends on such an argument, those beliefs will not be foundational after all. Their justification will then depend on the justification of other beliefs: the beliefs represented by the premises of the argument (a)–(c).

Sellars and BonJour must face a problem. It is too demanding to hold that the justification of one's belief that P requires one's being *justified in believing* premises (a) and (b). Given that requirement, you will be justified in believing that P only if you are justified in believing that your belief that P has feature F. In addition, given those requirements, you will be justified in believing that (i) your belief that P has F only if you are justified in believing an additional proposition: that (ii) your belief that (i) has F. Given the requirements in question, we have no nonarbitrary way to avoid the troublesome implication that similar requirements apply not only to this latter proposition—viz., (ii)—but also to each of the ensuing infinity of required justified beliefs. The problem is that we seem not to have the required infinity of increasingly complex justified beliefs.

One lesson here is that if justificational support for a belief must be accessible to the believer, that accessibility should not itself be regarded as requiring further justified belief. Current debates over *inter-*

nalism and *externalism* regarding epistemic justification concern what sort of access, if any, one must have to the support for one's justified beliefs. (For some indication of these debates, see BonJour (1985, chap. 3), Alston (1989, chaps. 8, 9), Moser (1985, chap. 4; 1989b, chap. 2), Audi (1993, chap. 11), and Fumerton (1995).) Internalism incorporates an accessibility requirement, of some sort, on what provides justification, whereas externalism does not. Debates over internalism and externalism are currently unresolved in contemporary epistemology.

Foundationalists must explain not only the conditions for noninferential justification, but also how justification transmits from foundational beliefs to inferentially justified, nonfoundational beliefs. Modest foundationalists, unlike radical foundationalists, allow for nondeductive, merely probabilistic connections that transfer justification. They have not, however, reached agreement on the exact nature of such connections. Some modest foundationalists hold that some kind of "inference to a best explanation" can account for transmission of justification in many cases. For example, the belief that there is a book before me can, in certain circumstances, provide a best explanation of various foundational beliefs about my perceptual inputs, or at least a best explanation of the perceptual inputs themselves. This, however, is a controversial matter among epistemologists. (For discussion, see Alan Goldman (1988), Lycan (1988), and Moser (1989b).)

A special problem troubles versions of foundationalism that restrict noninferential justification to subjective beliefs about what one *seems* to see, hear, feel, smell, and taste. Those versions must explain how such subjective beliefs can provide justification for beliefs about *conceiver-independent* physical objects. Such subjective beliefs do not logically entail beliefs about physical objects. Since extensive hallucination is always possible, it is always possible that one's subjective beliefs are true while the relevant beliefs about physical objects are false. This consideration challenges foundationalists who endorse *linguistic phenomenalism,* the view that statements about physical objects can be translated without loss of meaning into logically equivalent statements solely about subjective states characterized by subjective beliefs. (For discussion of this and other versions of phenomenalism, see Ayer (1940, 1947), Chisholm (1957, pp. 189–97), Cornman (1975), and Fumerton (1985).) Perhaps a foundationalist, following Chisholm (1977) and Cornman (1980), can invoke a set of nondeductive relations to explain how subjective beliefs can justify beliefs about physical objects. This remains a challenge, however, as no set of such relations has attracted widespread acceptance from foundationalists. We should note, though, that some versions of foundationalism allow for the noninferential justification of beliefs about physical objects, and thus avoid the problem at hand. (Cf. Swain 1981, Moser 1989b.)

A fourth nonskeptical reply to the regress problem is *epistemic contextualism,* a view suggested by Wittgenstein (1969) and formulated explicitly by Annis (1978, selection 7 below). Wittgenstein set forth a central tenet of contextualism with his claim that "at the foundation of well-founded belief lies belief that is not founded" (1969, §253). If we construe Wittgenstein's claim as stating that at the foundation of justified beliefs lie beliefs that are unjustified, we have an alternative to infinitism, coherentism, and foundationalism. (The interpretation of Wittgenstein's *On Certainty* is, however, a matter of controversy among philosophers; for efforts at interpretation, see Shiner (1977) and Morawetz (1978).)

In any context of inquiry, according to contextualism, people simply assume (the acceptability of) some propositions as starting points for inquiry. These "contextually basic" propositions, while themselves lacking evidential support, can support other propositions. Contextualists emphasize that contextually basic propositions can vary from social group to social group, and from context to context—for example, from theological inquiry to biological inquiry. So, what functions as an unjustified justifier in one context need not in another.

The main problem for contextualism comes from the view that unjustified beliefs can yield epistemic justification for other beliefs. If we grant that view, we need to avoid the implausible view that *any* unjustified belief, however obviously false or contradictory, can yield justification in certain contexts. If any unjustified proposition can serve as a justifier, we shall be able to justify anything we want—an intolerable result. Even if we do typically take certain things for granted in certain contexts of discussion, this does not support the view that there are unjustified justifiers. Perhaps the things typically taken for granted are actually supportable by good reasons. If they are not, we need some way to distinguish them from unjustified beliefs that cannot transmit justification to other beliefs. The contextualist must explain, then, how an unjustified belief—but not just *any* unjustified belief—can confer inferential justification on other beliefs. Contextualists have not reached agreement on the needed explanation.

In sum, then, the epistemic regress problem for inferential justification is troublesome and resilient. Infinitism, coherentism, foundationalism, or contextualism may offer a viable solution to the problem, but only after a resolution of the problems noted above. Let us turn briefly now to some complications facing the analysis of knowledge.

II. Conditions for Knowledge

Recently some epistemologists have proposed that we abandon the traditional justification condition for propositional knowledge. They recommend, following Alvin Goldman (1967), that we construe the justification condition as a *causal* condition. Put roughly, the idea is that you know that *P* if (a) you believe that *P*, (b) *P* is true, and (c) your believing that *P* is causally produced and sustained by the fact that makes *P* true. This is the basis of a *causal theory of knowing,* a theory that admits of various manifestations.

A causal theory of knowing faces serious problems from knowledge of universal propositions. We know, for instance, that all computers are produced by humans. Our believing that this is so seems not to be causally supported by the fact that all computers are humanly produced. It is not clear that the latter fact causally produces *any* beliefs. We need, then, an explanation of how a causal theory can account for knowledge of such universal propositions.

The analysis of knowledge as justified true belief, however elaborated, faces a serious challenge that initially gave rise to causal theories of knowledge: *the Gettier problem.* In 1963 Edmund Gettier published an influential challenge to the view that if you have a justified true belief that *P,* then you know that *P* (see selections 9–12 below). Here is one of Gettier's counterexamples to this view: Smith is justified in believing the false proposition that (i) Jones owns a Ford. On the basis of (i), Smith infers, and thus is justified in believing, that (ii) either Jones owns a Ford or Brown is in Barcelona. As it happens, Brown is in Barcelona, and so (ii) is true. Hence, although Smith is justified in believing the true proposition (ii), Smith does not know (ii).

Gettier-style counterexamples are cases where a person has justified true belief that *P* but lacks knowledge that *P.* The Gettier problem, stated generally, is the problem of finding a modification of, or an alternative to, the standard, justified-true-belief analysis that escapes difficulties from Gettier-style counterexamples. The controversy over the Gettier problem is complex and unsettled. Many epistemologists take the main lesson of Gettier-style counterexamples to be that propositional knowledge requires a *fourth* condition, beyond the justification, truth, and belief conditions. No specific fourth condition has received widespread acceptance by epistemologists, but some proposals are prominent.

The so-called "defeasibility condition" requires that the justification appropriate to knowledge be "undeafeated" in the general sense that some appropriate subjunctive conditional concerning defeaters of justification be true of that justification (see Swain (1981) and Shope (1983, chap. 2)). For instance, one simple defeasibility fourth condition requires of Smith's knowing that *P* that there be no true proposition, *Q*, such that if *Q* became justified for Smith, *P* would no longer be justified for Smith. So if Smith knows, on the basis of his visual perception, that Jeanne removed books from the library, then Smith's coming to believe the true proposition that Jeanne's identical twin removed books from the library would not undermine the justification for Smith's belief concerning Jeanne herself. A different approach avoids subjunctive conditionals of that sort, claiming that propositional knowledge requires justified true belief that is sustained by the collective totality of actual truths. This approach requires a detailed account of when justification is undermined and restored. (For some details on this general approach, see Pollock (selection 11 below) and Moser (1989b, chap. 6). On the history of the Gettier problem, prior to 1982, see Shope (1983).)

The Gettier problem, according to many epistemologists, is epistemologically important. One branch of epistemology seeks a precise understanding of the nature—e.g., the essential components—of propositional knowledge. Our having a precise understanding of propositional knowledge requires our having a Gettier-proof analysis of such knowledge. Epistemologists thus need a defensible solution to the Gettier problem, however complex that solution is (see selection 12 below). This conclusion is compatible with the view that various epistemologists employ different notions of knowledge at a level of specificity (on which see Moser (1989b, chap. 6, 1993)).

III. Skepticism and Naturalism

Epistemologists have long debated the limits, or scope, of human propositional knowledge (see selections 13–15 below). The more restricted we take the scope to be, the more skeptical we are. Two noteworthy types of skepticism are *knowledge*-skepticism and *justification*-skepticism. Unrestricted knowledge-skepticism implies that no one knows anything. Unrestricted justification-skepticism implies the more extreme view that no one is even justified in believing anything. Knowledge-skepticism in its strongest form implies that it is *impossible* for anyone to know anything. A weaker form denies the *actuality* of our having knowledge, but leaves open its possibility. Many skeptics have restricted their skepticism to a particular domain of supposed knowledge: e.g., knowledge of the external world, knowledge of other minds, knowledge of the past or the future, or knowledge of unperceived items. Limited skepticism has been more common than unrestricted skepticism in the history of philosophy.

One of the most difficult problems supporting skepticism is *the problem of the criterion,* a version of which comes from the sixteenth-century skeptic Michel de Montaigne:

> To adjudicate [between the true and the false] among the appearances of things, we need to have a distinguishing method; to validate this method, we need to have a justifying argument; but to validate this justifying argument, we need the very method at issue. And there we are, going round on the wheel.

This line of skeptical argument originated in ancient Greece, with epistemology itself. (See Sextus Empiricus, *Outlines of Pyrrhonism,* Book II; for relevant discussion, see Striker (1980, 1990), Barnes (1990), and Amico (1993).) The argument raises this question: How can we specify *what* we know without having specified *how* we know, and how can we specify *how* we know without having specified *what* we know? Is there any reasonable way out of this threatening circle? A cogent epistemology must provide a defensible solution to this difficult problem. Contemporary epistemology still lacks a widely accepted reply to this problem. One influential reply from Roderick Chisholm (1982) rules out skepticism from the start, with the assumption that we do know some specific propositions about the external world. Chisholm endorses a *particularist* reply that begins with an answer to the question of what we know. Such a reply seems, however, to beg a key question against the skeptic. A *methodist* reply to the problem of the criterion begins with an answer to the question of how we know. Such a reply risks, however, divorcing knowledge from our considered judgments about particular cases of knowledge. It also must avoid begging key questions raised by skeptics.

Let us briefly consider another skeptical argument (developed in detail in Moser (1993)). Suppose that you are a realist claiming to have knowledge that external, mind-independent objects exist, and that you take such knowledge to entail that external objects exist. Suppose also that you regard yourself as having a cogently sound argument, of the following form, for your claim to knowledge that external objects exist.

1. If one's belief that P has feature F, then one knows that P.
2. My belief that external objects exist has F.
3. Hence, I know that external objects exist.

Even critics of realism may grant premise 1, if only for the sake of argument. Premise 1 may be just an obvious implication of what a realist *means* by 'knows that P'.

Feature F can incorporate any of a number of familiar well-foundedness properties: either (a) suitable doxastic coherence (cf. Lehrer (1990, chap. 7)), (b) maximal explanatory efficacy (cf. Lycan (1988, chap. 7)), (c) undefeated self-evidentness (cf. Foley (1987, chap. 2)), (d) consistent predictive success (cf. Almeder (1992, chap. 4)), (e) uncontested communal acceptance (cf. Annis (1978, selection 7 below)), (f) causal sustenance by such a belief-forming process as perception, memory, introspection, or testimony (cf. Goldman (1986, chap. 5)), (g) adequate theoretical elegance in terms of such virtuous characteristics as simplicity and comprehensiveness (cf. Thagard (1988, chap. 5)), (h) survival value in the evolutionary scheme of things (cf. Carruthers (1992, chap. 12)), or (i) some combination of (a)–(h) (cf. Cornman (1980)). Such well-foundedness properties cannot conceptually exhaust F, because one's knowing that P (unlike those well-foundedness properties) *logically entails* its objectively being the case that P.

If knowledge consists of more than objectively true belief, as it does on standard conceptions since the time of Plato's *Theaetetus* (cf. 202b), then F will be a complex property—the property of being objectively true plus some additional property. The additional property, on standard conceptions of knowledge, incorporates a well-foundedness feature (and sometimes a no-defeaters restriction on that feature to handle the aforementioned Gettier problem). Call this *the well-foundedness component of F*. A well-foundedness feature can serve to distinguish knowledge from true belief due simply to such coincidental phenomena as lucky guesswork. In this respect, a well-foundedness feature may be regarded as making a belief "likely to be true" to some extent. As suggested earlier, an *internalist* well-foundedness feature is accessible—directly or indirectly—to the knower for whom it yields likelihood of truth; an *externalist* well-foundedness feature is not.

Since knowledge that *P* entails that it is objectively the case—or objectively true—that *P,* the relevant kind of likelihood of *truth* must entail likelihood of what is objectively true, or objectively the case. So, whether internalist or externalist, a well-foundedness feature must yield likelihood of what is objectively the case. It must, in other words, indicate with some degree of likelihood what is the case conceiver-independently. A well-foundedness component of *F* that violates this requirement will fail to distinguish knowledge from true belief due simply to such coincidental phenomena as lucky guesswork.

Premise 2 generates a problem motivating skepticism about realism concerning external objects. It affirms (i) that your belief that external objects exist is objectively true, and (if you hold that knowledge has a well-foundedness component) (ii) that a well-foundedness feature indicates with some degree of likelihood that external objects exist. A skeptic can plausibly raise this question:

Q1. What non-questionbegging reason, if any, have we to affirm that your belief that external objects exist is objectively true?

If you are a typical realist, committed to a well-foundedness component of knowledge, you will appeal to your preferred well-foundedness component of *F* to answer *Q1.* You will answer that the satisfaction of the conditions for that well-foundedness component yields the needed reason to affirm that your belief is true. This answer to *Q1* is predictable, given the aforementioned assumption that a well-foundedness component yields likelihood of truth.

For any well-foundedness component a realist offers, a skeptic can raise the following challenge:

Q2. What non-questionbegging reason, if any, have we to affirm that the satisfaction of the conditions for *that* well-foundedness component of *F* is actually indicative, to any extent, of what is objectively the case?

Realists might reply that it is true in virtue of what they *mean* by "indicative of what is objectively the case" that their preferred well-foundedness component is indicative of what is objectively the case. (Cf. Pollock (1986, chap. 5).) This move uses definitional (or, conceptual) fiat to try to answer skeptics, but it actually fails to answer their main concern.

We can put the main concern now as follows:

Q3. What non-questionbegging reason, if any, have we to affirm that the satisfaction of the conditions for a preferred well-foundedness component of *F*—including the satisfaction of conditions definitive of what a realist means by 'indicative of what is objectively the case'—is ever a genuinely reliable means of representational access to what is objectively the case?

In other words, what non-questionbegging reason, if any, have we to affirm that some claim satisfying the conditions for a preferred well-foundedness component (e.g., the claim that external objects exist) is actually objectively true? We can grant realists their preferred defini-tion of "indicative of what is the case," but then follow up with *Q3*. A skeptic may begin with *Q1*, but will naturally raise *Q3* once a realist appeals to a well-foundedness component of *F*.

Your invoking your preferred well-foundedness component to defend realism against *Q3* would be questionbegging. The realiability of your preferred well-foundedness component is under question now. Begging this question offers no cogent support for realism. If, for instance, you hold that coherent belief is indicative of objective truth, you cannot now simply presume that coherent belief is indicative of objective truth. The issue now is whether coherent belief (or any similar well-foundedness component) is actually indicative of objective truth. Per-haps *given* your preferred well-foundedness component, that well-foundedness component is itself well-founded. This consideration, however, does not answer *Q3*. *Q3* asks what *non-questionbegging* reason, if any, we have to regard your preferred well-foundedness component as ever being a reliable means to objective truth. *Apart from* appeal to (the reliability of) your preferred well-foundedness component, what reason have we to regard that component as ever being a reliable means to objective truth—e.g., objective truth regard-ing your belief that external objects exist?

A skeptic's use of *Q3* allows for fallibilism about well-foundedness: the view that a well-founded belief can be false. In addition, a skeptic's use of *Q3* need not assume that evidence on which a claim is well-founded must logically entail (or, deductively support) that claim; nor does it require that we take a controversial stand on purely conceptual disputes over the exact conditions for epistemic justification.

Suppose then that you are a realist wielding argument 1–3, along with the standard view that *F* has a well-foundedness component (say, coherent belief). You will then hold that your belief that external objects exist illustrates a case where a belief's meeting the conditions for your preferred well-foundedness component (such as coherent belief) is an objectively true belief. You will then hold, given premise 2, that your belief that external objects exist is objectively true, and that your preferred well-foundedness component of F (namely, coherent belief) is satisfied by an objectively true belief in this case. Still, you will owe the skeptic a non-questionbegging reason for thinking that your preferred well-foundedness component of F is, in this case, a genuinely reliable means to objectively true belief.

Realists might aim to silence the skeptic by claiming, following Pollock (1986), that our concept of an external object is actually constituted, or wholly determined, by certain well-foundedness conditions involving one or more of the well-foundedness properties noted above. The claim here is that certain conditions for well-founded ascription of our concept of an external object *fully* determine that concept. This claim entails a kind of verificationism about our notion of an external object, and is not equivalent to the previous view that appealed to considerations of meaning regarding 'indication of what is objectively true'.

Skeptics will note that our notion of an external object logically outstrips various standard well-foundedness conditions for that notion. Our concept of an external object involves, for instance, the condition that any object falling under it does not perish whenever one looks away from it, but would exist even when unperceived. The condition that an external object would exist even when unperceived is not logically entailed by various standard well-foundedness conditions for our notion of an external object. For example, maximally effective explanation of (the origin of) our common perceptual experiences does not *logically* require that there be objects that exist when unperceived. Further, even if well-foundedness conditions fully determined our notion of an external object, it would still be an open question whether one's having our notion of an external object involves one's *actually satisfying* those well-foundedness conditions with one's belief that external objects exist. If those conditions entail the subjunctive condition just noted, a skeptic will demand a non-questionbegging reason to think it is ever actually satisfied. In that case, the now familiar worries motivating skepticism will resurface. Realists, in any case, are not typically verificationists about our notion of an external object.

A non-questionbegging reason will not simply *presume* a realism-favoring answer to a skeptic's familiar questions about reliability. Some of these familiar questions concern the reliability, in any actual case, of our belief-forming processes (e.g., perception, introspection, judgment, memory, testimony) that sometimes produce belief in the existence of external objects. Some other familiar questions concern the reliability, in any actual case, of suitably coherent, explanatorily efficacious, or predictively successful belief regarding the existence of external objects. Each of the well-foundedness properties above will attract such a question about reliability from a skeptic. A skeptic will thus be unmoved by observations concerning the simplicity and comprehensiveness provided by realism about external objects. The application of *Q3* will ask for a non-questionbegging reason to affirm that the simplicity and comprehensiveness provided by such realism is ever a reliable means to objectively true belief. Any higher-order use of a well-foundedness component—to support a well-foundedness component—meets the same fate as first-order use. *Q3* applies equally to any higher-order use. Lacking answers to the skeptic's questions, we cannot cogently infer that realism about external objects has been substantiated.

A skeptic is not guilty of the following vacuous challenge: Give me a cogent argument, but do not use any premises. The challenge is rather: Give me a cogent non-questionbegging reason to hold that your belief that external objects exist is a case where a belief satisfying your preferred well-foundedness component is an objectively true belief. The demand is not that realists forgo the use of premises. It rather is that the realist forgo the use of *questionbegging* premises—premises that beg relevant questions about reliability motivating skepticism. (*Q1–Q3* illustrate some standard skeptical questions.) A questionbegging argument from a realist will not even begin to approach cogency for a skeptic. Mere soundness of argument, then, is not at issue; *cogent non-questionbegging soundness* is.

One consideration—perhaps employed just for the sake of argument by a skeptic—indicates that a skeptic offering *Q3* will *not* be successfully answered by a realist. Cognitively relevant access to *anything* by us humans depends on such belief-forming processes as perception, introspection, judgment, memory, testimony, intuition, and common sense. Such processes are open to question via *Q3,* and cannot themselves deliver non-questionbegging support for their own reliability. Clearly, we cannot assume a position independent of our own cognitively relevant processes to deliver a non-questionbegging indication of the reliability of those processes. This, for better or worse, is the human cognitive predicament, and no realist has yet shown how we can escape it. This consideration favors the conclusion that we must take skepticism seriously.

A non-questionbegging reason favoring realism would be a reason that provides *effective* discernment of conceiver-independent truth. We cannot effectively rely on the deliverances of our belief-forming processes (e.g., perception, introspection, judgment, memory, testimony, intuition, and common sense) to test the reliability of those processes regarding their accessing conceiver-independent facts. Appeal to the deliverances of those processes would beg the key question against an inquirer doubtful of the reliability of those deliverances and processes. The belief-forming processes in question need testing, with respect to their reliability, in order to yield non-questionbegging support for the realist's claim to objective truth.

A skeptic's use of *Q1–Q3* bears on a wide range of positions commonly called 'realism'. This range includes each of the following species of realism: *weak* realism (i.e., something objectively exists independently of conceivers); *common-sense* realism (i.e., tokens of most current observable common-sense, and scientific, physical types objectively exist independently of conceivers); and *scientific* realism (i.e., tokens of most current unobservable scientific physical types objectively exist independently of conceivers). Michael Devitt has claimed that weak realism is "so weak as to be uninteresting" (1984, p. 22). Even if it is, this is just mere biography. It does not excuse realists from delivering non-questionbegging reasons in support of weak realism. Realists cannot plausibly follow Devitt in appealing just to our current science to settle epistemological debates about realism. Realists must, to defend against the skeptic, explain how our current science gives us the needed effective reason in favor of realism. It settles no philosophical questions simply to say, with Devitt, that "scepticism [regarding knowledge of conceiver-independent facts] is simply uninteresting: it throws the baby out with the bath water" (1984, p. 63). One pressing question is whether realists actually have a real baby—i.e., effectively supportable knowledge—to throw out. We cannot simply beg this question, if we wish to make genuine philosophical progress.

Questions under dispute in a philosophical context cannot attract non-questionbegging answers from mere *presumption* of the correctness of a disputed answer. If we allow such questionbegging in general, we can support *any* disputed position we prefer: Simply beg the key question in any dispute regarding the preferred position. Given that strategy, argument becomes superfluous in the way circular argument is typically pointless. Questionbegging strategies promote an undesir-able arbitrariness in philosophical debate. They are thus rationally inconclusive regarding the questions under dispute. What is question-begging is always relative to a context of disputed issues, a context that is not necessarily universally shared. (For doubts about any purely formal criterion of vicious circularity in argument, see Sorensen (1991).)

A pragmatic defense of realism, in terms of a belief's overall utility, fares no better than the well-foundedness properties noted above. A variation on *Q3* applies straightforwardly: What non-questionbegging reason, if any, have we to affirm that a belief's overall pragmatic utility is ever a genuinely reliable means of access to what is objectively the case? It does no good here to note that it is pragmatically useful to regard pragmatic utility as a reliable means to objective truth. A skeptic, once again, seeks non-questionbegging reasons. Given the aforementioned human cognitive predicament, we can offer little hope for the needed non-questionbegging support on pragmatic grounds. Pragmatic support for realism is one thing; non-questionbegging sup-port, another. A skeptic demands, but despairs of achieving, the latter.

Even if realism has certain theoretical advantages over various species of idealism, it still must face a skeptic's worries about the absence of non-questionbegging supporting reasons. Skeptics doubtful of the correctness of realism need not be idealists holding that an individual's mental activity creates all objects. They rather can plausi-bly hold that we lack non-questionbegging reasons to endorse—to any degree—idealism as well as realism. Skepticism rejects *any* position that does not enjoy non-questionbegging reasons; for questionbegging support is really no support at all. The burden of cogent argument is now squarely on the realist's shoulders.

Some philosophers have tried to avoid the problems of skepticism and traditional epistemology by naturalizing epistemology. A recent tradition, stemming from the views of W. V. Quine (see selection 16), proposes that epistemology be "naturalized" in that it be treated as continuous with the natural sciences. (See selections 16–18 for various approaches to naturalized epistemology.) Quine himself (1951) rejects the analytic-synthetic distinction as philosophically irrelevant, and holds that there is no "first philosophy," no philosophy prior to the natural sciences. Philosophy, Quine maintains, is methodologically and doctrinally continuous with the natural sciences. Epistemology, for instance, is a branch of psychology; it is not a discipline that offers independent standards of assessment for the natural sciences. Quine denies, then, that philosophy is autonomous with respect to the natural sciences. He denies, accordingly, that philosophical truths are necessary or knowable a priori. Some other philosophers have offered less austere ways to naturalize philosophy (see selection 18), but Quine's approach has attracted the most attention in contemporary epistemology.

A pressing issue for Quine's approach is: In the absence of any "first philosophy," how are we to discern which of the various so-called sciences are genuinely reliable and thus regulative for purposes of theory formation? Our list of genuine sciences will perhaps include the dominant physics and chemistry, but exclude astrology and parapsychology. Such a list, regardless of its exact components, seemingly depends on some first philosophy, some philosophical commitments prior to the natural sciences in question. (See Moser (1993, chap. 3) for elaboration.) It is an open question, however, whether the latter philosophical commitments are analytic or synthetic. We cannot do justice, in any case, to skeptical worries by ignoring them.

A very recent approach to philosophy, including epistemology, emerges from the writings of Richard Rorty (see selection 8), who has acknowledged influence from Thomas Kuhn's social-political approach to the sciences. If Quine's approach exalts the natural sciences, Rorty's approach elevates the social sciences. Rorty proposes that we replace first philosophy with philosophy as the comparing and contrasting of cultural traditions. He offers a kind of pragmatism that aims to change the subject from platonic and positivistic questions about the nature of truth to intellectual history of a certain sort. Rorty endorses the Hegelian view that philosophy is "its own time apprehended in thoughts," and he understands talk of "our own time" as including the notion of "our view of previous times." Rorty's pragmatism apparently merges philosophy with intellectual history and literary criticism, leaving no special subject-matter for the discipline of philosophy, including epistemology.

Even if pragmatists wish to change the subject, we can still ask whether one approach to our intellectual history is more reliable—closer to the truth—than another. Indeed, we shall naturally raise such an issue when faced with incompatible lessons from alternative approaches to intellectual history. In raising such an issue, however, we shall also open questions about the nature of truth, questions that appear to be philosophical and epistemological, not merely historical or literary. Rorty's pragmatism owes us, at the least, an explanation of how philosophical issues about the nature of truth can actually be set aside. Such issues, contrary to Rorty's pragmatism, seem unavoidable in the end.

IV. The Selections

In selection 1, William Alston surveys some prominent notions of epistemic justification, distinguishing between deontological and non-deontological evaluative notions of justification. According to a deonto-logical notion, one is justified in believing a proposition, *P,* if and only if no intellectual obligations concerning belief formation or belief sustenance are violated in one's believing that *P.* According to a nondeontological evaluative notion, one is justified in believing that *P* if and only if one's believing that *P,* as one now does, is a good thing from the epistemic point of view. A nondeontological evaluative notion presupposes that there is a way of being good from the epistemic point of view that is not identical to one's being blameless regarding the violation of an intellectual obligation. Alston defines the epistemic point of view in terms of the aim to maximize true belief and minimize false belief. He proposes that epistemic goodness relative to this aim consists in a belief's being true as far as a believer can tell from what is available to him or her. Such epistemic goodness requires *adequate evidence* for a belief, that is, warrant sufficiently indicative of the truth of the belief in question. Alston argues that this nondeontological notion, rather than a deontological notion, is essential to our concept of empirical knowledge.

In selection 2, Roderick Chisholm outlines an influential foundation-alist version of empirical knowledge. He begins by clarifying the controversial doctrine of the "epistemological given" as a thesis about the two-tier structure of knowledge and about the self-justifying foundations of knowledge. Chisholm uses an eliminative regress argu-ment to support the claim that empirical knowledge has self-justified foundations. He opposes, however, the view that our empirical knowl-edge is an edifice supported by sense-impressions alone. Chisholm argues that we need to recognize the central role of certain "rules of evidence" in the derivation of empirical knowledge.

In selection 3, Laurence BonJour sets forth an argument against foundationalism. The gist of the argument is that foundationalist ac-counts of empirical knowledge face a serious dilemma: they commit one either to an implausible sort of "epistemic externalism" or to a dubious doctrine of "the given." BonJour argues against both epistemic externalism and the relevant doctrine of the given. In light of the aforementioned dilemma, BonJour concludes that foundationalism provides at best an ad hoc evasion of the epistemic regress problem, the problem of explaining the nature of inferential justification. If BonJour's argument is sound, the traditional regress argument is incapable of justifying epistemic foundationalism.

In selection 4, Keith Lehrer explains and defends a coherence theory of justification and of knowledge. His theory depends on a certain conception of ourselves as knowers. Our cognitive faculties convey information, but knowledge requires that we evaluate the trustworthiness of our faculties on particular occasions and that we assess the probability of beliefs formed on the basis of available information. Such evaluations constitute a higher order processing of lower order information, and they must be carried out against a background system of accepted statements. This evaluation against a background acceptance system is based on an assessment of coherence with the acceptance system. Lehrer's theory of epistemic justification distinguishes a subjective from an objective component of justification. Lehrer claims that accepting a statement is justified subjectively, or personally, if and only if it coheres with one's system of accepted statements. A statement is objectively, or verifically, justified if and only if it coheres with that subset of one's system of accepted statements that contains only the true members of the system. Accepting a statement is completely justified for one if it is both personally and verifically justified for one. Lehrer explains that a statement coheres with a system if it is more reasonable to accept it than to accept anything else with which it conflicts on the basis of the system. He explicates the notion of conflict, or competition, and explains what it is for one thing to beat or to neutralize a competitor. Coherence, then, consists of beating or neutralizing all competitors against a background acceptance system. Lehrer claims that his coherence theory is superior to foundation theories because it explains the foundational principles that the foundation theorist must take as primitive.

In selection 5, Robert Audi clarifies the distinction between a foundationalist and a coherentist theory of justification (or of knowledge) by distinguishing their responses to the epistemic regress problem. Foundationalism claims that epistemic chains of justification terminate in a belief constituting direct knowledge, knowledge not inferentially based on further belief. Audi develops and defends a fallibilist foundationalism stating that direct knowledge need not be indefeasible. That is, the foundations for knowledge claims can be overridden or defeated. Considerations of coherence are important for foundationalism of this sort, because a belief's incoherence with other beliefs can defeat its justification. Coherence can also contribute to justification, although it can never be the sole basis of justification for a belief. Audi argues that there is a negative epistemic dependence on coherence, in that incoherence is a defeater for justification, but that this does not imply a positive epistemic dependence on coherence for justification. The epistemic significance of incoherence does not imply that coherence alone can justify a belief. Holistic coherentism rejects a linear epistemic chain for justification of a belief. It claims that a belief can be justified solely on the basis of its coherence with others of one's beliefs. Although a belief may be psychologically direct (grounded directly in an experience rather than another belief), it must, given coherentism, be epistemically indirect, depending on its coherence with other beliefs for its justification. Although fallibilist foundationalism and holistic coherentism differ less than is commonly thought, Audi indicates three important respects in which they are distinct. Audi argues that neither position is committed to epistemological dogmatism, although both positions allow for dogmatism.

In selection 6, Ernest Sosa develops and defends "virtue perspectivism" as an improvement on various forms of reliabilism. His position incorporates some elements from reliabilism and coherentism. Sosa begins by characterizing "generic reliabilism" and explaining three of the main problems it faces. The problems addressed in this article are "the evil-demon problem" and "the meta-incoherence problem." The evil-demon scenario presents a problem for reliabilism because it raises the possibility of a person who has highly unreliable belief-forming mechanisms (and mostly false beliefs) but is still in some sense epistemically justified (blameless). The meta-incoherence problem poses the possibility of a person holding beliefs produced by a reliable mechanism yet being in some sense unjustified (blameworthy) in holding them. Sosa offers a notion of "meta-justification" to capture the characteristic we share with the evil-demon victim. A belief is meta-justified for one if and only if one neither takes it to be ill-formed (produced by an unreliable belief-forming mechanism) nor has any available way of determining it to be ill-formed. Sosa presents a new problem facing his characterization of meta-justification (presenting a stronger notion of meta-justification along the way), based on the idea of an epistemic agent having beliefs that are not connected in the right way with sensory experience (the connection is, perhaps, purely random). A random connection could produce a coherent and comprehensive set of beliefs that are weakly justified and meta-justified (in the weak and the strong sense), but in this case we want an epistemic notion to capture the sense in which this person is internally defective in a way that a normal human is not. Sosa develops a notion of internal justification based on his idea of an intellectual virtue or faculty. An intellectual virtue is a competence, based on an *inner nature,* that leads one mostly to attain the truth and avoid error (i.e., to be reliable) relative to a certain *field* of propositions and relative to *conditions* the subject is in and relative to the subject's *environment.*

In selection 7, David Annis develops a theory of justification that purports to differ from foundationalism, coherentism, and reliabilism. The theory emphasizes the pragmatic and social (i.e., contextual) factors in the acquisition of empirical knowledge. Specifically, Annis bases his contextualism on the notions of an *issue-context* and an appropriate *objector-group.* The issue-context concerning a belief is merely the specific issue someone raises about the belief. The appropriate objector-group is simply the group of people qualified to raise objections about the belief. According to Annis, a belief is *contextually basic* for a person at a particular time, relative to an appropriate objector-group, if and only if at that time the objector-group does not require the person in question to have any reasons for the belief. If, however, some member of the objector-group raises an objection to a belief, one will have to provide an appropriate response to support the belief. Given such a reason, the belief in question will not be contextually basic. On Annis's contextualism, however, a belief is epistemically justified if and only if it is itself contextually basic or is supported by reasons that are contextually basic beliefs. One of the key implications of Annis's account is that epistemology must be *naturalized,* for given contextualism, we cannot neglect the *actual* social practices and standards of justification of a group of inquirers.

In selection 8, Richard Rorty claims that certain pragmatists should be read as breaking with the Kantian epistemological tradition that attempts to "ground" all areas of human intellectual endeavor with foundational philosophical theories. James and Dewey criticized the assumption that we need philosophical theories about knowledge, truth, morality, and the like. They rejected the attempt to find an all-embracing ahistorical context in which to organize human discourse. Pragmatism is anti-essentialism about the standard subjects of philosophical theorizing, a rejection of the fact-value distinction, and an endorsement of the claim that there are no constraints on inquiry other than conversational constraints. The pragmatist's anti-essentialism emphasizes the practical pay-off for philosophical reflection. Rejection of the fact-value dichotomy is a distinctively philosophical rejection of a metaphysical distinction between what is and what ought to be, as well as a rejection of any fundamental distinction between science and ethics. This position is based on a deeper rejection of the idea that scientific rationality is governed by an algorithmic and ahistorical method. The pragmatist's denial of constraints on inquiry other than conversational constraints is meant to persuade us to give up the notion of some universal and objective source of constraint and guidance toward the Truth in our theoretical pursuits. Pragmatists have been accused of being relativists, but Rorty explains that they are relativists not about "real theories," only about "philosophical theories."

In selection 9, Edmund Gettier presents two examples showing that justified true belief is not sufficient for one's having propositional knowledge. Here is one of Gettier's examples: Smith is justified in believing the false proposition that (a) Jones owns a Ford. On the basis of (a), Smith infers, and thus is justified in believing, that (b) either Jones owns a Ford or Brown is in Barcelona. As it happens, Brown is in Barcelona; so (b) is true. Although Smith is justified in believing the true proposition (b), Smith does not know (b). Gettier-style counterexamples are cases where a person has justified true belief that *P* but lacks knowledge that *P*. In *The Analysis of Knowing* (Princeton, 1983), Robert Shope suggests the following abstract description of Gettier-style counterexamples to the traditional analysis of knowledge:

> (G) In a Gettier-style counterexample concerning a person, *S,* and a proposition, *P;* (1) the truth condition holds regarding *P;* (2) the belief condition holds regarding *P;* (3) the justification condition holds regarding *P;* (4) some proposition, *O,* is false; (5) either the justification condition holds regarding *O,* or at least *S* would be justified in believing *O*; (6) S does not know *P*.

The Gettier problem is just the problem of providing an alternative to, or a modification of, the traditional justified-true-belief analysis that avoids difficulties from Gettier-style counterexamples. In solving this problem, one would go a long way toward identifying the logically sufficient conditions for one's having knowledge.

In selection 10, Richard Feldman shows that we cannot fault Gettier-style counterexamples on the ground that they rely on the allegedly false principle that false propositions can justify one's belief in other propositions. Feldman offers a counterexample to the justified-true-belief analysis that resembles Gettier's examples but does not rely on the allegedly false principle. Such a counterexample is perhaps the most difficult for the traditional analysis of knowledge.

In selection 11, John Pollock begins by recounting the history of responses to the Gettier problem, illustrating some failed attempts at a solution. Pollock then develops his solution to the Gettier problem, offering "objective epistemic justification" as a necessary condition for knowledge. Objective epistemic justification entails justified true belief, but requires also that one arrive at true belief "while doing everything right." His explanation of this claim fills in some of the details of epistemic justification generally, within his own version of direct realism. Using his general characterization of epistemic justification, Pollock defines objective epistemic justification: "S is objectively justified in believing P if and only if S instantiates some argument A supporting P which is ultimately undefeated relative to the set of all truths." Pollock takes this characterization to capture the way in which knowledge requires justification that is ultimately undefeated by true defeaters. In the last section of his paper, Pollock expands on this definition to accommodate the social aspects of knowledge emphasized by Gilbert Harman and others. He does this by including reference to the set of all propositions one is socially expected to believe when true.

In selection 12, Earl Conee defends the philosophical importance of solving the Gettier problem. He responds directly to recent charges that nothing important turns on solving the Gettier problem. Conee claims that solving the Gettier problem would provide independent conditions that are individually necessary and jointly sufficient for factual knowledge. This result, he claims, would be philosophically important, because the nature of factual knowledge is itself an important philosophical topic. Recent criticisms of the Gettier Problem show, at most, that solving the Gettier problem (a) would not contribute to refuting or supporting radical skepticism, (b) would not contribute to a theory of reasoning, (c) would not improve some historically important account of knowledge, and (d) would not improve our understanding of rational inquiry. There are, in fact, very good reasons for doubting some of these conclusions. Conee contends, however, that, even granting all these points, recent criticisms have not shown the Gettier problem to be philosophically unimportant. Conee explains why it is in fact important.

In selection 13, William Alston responds to the challenge of skepticism, arguing that we can reasonably take some of our "doxastic practices" to be reliable. His idea of a doxastic practice is based loosely on Wittgenstein's idea of a language-game, although Alston rejects verificationism. The main problem posed by skepticism is the charge of circularity in any philosophical defense of a belief-forming process, such as sense perception. We seem trapped in epistemic circularity, because any attempt to defend some source of beliefs will have to presuppose the reliability of that very source of beliefs. Such self-justification can validate *any* process of belief-formation. Alston grants that we are trapped in such epistemic circularity, but he claims that we can nonetheless reasonably defend reliance on some doxastic practices, such as the sense perception doxastic practice, as superior to others (e.g., crystal-ball gazing). Selection of doxastic practices is not dictated by rules but must be based on sensitivity, experience, trained judgment, and intuition. This latter sort of judgment frees us from objectionably using the standards of one doxastic practice to judge others. In this sort of judgment, we take all familiar doxastic practices as *prima facie* rational, and we only eliminate practices as they come to exhibit some deficiency such as inconsistency or conflict with other *well-established* practices. One task, then, of the epistemologist is to clarify the structure of various doxastic practices to aid judgments concerning what practices can rationally be taken as reliable. Alston ends by arguing that the *meta-epistemological* position he has developed in this article recommends certain positions in first-order epistemology, positions consistent with his prior work along the lines of reliabilism.

In selection 14, Ernest Sosa defends an externalist, reliabilist theory of knowledge as the only way to avoid the skeptical conclusion that a fully general theory of knowledge is impossible. He rejects "internalism," the view that a belief can be justified only through the backing of reasons or arguments, for it gives us no way to refute the skepticism he mentions. He identifies three version of "externalism," in the special sense of the word he uses. He finds that coherentism and foundationalism of the given both fail to provide a satisfactory general theory of knowledge that could refute skepticism. He defends a version of reliabilism while giving particular attention to the problem of circularity in epistemological theorizing. Sosa claims to be able to avoid the skeptical conclusion by rejecting the demand for a fully general and *legitimating* philosophical understandings of all our knowledge. We must resign ourselves to a kind of circularity, but it is not a philosophically destructive kind of circularity. Some epistemically circular arguments, he claims, can discriminate between reliable and unreliable doxastic practices. He concludes that philosophers such as Alston and Stroud have presented no good reason to accept philosophical scepticism.

In selection 15, Barry Stroud responds to selection 14 by Sosa. He claims that scepticism is inescapable, but he uses 'skepticism' to refer to a view different from the view Sosa intends with this term. Stroud calls "skepticism" the epistemological position that "nobody knows anything, or that nobody has any good reason to believe anything." He argues that the traditional epistemological project of explaining how we do in fact know the things we think we know should be rejected. Stroud adds: "I find the force and resilience of scepticism in the theory of knowledge to be so great, once the epistemological project is accepted, and I find its consequences to be so paradoxical, that I think the best thing to do now is to look much more closely and critically at the very enterprise of which scepticism or one of its rivals is the outcome: the task of the philosophical theory of knowledge itself." He claims that the circularity Sosa found to be unavoidable but philosophically acceptable renders the epistemological project itself and any possible answer to it thoroughly "dissatisfying." It is, he claims, the complete generality in a theory of knowledge demanded by the traditional epistemological project that compels us ultimately to be dissatisfied.

In selection 16, W. V. Quine argues that we should dispense with traditional epistemology and settle for empirical psychology instead. We must settle for psychology, according to Quine, because of the impossibility of grounding natural science on immediate experience in a strictly logical way. Epistemologists have traditionally avoided a reliance on empirical psychology, because the goal had been the grounding of empirical science, and it would be clearly circular to rely on empirical psychology in that enterprise. Quine claims that such worries about circularity have little point once we abandon the goal of deriving all of our knowledge from observations. We aim to understand the actual connection between observation and knowledge (especially science), and psychology can clarify that connection. Quine attacks not only the thesis that all knowledge can be justified on a strict observational basis but also the thesis that all meaning can be reduced to observation statements. He defends this latter attack on the basis of his account of theoretical statements as lacking empirical meaning in isolation. Quine concludes with a brief account of observation sentences, as they are basic both to considerations of meaning and to considerations of truth. He rejects an extreme relativism based on giving too much weight to the indeterminacy of translation. He explains why the indeterminacy of translation has little significance for observation sentences.

In selection 17, Louise Antony argues that we need epistemology to be sensitive to the insights of feminism. Feminist epistemology need not, she adds, reject all existing epistemological paradigms. Naturalized epistemology can contribute significantly to the goals of a feminist epistemology. She claims that feminists, in criticizing traditional epistemology and contemporary analytic epistemology, have seriously misunderstood both. Naturalized epistemology benefits feminism by providing criticisms of the epistemic ideal of complete impartiality while also showing how to condemn bad, or harmful, epistemic biases. Its criticisms of the ideal of impartiality are, in fact, more radical than many of the criticisms coming from outside contemporary analytic epistemology, according to Antony. Naturalized epistemology supports the claim that all human knowledge is "biased," in the sense that it must start from some perspective containing presuppositions, but it also allows us to identify the "good biases" as those that facilitate the search for truth (in a realist sense) and the "bad biases" as those that impede the search for truth. Which biases facilitate the search for truth is an empirical matter. Antony argues that many feminists have misunderstood traditional epistemology because they have failed to appreciate the importance of the distinction between empiricist and rationalist theories of knowledge and of the mind. They have focussed on a notion of objectivity as impartiality that received support only from empiricist philosophers. They have also failed to represent accurately recent analytic epistemology, for they have failed to appreciate its radical critique of neo-empiricism, especially the empiricism of the Logical Positivists. They have neglected the *rationalism* in naturalized epistemology that is associated mostly with Chomsky's *empirically based* arguments about the nature of human knowers. Feminists' wholesale rejection of analytic epistemology, including naturalized epistemology, harms them in many ways, according to Antony.

In selection 18, Alvin Goldman claims that our commonsense epistemic concepts and norms, our *epistemic folkways,* should receive both clarification and improvement, or refinement, from epistemology. Lessons from cognitive science are crucial for both of these projects, Goldman claims. Cognitive science helps us understand epistemic folkways and guides us in "transcending" our epistemic folkways by offering a detailed, empirically based characterization of psychological mechanisms. He recommends a kind of *scientific epistemology,* informed by empirical psychology, that has descriptive and normative branches. His own approach to the concept of justified belief identifies it with the concept of belief obtained through the exercise of intellectual virtues. His account of intellectual virtues and vices is based on our use of exemplars for identifying virtuous psychological processes. Goldman uses his notion of justified belief to solve some epistemological problems. His descriptive scientific epistemology is based primarily on considerations of reliability, and it includes a crucial function for the judgments of an epistemic *evaluator.* Cognitive science can help to clarify an evaluator's judgments about intellectual virtues and vices. In normative scientific epistemology, cognitive science can help us primarily in our identification of the psychological mechanisms that are objects of evaluation. It highlights also the importance of question-answering power and question-answering speed, among other things, as standards for epistemic evaluation.

References

Almeder, Robert. 1992. *Blind Realism.* Lanham, Md.: Rowman & Littlefield.

Alston, William. 1989. *Epistemic Justification.* Ithaca: Cornell University Press.

Amico, Robert. 1993. *The Problem of the Criterion.* Lanham, Md.: Rowman & Littlefield.

Annis, David. 1978. "A Contextualist Theory of Epistemic Justification." *American Philosophical Quarterly* 15, 213–19. Selection 7, this volume.

Audi, Robert. 1988. *Belief, Justification, and Knowledge.* Belmont, Calif.: Wadsworth.

———. 1993. *The Structure of Justification.* Cambridge: Cambridge University Press.

Ayer, A. J. 1940. *The Foundations of Empirical Knowledge.* London: Macmillan.

———. 1947. "Phenomenalism." In Ayer, *Philosophical Essays.* London: Macmillan, 1954.

Barnes, Jonathan. 1990. *The Toils of Scepticism.* Cambridge: Cambridge University Press.

Blanshard, Brand. 1939. *The Nature of Thought,* Vol. 2. London: Allen & Unwin.

———. 1980. "Reply to Nicholas Rescher." In P. A. Schilpp. ed., *The Philosophy of Brand Blanshard,* pp. 589–600. La Salle, Ill.: Open Court.

BonJour, Laurence. 1985. *The Structure of Empirical Knowledge.* Cambridge, Mass.: Harvard University Press.

Paul K. Moser

Carruthers, Peter. 1992. *Human Knowledge and Human Nature*. New York: Oxford University Press.

Chisholm, Roderick. 1957. *Perceiving: A Philosophical Study*. Ithaca: Cornell University Press.

———. 1964. "Theory of Knowledge in America." In Chisholm, *The Foundations of Knowing*, pp. 109–93. Minneapolis: University of Minnesota Press, 1982.

———. 1977. *Theory of Knowledge, 2d ed.* Englewood Cliffs, N.J.: Prentice Hall.

———. 1982. "The Problem of the Criterion." In Chisholm, *The Foundations of Knowing*, pp. 61–75. Minneapolis: University of Minnesota Press, 1982.

———. 1989. *Theory of Knowledge, 3d ed.* Englewood Cliffs, N.J.: Prentice Hall.

Cornman, James. 1975. *Perception, Common Sense, and Science*. New Haven, Conn.: Yale University Press.

———. 1980. *Skepticism, Justification, and Explanation*. Dordrecht: Reidel.

Devitt, Michael. 1984. *Realism and Truth*. Oxford: Basil Blackwell.

Ducasse, C. J. 1968. "Propositions, Truth, and the Ultimate Criterion of Truth." In Ducasse, *Truth, Knowledge, and Causation*. London: Routledge & Kegan Paul.

Firth, Roderick. 1969. "Lewis on the Given." In P. A. Schilpp, ed., *The Philosophy of C. I. Lewis*, pp. 329–50. LaSalle, Ill.: Open Court.

Foley, Richard. 1987. *The Theory of Epistemic Rationality*. Cambridge, Mass.: Harvard University Press.

Fumerton, Richard. 1985. *Metaphysical and Epistemological Problems of Perception*. Lincoln, Neb.: University of Nebraska Press.

———. 1995. *Metaepistemology and Skepticism*. Lanham, Md.: Rowman & Littlefield.

Goldman, Alan. 1988. *Empirical Knowledge*. Berkeley: University of California Press.

Goldman, Alvin. 1967. "A Causal Theory of Knowing." *The Journal of Philosophy* 64, 357–72.

———. 1986. *Epistemology and Cognition*. Cambridge, Mass.: Harvard University Press.

Harman, Gilbert. 1973. *Thought*. Princeton: Princeton University Press.

———. 1986. *Change In View*. Cambridge, Mass.: The MIT Press.

Lehrer, Keith. 1974. *Knowledge*. Oxford: Clarendon Press.

———. 1990. *Theory of Knowledge*. Boulder, Colo.: Westview.

Lewis, C. I. 1929. *Mind and the World-Order*. New York: Scribner's.

———. 1946. *An Analysis of Knowledge and Valuation*. LaSalle, Ill.: Open Court.

Lycan, William. 1988. *Judgement and Justification*. Cambridge: Cambridge University Press.

Meyers, Robert. 1988. *The Likelihood of Knowledge*. Dordrecht: Reidel.

Morawetz, Thomas. 1978. *Wittgenstein and Knowledge*. Amherst, Mass.: University of Massachusetts Press.
Moser, Paul. 1985. *Empirical Justification*. Dordrecht: Reidel.
———. 1988. "Foundationalism, the Given, and C. I. Lewis." *History of Philosophy Quarterly* 5, 189–204.
———. 1989a. "Lehrer's Coherentism and the Isolation Objection." In J. W. Bender, ed., *The Current State of the Coherence Theory*, pp. 29–37. Dordrecht: Reidel.
———. 1989b. *Knowledge and Evidence*. Cambridge: Cambridge University Press.
———. 1993. *Philosophy After Objectivity*. New York: Oxford University Press.
Oakley, I. T. 1976. "An Argument for Scepticism Concerning Justified Belief." *American Philosophical Quarterly* 13, 221–28.
Peirce, C. S. 1868. "Questions Concerning Certain Faculties Claimed for Man." In Philip Wiener, ed., *Charles S. Peirce: Selected Writings*, pp. 15–38. New York: Dover, 1966.
Pollock, John. 1974. *Knowledge and Justification*. Princeton: Princeton University Press.
———. 1986. *Contemporary Theories of Knowledge*. Lanham, Md.: Rowman & Littlefield.
Quine, W. V. 1951. "Two Dogmas of Empiricism." In Quine, *From A Logical Point of View, 2d ed.*, pp. 20–46. New York: Harper & Row, 1963.
Rescher, Nicholas. 1973. *The Coherence Theory of Truth*. Oxford: Clarendon Press.
———. 1979. *Cognitive Systematization*. Oxford: Basil Blackwell.
Russell, Bertrand. 1940. *An Inquiry Into Meaning and Truth*. London: Allen & Unwin.
Sellars, Wilfrid. 1956. "Empiricism and the Philosophy of Mind." In Sellars, *Science, Perception, and Reality*. London: Routledge & Kegan Paul, 1963.
———. 1975. "Epistemic Principles." In H.-N. Castañeda, ed., *Action, Knowledge, and Reality*, pp. 332–48. Indianapolis: Bobbs-Merrill.
———. 1979. "More on Givenness and Explanatory Coherence." In G. S. Pappas, ed., *Justification and Knowledge*, pp. 169–81. Dordrecht: Reidel.
Shiner, Roger. 1977. "Wittgenstein and the Foundations of Knowledge." *Proceedings of the Aristotelian Society* 78, 102–24.
Shope, Robert. 1983. *The Analysis of Knowing*. Princeton: Princeton University Press.
Sorensen, Roy. 1991. " '*P,* Therefore, *P*' Without Circularity." *The Journal of Philosophy* 88, 245–66.
Sosa, Ernest. 1991. *Knowledge In Perspective*. Cambridge: Cambridge University Press.
Striker, Gisela. 1980. "Sceptical Strategies." In Malcolm Schofield, Myles Burnyeat, and Jonathan Barnes, eds., *Doubt and Dogmatism*, pp. 54–83. Oxford: Clarendon Press.

————. 1990. "The Problem of the Criterion." In Stephen Everson, ed., *Epistemology,* pp. 143–60. Cambridge: Cambridge University Press.

Swain, Marshall. 1981. *Reasons and Knowledge.* Ithaca: Cornell University Press.

Thagard, Paul. 1988. *Computational Philosophy of Science.* Cambridge, Mass.: MIT Press.

Part I

Justification and Knowledge

1

Concepts of Epistemic Justification

William P. Alston

I

Justification, or at least "justification," bulks large in recent epistemology. The view that knowledge consists of true-justified-belief (+ . . .) has been prominent in this century, and the justification of belief has attracted considerable attention in its own right. But it is usually not at all clear just what an epistemologist means by "justified," just what concept the term is used to express. An enormous amount of energy has gone into the attempt to specify conditions under which beliefs of one or another sort are justified; but relatively little has been done to explain *what it is* for a belief to be justified, what that is for which conditions are being sought.[1] The most common procedure has been to proceed on the basis of a number of (supposedly) obvious cases of justified belief, without pausing to determine what property it is of which these cases are instances. Now even if there were some single determinate concept that all these theorists have implicitly in mind, this procedure would be less than wholly satisfactory. For in the absence of an explicit account of the concept being applied, we lack the most fundamental basis for deciding between supposed intuitions and for evaluating proposed conditions of justification. And in any event, as philosophers we do not seek merely to speak the truth, but also to gain an explicit, reflective understanding of the matters with which we deal. We want to know not only when our beliefs are justified, but also what it is to enjoy that status. True, not every fundamental concept can be explicated, but we shall find that much can be done with this one.

Reprinted from *The Monist* 68 (1985): 57–89, by permission of the author and the publisher. Copyright 1985, *The Monist*.

And since, as we shall see in this essay, there are several distinct concepts that are plausibly termed "concepts of epistemic justification," the need for analysis is even greater. By simply using "justified" in an unexamined, intuitive fashion the epistemologist is covering up differences that make important differences to the shape of a theory of justification. We cannot fully understand the stresses and strains in thought about justification until we uncover the most crucial differences between concepts of epistemic justification.

Not all contemporary theorists of justification fall under these strictures. Some have undertaken to give an account of the concept of justification they are using.[2] But none of them provides a map of this entire conceptual territory.

In this essay I am going to elaborate and interrelate several distinct concepts of epistemic justification, bringing out some crucial issues involved in choosing between them. I shall give reasons for disqualifying some of the contenders, and I shall explain my choice of a winner. Finally I shall vouchsafe a glimpse of the enterprise for which this essay is a propaedeutic, that of showing how the differences between these concepts make a difference in what it takes for the justification of belief, and other fundamental issues in epistemology.

Before launching this enterprise we must clear out of the way a confusion between one's *being* justified in believing that p, and one's *justifying* one's belief that p, where the latter involves one's *doing* something to show that p, or to show that one's belief was justified, or to exhibit one's justification. The first side of this distinction, on the other hand, is a state or condition one is in, not anything one does or any upshot thereof. I might *be* justified in believing that there is milk on the table because I see it there, even though I have done nothing to show that there is milk on the table or to show that I am justified in believing there to be. It is amazing how often these matters are confused in the literature. We will be concentrating on the "be justified" side of this distinction, since that is of more fundamental epistemological interest. If epistemic justification were restricted to those cases in which the subject carries out a "justification," it would *obviously* not be a necessary condition of knowledge or even of being in a strong position to acquire knowledge. Most cases of perceptual knowledge, for example, involve no such activity.[3]

II

Let's begin our exploration of this stretch of conceptual territory by listing a few basic features of the concept that would seem to be common ground.

1. It applies to beliefs, or alternatively to a cognitive subject's having a belief. I shall speak indifferently of S's belief that *p* being justified and of S's being justified in believing that *p*. This is the common philosophical concept of belief, in which S's believing that *p* entails neither that S knows that *p* nor that S does not know that *p*. It is not restricted to conscious or occurrent beliefs.
2. It is an evaluative concept, in a broad sense in which this is contrasted with "factual." To say that S is justified in believing that *p* is to imply that there is something all right, satisfactory, in accord with the way things should be, about the fact that S believes that *p*. It is to accord S's believing a positive evaluative status.
3. It has to do with a specifically *epistemic* dimension of evaluation. Beliefs can be evaluated in different ways. One may be more or less prudent, fortunate, or faithful in holding a certain belief. Epistemic justification is different from all that. Epistemic evaluation is undertaken from what we might call the "epistemic point of view." That point of view is defined by the aim at maximizing truth and minimizing falsity in a large body of beliefs. The qualification "in a large body of beliefs" is needed because otherwise one could best achieve the aim by restricting one's belief to those that are obviously true. That is a rough formulation. How large a body of beliefs should we aim at? Is any body of beliefs of a given size, with the same truth-falsity ratio, equally desirable, or is it more important, epistemically, to form beliefs on some matters than others? And what relative weights should be assigned to the two aims at maximizing truth and minimizing falsity? We can't go into all that here; in any event, however these issues are settled it remains true that our central cognitive aim is to amass a large body of beliefs with a favorable truth-falsity ratio. For a belief to be epistemically justified is for it, somehow, to be awarded high marks relative to that aim.
4. It is a matter of degree. One can be more or less justified in believing that *p*. If, e.g., what justifies one is some evidence one has, one will be more or less justified depending on the amount and strength of the evidence. However, in this essay I shall, for the sake of simplicity, treat justification as absolute. You may, if you like, think of this as the degree of justification required for some standard of acceptability.

III

Since any concept of epistemic justification is a concept of some condition that is desirable or commendable from the standpoint of the aim at maximizing truth and minimizing falsity, in distinguishing different concepts of justification we will be distinguishing different ways in which conditions can be desirable from this standpoint. As I see it, the major divide in this terrain has to do with whether believing and refraining from believing are subject to obligation, duty, and the like. If they are, we can think of the favorable evaluative status of a certain belief as consisting in the fact that in holding that belief one has fulfilled one's obligations, or refrained from violating one's obligations, to achieve the fundamental aim in question. If they are not so subject, the favorable status will have to be thought of in some other way.

I shall first explore concepts of the first sort, which I shall term "deontological,"[4] since they have to do with how one stands in believing that *p,* vis-à-vis duties or obligations. Most epistemologists who have attempted to explicate justification have set out a concept of this sort.[5] It is natural to set out a deontological concept on the model of the justification of behavior. Something I *did* was justified just in case it was *not in violation* of any relevant duties, obligations, rules, or regulations, and hence was not something for which I could rightfully be blamed. To say that my expenditures on the trip were justified is not to say that I was obliged to make those expenditures (e.g., for taxis), but only that it was all right for me to do so, that in doing so I was not in violation of any relevant rules or regulations. And to say that I was justified in making that decision on my own, without consulting the executive committee, is not to say that I was required to do it on my own (though that *may* also be true); it is only to say that the departmental by-laws permit the chairman to use his own discretion in matters of this kind. Similarly, to say that a belief was deontologically justified is not to say that the subject was obligated to believe this, but only that he was permitted to do so, that believing this did not involve any violation of relevant obligations. To say that I am justified in believing that salt is composed of sodium and chlorine, since I have been assured of this by an expert, is not to say that I am obligated to believe this, though this might also be true. It is to say that I am permitted to believe it, that believing it would not be a violation of any relevant obligation, e.g., the obligation to refrain from believing that *p* in the absence of adequate reasons for doing so. As Carl Ginet puts it, "One is *justified* in being confident that *p* if and only if it is not the case that one ought not to be confident that *p;* one could not be justly reproached for being confident that *p*."[6]

Since we are concerned specifically with the *epistemic* justification of belief, the concept in which we are interested is not that of *not violating obligations of any sort in believing*, but rather the more specific concept of *not violating "epistemic," "cognitive," or "intellectual" obligations in believing*. Where are such obligations to be found? If we follow out our earlier specification of the "epistemic point of view," we will think of our basic epistemic obligation as that of doing what we can to achieve the aim at maximizing truth and minimizing falsity within a large body of beliefs. There will then be numerous more specific obligations that owe their status to the fact that fulfilling them will tend to the achievement of that central aim. Such obligations might include *to refrain from believing that p in the absence of sufficient evidence* and *to accept whatever one sees to be clearly implied by something one already believes (or, perhaps, is already justified in believing.).*[7] Of course other positions might be taken on this point.[8] One might suppose that there are a number of ultimate, irreducible intellectual duties that cannot be derived from any basic goal of our cognitive life. Or alternative versions of the central aim might be proposed. Here we shall think of terms of the basic aim we have specified, with more specific obligations derived from that.

Against this background we can set out our first concept of epistemic justification as follows, using "d" for "deontological":

I. S is J_d in believing that *p iff* in believing that *p* S is not violating any epistemic obligations.

There are important distinctions between what we may call "modes" of obligation, justification, and other normative statuses. These distinctions are by no means confined to the epistemic realm. Let's introduce them in connection with moral norms for behavior. Begin with a statement of obligation in "objective" terms, a statement of the objective state of affairs I might be said to be obliged to bring about. For example, it is my obligation as a host to make *my guest, G, feel welcome*. Call that underlined state of affairs "A." We may think of this as an *objective* conception of my obligation as a host. I have fulfilled that obligation *iff* G feels welcome.[9] But suppose I did what I sincerely believed would bring about A? In that case surely no one could blame me for dereliction of duty. That suggests a more *subjective* conception of my obligation as *doing what I believed was likely to bring about A*.[10] But perhaps I should not be let off so easily as that. "You should have realized that what you did was not calculated to make G feel welcome." This retort suggests a somewhat more stringent formulation of my obligation than the very permissive subjective conception just specified. It suggests that I can't fulfill my obligation by doing just anything I happen to believe will bring about A? I am not off the hook unless *I did what the facts available to me indicate will have a good chance of leading to A*. This is still a subjective conception in that what it takes to fulfill my obligation is specified from my point of view; but it takes my point of view to range over not all my beliefs, but only my justified beliefs. This we might call a *cognitive* conception of my obligation.[11] Finally, suppose that I did what I had adequate reason to suppose would produce A, and I did produce A, but I didn't do it for that reason. I was just amusing myself, and I would have done what I did even if I had known it would not make G feel welcome. In that case I might be faulted for moral irresponsibility, however well I rate in the other modes. This suggests what we may call a motivational conception of my obligation as *doing what I believed (or was justified in believing) would bring about A, in order to bring about A*.

We may sum up these distinctions as follows:

II. S has fulfilled his *objective* obligation *iff* S has brought about A.
III. S has fulfilled his *subjective* obligation *iff* S has done what he believed to be most likely to bring about A.
IV. S has fulfilled his *cognitive* obligation *iff* S did what he was justified in believing to be most likely to bring about A.
V. S has fulfilled his *motivational* obligation *iff* S has done what he did because he supposed it would be most likely to bring about A.

We can make analogous distinctions with respect to the justification of behavior or belief, construed as the absence of any violation of obligations.[12] Let's indicate how this works out for the justification of belief.

VI. S is *objectively* justified in believing that *p iff* S is not violating any objective obligation in believing that *p*.
VII. S is *subjectively* justified in believing that *p iff* S is not violating any subjective obligation in believing that *p*.
VII. S is *cognitvely* justified in believing that *p iff* S is not violating any cognitive obligation in believing that *p*.
IX. S is *motivationally* justified in believing that *p iff* S is not violating any motivational obligation in believing that *p*.

If we assume that only one intellectual obligation is relevant to the belief in question, viz., the obligation to believe that *p* only if one has adequate evidence for *p*, we can be a bit more concrete about this.

> X. S is objectively justified in believing that *p* *iff* S has adequate evidence for *p*.[13]
>
> XI. S is subjectively justified in believing that *p* *iff* S believes that he possesses adequate evidence for *p*.
>
> XII. S is cognitively justified in believing that *p* *iff* S is justified in believing that he possesses adequate evidence for *p*.[14]
>
> XIII. S is motivationally justified in believing that *p* *iff* S believes that *p* on the basis of adequate evidence, or, alternatively, on the basis of what he believed, or was justified in believing, was adequate evidence.

I believe that we can safely neglect XI. To explain why, I will need to make explicit what it is to have adequate evidence for *p*. First a proposition, *q*, is adequate evidence for *p* provided they are related in such a way that if *q* is true then *p* is at least probably true. But I *have* that evidence only if I believe that *q*. Furthermore I don't "have" it in such a way as to thereby render my belief that *p* justified unless I know or am justified in believing that *q*. An unjustified belief that *q* wouldn't do it. If I believe that Begin has told the cabinet that he will resign, but only because I credited an unsubstantiated rumor, then even if Begin's having told the cabinet that he would resign is an adequate indication that he will resign, I will not thereby be justified in believing that he will resign.

Now I might very well *believe* that I have adequate evidence for *q* even though one or more of these conditions is not satisfied. This is an especially live possibility with respect to the first and third conditions. I might mistakenly believe that my evidence is adequate support, and I might mistakenly suppose that I am justified in accepting it. But, as we have just seen, if I am not justified in accepting the evidence for *p*, then my believing it cannot render me justified in believing that *p*, however adequate that evidence. I would also hold, though this is perhaps more controversial, that if the evidence is not in fact adequate, my having that evidence cannot justify me in believing that *p*. Thus, since my believing that I had adequate evidence is compatible with these non-justifying states of affairs, we cannot take subjective justification, as defined in XI, to constitute epistemic justification.

That leaves us with three contenders. Here I will confine myself to pointing out that there is a strong tendency for J_d to be used in a cognitive rather than a purely objective form. J_d is, most centrally, a concept of freedom from blameworthiness, a concept of being "in the clear" so far as one's intellectual obligations are concerned. But even if I don't have adequate evidence for p, I could hardly be blamed for believing that p (even assuming, as we are in this discussion, that there is something wrong with believing in the absence of adequate evidence), provided I am justified in supposing that I have adequate evidence. So long as that condition holds I have done the right thing, or refrained from doing the wrong thing, so far as I am able to tell; and what more could be required of me? But this means that it is XII, rather than X, that brings out what it takes for freedom from blame, and so brings out what it takes for being J_d.[15]

What about the motivational form? We can have J_d in any of the first three forms with or without the motivational form. I can have adequate evidence for p, and believe that p (XI), whether or not my belief is based on that evidence; and so for the other two. But the motivational mode is parasitic on the other modes, in that the precise form taken by the motivational mode depends on the status of the (supposed) evidence on which the belief is based. This "unsaturated" character of the motivational mode is reflected in the threefold alternative that appears in our formulation of XIII. If S bases his belief that p on actually possessed adequate evidence, then XIII combines with X. If the evidence on which it is based is only believed to be adequate evidence, or only justifiably believed to be adequate evidence, then XIII combines with XI or XII. Of course, it may be based on actually possessed adequate evidence, which is justifiably believed to be such; in which case S is justified in all four modes. Thus the remaining question concerning J_d is whether a "motivational rider" should be put on XII. Is it enough for J_d that S be justified in believing that he has adequate evidence for p, or should it also be required that S's belief that p be based on that evidence? We will address this question in section V in the form it assumes for a quite different concept of justification.[16]

IV

We have explained *being J_d in believing that p as not violating any intellectual obligations in believing that p*. And, in parallel fashion, being J_d in refraining from believing that p would consist in not having violated any intellectual obligations in so doing. But if it is possible for me to violate an obligation in refraining from believing that p, it must be that I can be obliged to refrain from believing that p. And this is the way we have been thinking of it. Our example of an intellectual obligation has been the obligation to refrain from believing that p in the absence of adequate evidence. On the other side, we might think of a person as being obliged to believe that p if confronted with conclusive evidence that p (where that includes the absence of sufficient overriding evidence to the contrary).

Now it certainly looks as if I can be obliged to believe or to refrain from believing, only if this is in my direct voluntary control; only if I can, here and now, believe that p or no just by willing (deciding, choosing . . .). And that is the way many epistemologists seem to construe the matter. At least many formulations are most naturally interpreted in this way. Think back, e.g., on Chisholm's formulation of our intellectual obligation (1977, p. 14), cited in n16. Chisholm envisages a person thinking of a certain proposition as a candidate for belief, and then effectively choosing belief or abstention on the basis of those considerations.[17] Let's call the version of J_d that presupposes direct voluntary control over belief (and thus thinks of an obligation to believe as an obligation to bring about belief here and now) "J_{dv}" ("v" for "voluntary").

I find this assumption of direct voluntary control over belief quite unrealistic. There are strong reasons for doubting that belief is usually, or perhaps ever, under direct voluntary control. First, think of the beliefs I acquire about myself and the world about me through experience—through perception, self-consciousness, testimony, and simple reasoning based on these data. When I see a car coming down the street I am not capable of believing or disbelieving this at will. In such familiar situations the belief-acquisition mechanism is isolated from the direct influence of the will and under the control of more purely cognitive factors.

Partisans of a voluntary control thesis will counter by calling attention to cases in which things don't appear to be so cut and dried: cases of radical underdetermination by evidence, as when a general has to dispose his forces in the absence of sufficient information about the position of enemy forces; or cases of the acceptance of a religious or philosophical position where there seem to be a number of equally viable alternatives. In such cases it can appear that one makes a decision as to what to believe and what not to believe. My view on these matters is that insofar as something is chosen voluntarily it is something other than a belief or abstention from belief. The general chooses to proceed on the working assumption that the enemy forces are disposed in such-and-such a way. The religious convert to whom it is not clear that the beliefs are correct has chosen to live a certain kind of life, or to selectively subject himself to certain influences. And so on. But even if I am mistaken about these kinds of cases, it is clear that for the vast majority of beliefs nothing like direct voluntary control is involved. And so J_{dv} could not possibly be a generally applicable concept of epistemic justification.

If I am right in rejecting the view that belief is, in general or ever, under direct voluntary control, are we foreclosed from construing epistemic justification as freedom from blameworthiness? Not necessarily. We aren't even prevented from construing epistemic justification as the absence of obligation-violations. We *will* have to avoid thinking of the relevant obligations as obligations to believe or refrain from believing, on the model of obligations to answer a question or to open a door, or to do anything else over which we have immediate voluntary control.[18] If we are to continue to think of intellectual obligations as having to do with believing, it will have to be more on the model of the way in which obligations bear on various other conditions over which one lacks direct voluntary control but which one can influence by voluntary actions, such conditions as being overweight, being irritable, being in poor health, or having friends. I can't institute, nullify, or alter any of those conditions here and now just by deciding to do so. But I can do things at will that will influence those conditions, and in that way they may be to some extent under my indirect control. One might speak of my being obliged to be in good health or to have a good disposition, meaning that I am obliged to do what I can (or as much as could reasonably be expected of me) to institute and preserve those states of affairs. However, since I think it less misleading to say exactly what I mean, I will not speak of our being obliged to weigh a certain amount or to have a good disposition, or to believe a proposition; I will rather speak of our having obligations to do what we can, or as much as can reasonably be expected of us, to influence those conditions.[19]

The things we can do to affect our believings can be divided into (a) activities that bring influences to bear on, or withhold influences from, a particular situation, and (b) activities that affect our belief-forming habits. (a) includes such activities as checking to see whether I have considered all the relevant evidence, getting a second opinion, searching my memory for analogous cases, and looking into the question of whether there is anything markedly abnormal about my current perceptual situation. (b) includes training myself to be more critical of gossip, talking myself into being either more or less subservient to authority, and practicing greater sensitivity to the condition of other people. Moreover, it is plausible to think of these belief-influencing activities as being subject to intellectual obligations. We might, e.g., think of ourselves as being under an obligation to do what we can (or what could reasonably be expected of us) to make our belief-forming processes as reliable as possible.

All this suggests that we might frame a deontological conception of being epistemically justified in believing that p, in the sense that one's believing that p is not the result of one's failure to fulfill one's intellectual obligations vis-à-vis one's belief-forming and -maintaining activities. It would, again, be like the way in which one is or isn't to blame for other conditions that are not under direct voluntary control but which one can influence by one's voluntary activities. I am to blame for being overweight (being irritable, being in poor health, being without friends) only if that condition is in some way due to my own past failures to do what I should to limit my intake or to exercise or whatever. If I would still be overweight even if I had done everything I could and should have done about it, then I can hardly be blamed for it. Similarly, we may say that I am subject to reproach for believing that p, provided that I am to blame for being in the doxastic condition, in the sense that there are things I could and should have done, such that if I had done them I would not be believing that p. If that is the case I am unjustified in that belief. And if it is *not* the case, if there are no unfulfilled obligations the fulfilling of which would have inhibited that belief-formation, then I am justified in the belief.

Thus we have arrived at a deontological concept of epistemic justification that does not require belief to be under direct voluntary control. We may label this concept "J_{di}" ("i" for "involuntary"). It may be more formally defined as follows:

XIV. S is J_{di} in believing that p at t *iff* there are no intellectual obligations that (a) have to do with the kind of belief-forming or -sustaining habit the activation of which resulted in S's believing that p at t, or with the particular process of belief-formation or -sustenance that was involved in S's believing that p at t, and which (b) are such that:
  A. S had those obligations prior to t.
  B. S did not fulfill those obligations.
  C. If S had fulfilled those obligations, S would not have believed that p at t.[20]

As it stands, this account will brand too many beliefs as unjustified, just because it is too undiscriminating in the counter-factual condition C. There are ways in which the non-fulfillment of intellectual obligations can contribute to a belief-acquisition without rendering the belief unjustified. Suppose that I fail to carry out my obligation to spend a certain period in training myself to observe things more carefully. I use the time thus freed up to take a walk around the neighborhood. In the course of this stroll I see two dogs fighting, thereby acquiring the belief that they are fighting. There was a relevant intellectual obligation I didn't fulfill, which is such that if I had fulfilled it I wouldn't have acquired that belief. But if that is a perfectly normal perceptual belief, it is surely not thereby rendered unjustified.

Here the dereliction of duty contributed to belief-formation simply by facilitating access to the data. That's not the kind of contribution we had in mind. The sorts of cases we were thinking of were those most directly suggested by the two sorts of intellectual obligations we distinguished: (a) cases in which the belief was acquired by the activation of a habit that we would not have possessed had we fulfilled our intellectual obligations; (b) cases in which we acquire, or retain, the belief only because we are sheltered from adverse considerations in a way we wouldn't be if we had done what we should have done. Thus we can avoid counterexamples like the above by reformulating C as follows:

> C. If S had fulfilled those obligations, then S's belief-forming habits would have changed, or S's access to relevant adverse considerations would have changed, in such a way that S would not have believed that p at t.

But even with this refinement J_{di} does not give us what we expect of epistemic justification. The most serious defect is that it does not hook up in the right way with an adequate, truth-conducive ground. I may have done what could reasonably be expected of me in the management and cultivation of my doxastic life, and still hold a belief on outrageously inadequate grounds. There are several possible sources of such a discrepancy. First there is what we might call "cultural isolation." If I have grown up in an isolated community in which everyone unhesitatingly accepts the traditions of the tribe as authoritative, then if I have never encountered anything that seems to cast doubt on the traditions and have never thought to question them, I can hardly be blamed for taking them as authoritative. There is nothing I could reasonably be expected to do that would alter that belief-forming tendency. And there is nothing I could be expected to do that would render me more exposed to counterevidence. (We can suppose that the traditions all have to do with events distant in time and/or space, matters on which I could not be expected to gather evidence on my own.) I am J_{di} in believing these things. And yet the fact that it is the tradition of the tribe that p may be a very poor reason for believing that p.

Then there is deficiency in cognitive powers. Rather than looking at the extremer forms of this, let's consider a college student who just doesn't have what it takes to follow abstract philosophical reasoning, or exposition for that matter. Having read Bk. IV of Locke's *Essay,* he believes that it is Locke's view that everything is a matter of opinion, that one person's opinion is just as good as another's, and that what is true for me may not be true for you. And it's not just that he didn't work hard enough on this particular point, or on the general abilities involved. There is nothing that he could and should have done such that had he done so, he would have gotten this straight. He is simply incapable of appreciating the distinction between "One's knowledge is restricted to one's own ideas" and "Everything is a matter of opinion." No doubt teachers of philosophy tend to assume too quickly that this description applies to some of their students, but surely there can be such cases—cases in which either no amount of time and effort would enable the student to get straight on the matter, or it would be unreasonable to expect the person to expand that amount of time or effort. And yet we would hardly wish to say that the student is justified in believing what he does about Locke.

Other possible sources of a discrepancy between J_{di} and epistemic justification are poor training that the person lacks the time or resources to overcome, and an incorrigible doxastic incontinence. ("When he talks like that I just can't help believing what he says.") What this spread of cases brings out is that J_{di} is not sufficient for epistemic justification; we may have done the best we can, or at least the best that could reasonably be expected of us, and still be in a very poor epistemic position in believing that p; we could, blamelessly, be believing p for outrageously bad reasons. Even though J_{di} is the closest we can come to a deontological concept of epistemic justification, if belief is not under direct voluntary control, it still does not give us what we are looking for.

V

Thus neither version of J_d is satisfactory. Perhaps it was misguided all along to think of epistemic justification as freedom from blameworthiness. Is there any alternative, given the non-negotiable point that we are looking for a concept of epistemic evaluation? Of course there is. By no means all evaluation, even all evaluation of activities, states, and aspects of human beings, involves the circle of terms that includes "obligation," "permission," "right," "wrong," and "blame." We can evaluate a person's abilities, personal appearance, temperament, or state of health as more or less desirable, favorable, or worthwhile, without taking these to be within the person's direct voluntary control and so subject to obligation in a direct fashion (as with J_{dv}, and without making the valuation depend on whether the person has done what she should to influence these states (as with J_{di}). Obligation and blame need not come into it at all. This is most obvious when we are dealing with matters that are not even under indirect voluntary control, like one's basic capacities or bodily build. Here when we use positively evaluative terms like "gifted" or "superb," we are clearly not saying that the person has done all she could to foster or encourage the condition in question. But even where the condition is at least partly under indirect voluntary control, as with personal appearance or state of health, we need not be thinking in those terms when we take someone to present a pleasing appearance or to be in splendid health. Moreover, we can carry out these evaluations from a certain point of view. We can judge that someone has a fine bodily constitution from an athletic or from an aesthetic point of view, or that someone's manner is a good one from a professional or from a social point of view.

In like fashion one can evaluate S's believing that p as a good, favorable, desirable, or appropriate thing, without thinking of it as fulfilling or not violating an obligation, and without making this evaluation depend on whether the person has done what she could to carry out belief-influencing activities. As in the other cases, it could simply be a matter of the possession of certain good-making characteristics. Furthermore, believings can be evaluated from various points of view, including the epistemic, which, as we have noted, is defined by the aim of maximizing truth and minimizing falsity. It may be a good thing that S believes that p for his peace of mind, or from the standpoint of loyalty to the cause, or as an encouragement to the redoubling of his efforts. But none of this would render it a good thing for S to believe that p from the epistemic point of view. To believe that p because it gives peace of mind or because it stimulates effort may not be conducive to the attainment of truth and the avoidance of error.

All of this suggests that we can frame a concept of epistemic justification that is "evaluative," in a narrow sense of that term in which it contrasts with "deontological," with the assessment of conduct in terms of obligation, blame, right, and wrong. Let's specify an "evaluative" sense of epistemic justification as follows:

> XV. S is J_e in believing that p *iff* S's believing that p, as S does, is a good thing from the epistemic point of view.

This is a way of being commendable from the epistemic point of view that is quite different from the subject's not being to blame for any violation of intellectual obligations.[21] The qualification "as S does" is inserted to make it explicit that in order for S to be J_e in believing that p, it need not be the case that any believing of p by S would be a good thing epistemically, much less any believing of p by anyone. It is rather that there are aspects of *this* believing of p by S that make it a good thing epistemically. There could conceivably be person-proposition pairs such that any belief in that proposition by that person would be a good thing epistemically, but this would be a limiting case and not typical of our epistemic condition.

Is there anything further to be said about this concept? Of course we should avoid building anything very substantive into the constitution of the concept. After all, it is possible for epistemologists to differ radically as to the conditions under which one or another sort of belief is justified. When this happens they are at least sometimes using the same concept of justification; otherwise they wouldn't be disagreeing over what is required for justification, though they could still disagree over which concept of justification is most fundamental or most useful. Both our versions of J_d are quite neutral in this way. Both leave it completely open as to what intellectual obligations we have, and hence as to what obligations must not be violated if one is to be justified. But while maintaining due regard for the importance of neutrality, I believe that we can go beyond XV in fleshing out the concept.

We can get a start on this by considering the following question. If goodness from an epistemic point of view is what we are interested in, why shouldn't we identify justification with truth, at least extensionally? What could be better from that point of view than truth? If the name of the game is the maximization of truth and the minimization of falsity in our beliefs, then plain unvarnished truth is hard to beat.

However, this consideration has not moved epistemologists to identify justification with truth, or even to take truth as a necessary and sufficient condition for justification. The logical independence of truth and justification is a staple of the epistemological literature. But why should this be? It is obvious that a belief might be J_d without being true and vice versa, but what reason is there for taking J_e to be independent of truth?

I think the answer to this has to be in terms of the "internalist" character of justification. When we ask whether S is justified in believing that p, we are, as we have repeatedly been insisting, asking a question from the standpoint of an aim at truth; but we are not asking whether things are in fact as S believes. We are getting at something more "internal" to S's "perspective on the world." This internalist feature of justification made itself felt in our discussion of J_d when we pointed out that to be J_{dv} is to fail to violate any relevant intellectual obligations, *so far as one can tell,* to be J_{dv} in what we call the "cognitive" mode. With respect to J_e the analogous point is that although this is goodness vis-à-vis the aim at truth, it consists not in the beliefs fitting the way the facts actually are, but something more like the belief's being true "so far as the subject can tell from what is available to the subject." In asking whether S is J_e in believing that p, we are asking whether the truth of p is strongly indicated by what S has to go on; whether, given what S had to go on, it is at least quite likely that p is true. We want to know whether S had *adequate* grounds for believing that p, where *adequate* grounds are those sufficiently indicative to the truth of p.

If we are to make the notion of *adequate grounds* central for J_e we must say more about it. A belief has a certain ground, G, when it is "based on" G. What is it for a belief, B, to be *based on* G? That is a difficult question. So far as I know, there is no fully satisfactory general account in the literature, nor am I able to supply one. But we are not wholly at a loss. We do have a variety of paradigm cases; the difficulty concerns just how to generalize from them and just where to draw the line. When one infers p from q and *thereby* comes to accept p, this is a clear case of basing one belief on another. Again, when I come to believe that that is a tree because this visually appears to me to be the case, that is another paradigm; here my belief that that is a tree is based on my visual experience, or, if you prefer, on certain aspects of that experience. The main difficulties arise with respect to cases in which no conscious inference takes place but in which we are still inclined to say that one belief is based on another. Consider, e.g., my forming the belief that you are angry on seeing you look and act in a certain way. I perform no conscious inference from a proposition about your demeanor and behavior to a proposition about your emotional state. Nevertheless it seems plausible to hold that I did learn about your demeanor and behavior through seeing it, and that the beliefs I thereby formed played a crucial role in my coming to believe that you are angry. More specifically it seems that the former beliefs gave rise to the latter belief; that if I hadn't acquired the former I would not have acquired the latter; and, finally, that if I am asked why I suppose that you are angry I would cite the behavior and demeanor as my reason (perhaps only as "the way he looked and acted"). How can we get this kind of case together with the conscious-inference cases into a general account? We might claim that they are all cases of inference, some of them being unconscious. But there are problems as to when we are justified in imputing unconscious inferences. We might take it that what lets in our problem cases is the subject's disposition to cite the one belief(s) as his reason for the other belief, and then make our general condition a disjunction of conscious inference from *q* and a tendency to cite *q* as the reason. But then what about subjects (small children and lower animals) that are too unsophisticated to be able to answer questions as to what their reasons are? Can't their beliefs be based on something when no conscious inference is performed? Moreover this disjunctive criterion will not include cases in which a belief is based on an experience, rather than on other beliefs. A third suggestion concerns causality. In all cases mentioned thus far it is plausible to suppose that the belief that *q* was among the causes of the belief that *p*. This suggests that we might try to cut the Gordian knot by boldly identifying "based on" with "caused by." But this runs into the usual difficulties of simple causal theories. Many items enter into the causation of a belief, e.g., various neuro-physiological happenings, that clearly don't qualify as even part of what the belief is based on. To make a causal account work we would have to beef it up into "caused by *q* in a certain way." And what way is that? Some way that is paradigmatically exemplified by our paradigms? But how are we to state this way in such a fashion that it applies equally to the non-paradigmatic cases?[22]

In the face of these perplexities our only recourse is to keep a firm hold on our paradigms, and work with a less than ideally determinate concept of a relationship that holds in cases that are "sufficiently like" the paradigms. That will be sufficient to do the job over most of the territory.[23]

Let's return to "grounds." What a belief is based on we may term the ground of the belief. A ground, in a more dispositional sense of the term, is the sort of item on which a belief can be based. We have already cited beliefs and experiences as possible grounds, and these would seem to exhaust the possibilities. Indeed, some epistemologists would find this too generous already, maintaining that beliefs can be based only on other beliefs. They would treat perceptual cases by holding that the belief that a tree is over there is based on the *belief that* there visually appears to me to be a tree over there, rather than, as we are suggesting, on the visual appearance itself. I can't accept that, largely because I doubt that all perceptual believers have such beliefs about their visual experience,[24] but I can't pause to argue the point. Suffice it to say that since my opponents' position is, to be as generous as possible, controversial, we do not want to build a position on this issue into the *concept* of epistemic justification. We want to leave open at least the *conceptual* possibility of *direct* or *immediate* justification by experience (and perhaps in other ways also), as well as *indirect* or *mediate* justification by relation to other beliefs (inferentially in the most explicit cases). Finally, to say that a subject *has adequate* grounds for her belief that *p* is to say that she has other justified beliefs, or experiences, on which the belief could be based and which are strongly indicative of the truth of the belief. The reason for the restriction to *justified* beliefs is that a ground shouldn't be termed adequate unless it can confer justification on the belief it grounds. But we noted earlier that if I infer my belief that *p*, by even impeccable logic, from an *unjustified* belief that *q*, the former belief is not thereby justified.[25]

To return to the main thread of the discussion, we are thinking of S's being J_e in believing that *p* as involving S's having adequate grounds for that belief. That is, we are thinking of the possession of those adequate grounds as constituting the goodness of the belief from the epistemic point of view. The next thing to note is that the various "modes" of J_d apply here as well.

Let's begin by noting an objective-subjective distinction. To be sure, in thinking of J_e as *having truth-indicative grounds within one's "perspective on the world,"* we are already thinking of it as more subjective than flat-out truth. But within that perspectival conception we can set the requirements as more objective or more subjective. There is more than one respect in which the possession of adequate grounds could be "subjectivized." First, there is the distinction between the existence of the ground and its adequacy. S is *objectively* J_e

in believing that *p* if S does in fact have grounds that are in fact adequate grounds for that belief. A subjective version would require only that S *believe* one or the other part of this, or both; either (a) that there are (possible) grounds that are in fact adequate and he believes of those grounds that he has them; or (b) that he has grounds that he believes to be adequate; or the combination, (c), that he believes himself to have adequate grounds. Moreover, there are two ways in which the possession-of-grounds belief could go wrong. Confining ourselves to beliefs, one could mistakenly suppose oneself to believe that *p,* or one could mistakenly suppose one's belief that *p* to be justified. Lacking time to go into all these variations, I shall confine this discussion to the subjectivization of adequacy. So our first two modes will be:

XVI. Objective—S does have adequate grounds for believing that *p*.
XVII. Subjective—S has grounds for believing that *p* and he believes them to be adequate.

And here too we have a "justified belief," or "cognitive" variant on the subjective version.

XVIII. Cognitive—S has grounds for believing that *p* and he is justified in believing them to be adequate.

We can dismiss XVII by the same arguments we brought against the subjective version of J_d. The mere fact that I believe, however unjustifiably or irresponsibly, that my grounds for believing that *p* are adequate could scarcely render me justified in believing that *p*. If I believe them to be adequate just because I have an egotistical penchant to overestimate my powers, that could hardly make it rational for me to believe that *p*. But here we will not find the same reason to favor XVIII over XVI. With J_d the cognitive version won out because of what it takes for blameworthiness. But whether one is J_e in believing that *p* has nothing to do with whether he is subject to blame. It depends rather on whether his believing that *p* is a *good thing* from the epistemic point of view. And however justifiably S believes that his grounds are adequate, if they are not then his believing that *p* on those grounds is not a good move in the truth-seeking game. Even if he isn't to blame for making that move it is a bad move nonetheless. Thus J_e is properly construed in the objective mode.

We are also confronted with the question of whether J_e should be construed "motivationally." Since we have already opted for an objective reading, the motivational version will take the following form:

XIX. Motivational—S's belief that *p* is based on adequate grounds.

So our question is whether it is enough for justification that S *have* adequate grounds for his belief, whether used or not, or whether it is also required that the belief be based on those grounds. We cannot settle this question on the grounds that were available for J_{dv}, since with J_e we are not thinking of the subject as being obliged to take relevant considerations into account in *choosing* whether to believe that *p*.

There is something to be said on both sides of this issue. In support of the first, source-irrelevant position (XVI without XIX), it can be pointed out that S's *having a justification* for believing that *p* is independent of whether S does believe that *p*; I can have adequate grounds for believing that *p,* and so *have* a justification, even though I do not in fact believe that *p*. Hence it can hardly be a requirement for having a justification for *p* that my non-existent belief have a certain kind of basis. Likewise my having adequate grounds for believing that *p* is sufficient for this being *a rational thing for me to believe*. But, says the opponent, suppose that S does believe that *p*. If simply having adequate grounds were sufficient for this belief to be justified, then, provided S does have the grounds, her belief that *p* would be justified however frivolous the source. But surely a belief that stems from wishful thinking would not be justified, however strong one's (unutilized) grounds for it.[26]

Now the first thing to say about this controversy is that both antagonists win, at least to the extent that each of them is putting forward a viable concept, and one that is actually used in epistemic assessment. There certainly is the concept of *having* adequate grounds for the belief that *p,* whether or not one does believe that *p,* and there equally certainly is the concept of one's belief being based on adequate grounds. Both concepts represent favorable epistemic statuses. *Ceteris paribus,* one is better off believing something for which one has adequate grounds than believing something for which one doesn't. And the same can be said for the contrast between having a belief that is based on adequate grounds and having one that isn't. Hence I will recognize that these are both concepts of epistemic justification, and I will resist the pressure to decide which is *the* concept.

Nevertheless we can seek to determine which concept is more fundamental to epistemology. On this issue it seems clear that the motivational concept is the richer one and thereby embodies a more complete account of a belief's being a good thing from the epistemic point of view. Surely there is something epistemically undesirable about a belief that is generated in an intellectually disreputable way, however adequate the unutilized grounds possessed by the subject. If, possessing excellent reasons for supposing that you are trying to discredit me professionally, I nevertheless believe this, not for those reasons but out of paranoia, in such a way that even if I didn't have those reasons I would have believed this just as firmly, it was undesirable from the point of view of the aim at truth for me to form that belief as I did. So if we are seeking the most inclusive concept of what makes a belief a good thing epistemically, we will want to include a consideration of what the belief is based on. Hence I will take XIX as the favored formulation of what makes a belief a good thing from the epistemic point of view.

I may add that XVI can be seen as derivative from XIX. To simply *have* adequate grounds is to be in such a position that *if* I make use of that position as a basis for believing that *p* I will thereby be justified in that belief. Thus XVI gives us a concept of a potential for XIX; it is a concept of having resources that are sufficient for believing justifiably, leaving open the question of whether those resources are used.

The next point to be noted is that XIX guarantees only *prima facie* justification. As often noted, it is quite possible for my belief that *p* to have been formed on the basis of evidence that in itself adequately supports *p,* even though the totality of the evidence at my disposal does not. Thus the evidence on which I came to believe that the butler committed the murder might strongly support that hypothesis, but when arriving at that belief I was ignoring other things I know or justifiably believe that tend to exculpate the butler; the total evidence at my disposal is not sufficient support for my belief. In that case we will not want to count my belief as justified all things considered, even though the grounds *on the basis of which* it was formed were themselves adequate. Their adequacy is, so to say, *overridden* by the larger perspectival context in which they are set. Thus XIX gives us *prima facie* justification, what will be justification provided it is not cancelled by further relevant factors. Unqualified justification requires an additional condition to the effect that S does not also have reasons that suffice to override the justification provided by the grounds on which the belief is based. Building that into XIX we get:

XX. Motivational—S's belief that p is based on adequate grounds,
                   and S lacks overriding reasons to the contrary.

Even though XX requires us to bring in the unused portions of the
perspective, we cannot simplify the condition by ignoring the distinction between what provides the basis and what doesn't, and make
the crucial condition something like "The totality of S's perspective
provides adequate support." For then we would run up against the
considerations that led us to prefer XIX to XVI.

We have distinguished two aspects of our evaluative concept of
justification, the strictly evaluative portion—goodness from the epistemic point of view—and the very general statement of the relevant
good-making characteristic, *based on adequate grounds in the absence
of overriding reasons to the contrary*. In taking the concept to include
this second component we are opting for the view that this concept,
though unmistakably evaluative rather than "purely factual" in character, is not so purely evaluative as to leave completely open the basis
on which this evaluative status supervenes. I do not see how to justify
this judgment by reference to any more fundamental considerations. It
is just that in reflecting on epistemic justification, thought of in evaluative (as contrasted with deontological) terms, it seems clear to me that
the range of possible bases for epistemic goodness is not left completely open by the concept, that it is part of what we mean in terming
a belief justified, that the belief was based on adequate grounds (or, at
least, that the subject had adequate grounds for it).[27] Though this
means that J_e is not maximally neutral on the question of what it takes
for justification, it is still quite close to that. It still leaves open whether
there is immediate justification and if so on the basis of what, how
strong a ground is needed for justification, what dimensions of strength
there are for various kinds of grounds, and so on.

Let's codify our evaluative concept of justification as follows:

XXI. S is J_{eg} in believing that p *iff* S's believing that p, as S did, was
         a good thing from the epistemic point of view, in that S's belief
         that p was based on adequate grounds and S lacked sufficient
         overriding reasons to the contrary.

In the subscript "g" stands for "grounds."

My supposition that all justification of belief involves adequate
grounds may be contested. This does seem incontrovertible for beliefs
based on other beliefs and for perceptual beliefs based on experience.

But what about beliefs in self-evident propositions where the self-evidence is what justifies me in the belief.[28] On considering the proposition that two quantities equal to the same quantity are equal to each other, this seems obviously true to me; and I shall suppose, though this is hardly uncontroversial, that in those circumstances I am justified in believing it. But where are the adequate grounds on which my belief is based? It is not that there are grounds here about whose adequacy we might well have doubts; it is rather that there seems to be nothing indentifiable as grounds. There is nothing here that is distinguishable from my belief and the proposition believed, in the way evidence or reasons are distinct from that for which they are evidence or reasons, or in the way my sensory experience is distinct from the beliefs about the physical world that are based on it. Here I simply consider the proposition and straightaway accept it. A similar problem can be raised for normal beliefs about one's own conscious states. What is the ground for a typical belief that one feels sleepy?[29] If one replies "One's being conscious of one's feeling of sleepiness," then it may be insisted, with some show of plausibility, that where one is consciously feeling sleepy there is no difference between one's feeling sleepy and one's being conscious that one is feeling sleepy.

This is a very large issue that I will not have time to consider properly. Suffice it to say that one may treat these as limiting cases in which the ground, though real enough, is minimally distinguishable either from the belief it is grounding or from the fact that makes the belief true. In the first person belief about one's own conscious state, the ground coincides with the fact that makes the belief true. Since the face believed is itself an experience of the subject, there need be nothing "between" the subject and the fact that serves as an indication of the latter's presence. The fact "reveals itself" directly. Self-evident propositions require separate treatment. Here I think that we can take the *way* the proposition appears to one, variously described as "obviously true," "self-evident," and "clear and distinct," as the ground on which the belief is based. I accept the proposition because it *seems* to be so obviously true. This is less distinct from the belief than an inferential or sensory experiential ground, since it has to do with how I am aware of the proposition. Nevertheless there is at least a minimal distinctness. I can form an intelligible conception of someone's failing to believe that *p,* where *p* seems obviously true. Perhaps this person has been rendered unduly sceptical by over-exposure to the logical paradoxes.

VI

Let's go back to the idea that the "based on adequate grounds" part of J_{eg} is there because of the "internalist" character of justification. Contrasts between internalism and externalism have been popular in epistemology lately, but the contrast is not always drawn in the same way. There are two popular ways, both of which are distinct from what I have in mind. First there is the idea that justification is internal in that it depends on what support is available for the belief from "within the subject's perspective," in the sense of what the subject knows or justifiably believes about the world.[30] This kind of internalism restricts justification to mediate or discursive justification, justification by reasons. Another version takes "the subject's perspective" to include whatever is "directly accessible" to the subject, accessible just on the basis of reflection; internalism on this version restricts justifiers to what is directly accessible to the subject.[31] This, unlike the first version, does not limit us to mediate justification, since experience can be taken to be at least as directly accessible as beliefs and knowledge.

In contrast to both these ways of drawing the distinction, what I take to be internal about justification is that whether a belief is justified depends on what it is based on (grounds), and grounds must be other psychological state(s) of the same subject. I am not absolutely certain that grounds are confined to beliefs and experiences, even if experiences are not confined to sensations and feelings but also include, e.g., the way a proposition seems obvious to one, and religious and aesthetic experiences; but these are the prime candidates, and any other examples must belong to some kind of which these are the paradigms. So in taking it to be conceptually true that one is justified in believing that *p iff* one's belief that *p* is based on an adequate ground, I take justification to be "internal" in that it depends on the way in which the belief stems from the believer's psychological states, which are "internal" to the subject in an obvious sense. What would be an externalist contrast with this kind of internalism? We shall see one such contrast in a moment, in discussing the relation of J_{eg} to reliabilism. Moreover, it contrasts with the idea that one can be justified in a certain belief just because of the status of the proposition believed (necessary, infallible). My sort of internalism is different from the first one mentioned above, in that experiences as well as beliefs can figure as grounds. And it is different from the second if, as I believe, what a belief is based on may not be directly accessible. This will be the case if, as seems plausible, much belief-formation goes on below the conscious level. It would seem, e.g., that, as we move about the environment, we are constantly forming short-term perceptual beliefs without any conscious monitoring of this activity.

The most prominent exponents of an explicitly non-deontological conception of epistemic justification have been reliabilists, who have either identified justification with reliability[32] or have taken reliability to be an adequate criterion of justification.[33] The reliability that is in question here is the reliability of belief-formation and -sustenance.[34] To say that a belief was formed in a reliable way is, roughly, to say that it was formed in a way that can be depended on generally to form true rather than false beliefs, at least from inputs like the present one, and at least in the sorts of circumstances in which we normally find ourselves.[35] Thus if my visual system, when functioning as it is at present in yielding my belief that there is a tree in front of me, generally yields true beliefs about objects that are fairly close to me and directly in front of me, then my present belief that there is a tree in front of me was formed in a reliable manner.

Now it may be supposed that J_{eg}, as we have explained it, is just reliability of belief-formation with an evaluative frosting. For where a belief is based on adequate grounds, that belief has been formed in a reliable fashion. In fact, it is plausible to take reliability as a *criterion* for adequacy of grounds. If my grounds for believing that *p* are not such that it is generally true that beliefs like that formed on grounds like that are true, they cannot be termed "adequate." Why do we think that wanting State to win the game is not an adequate reason for supposing that it has won, whereas the fact that a victory has been reported by several newspapers is an adequate reason? Surely it has something to do with the fact that beliefs like that when formed on the first sort of grounds are not *generally* true, while they are *generally* true when formed on grounds of the second sort. Considerations like this may lead us to suppose that J_{eg}, in effect, identifies justification with reliability.[36]

Nevertheless the internalist character of justification prevents it from being identified with reliability, and even blocks an extensional equivalence. Unlike justification, reliability of belief-formation is not limited to cases in which a belief is based on adequate grounds within the subject's psychological states. A reliable mode of belief-formation *may* work through the subject's own knowledge and experience. Indeed, it is plausible to suppose that all of the reliable modes of belief-formation available to human beings are of this sort. But it is quite conceivable that there should be others. I might be so constituted that beliefs about the weather tomorrow which apparently just "pop into my mind" out of nowhere are in fact reliably produced by a mechanism of which we know nothing, and which does not involve the belief being based on anything. Here we would have reliably formed beliefs that are not based on adequate grounds from within my perspective, and so are not J_{eg}.

Moreover, even within the sphere of beliefs based on grounds, reliability and justification do not necessarily go together. The possibility of divergence here stems from another feature of justification embodied in our account, the way in which unqualified justification requires not only an adequate ground but also the absence of sufficient overriding reasons. This opens up the possibility of a case in which a belief is formed on the basis of grounds in a way that is in fact highly reliable, even though the subject has strong reasons for supposing the way to be unreliable. These reasons will (or may) override the *prima facie* justification provided by the grounds on which the belief was based. And so S will not be justified in the belief, even though it was reliably generated.

Consider, in this connection, a case presented by Alvin Goldman.[37]

> Suppose that Jones is told on fully reliable authority that a certain class of his memory beliefs are almost all mistaken. His parents fabricate a wholly false story that Jones suffered from amnesia when he was seven but later developed *pseudo*-memories of that period. Though Jones listens to what his parents say and has excellent reasons to trust them, he persists in believing the ostensible memories from his seven-year-old past.

Suppose that Jones, upon recalling his fifth birthday party, believes that he was given an electric train for his fifth birthday because, as it seems to him, he remembers being given it.[38] By hypothesis, his memory mechanism is highly reliable, and so his belief about his fifth birthday was reliably formed. But this belief is not adequately supported by the *totality* of what he justifiably believes. His justifiable belief that he has no real memory of his first seven years overrides the support from his ostensible memory. Thus Jones is not J_{eg} in his memory belief, because the "lack of overriding reasons to the contrary" requirement is not satisfied. But reliability is subject to no such constraint. Just as reliable mechanisms are not restricted to those that work through the subject's perspective, so it is not a requirement on the reliability of belief-formation that the belief be adequately supported by the totality of the subject's perspective. However many and however strong the reasons Jones has for distrusting his memory, the fact remains that his memory beliefs are still reliably formed. Here is another way in which the class of beliefs that are J_{eg} and the class of reliably formed beliefs can fail to coincide.[39]

I would suggest that, of our candidates, J_{eg} most fully embodies what we are looking for under the heading of "epistemic justification." (a) Like its deontological competitors it is an evaluative concept, in a broad sense, a concept of a favorable status from an epistemic point of view. (b) Unlike J_{dv} it does not presuppose that belief is under direct voluntary control. (c) Unlike J_{di}, it implies that the believer is in a strong epistemic position in believing that *p,* i.e., that there is something about the way in which he believes that *p* that renders it at least likely that the belief is true. Thus it renders it intelligible that justification is something we should prize from an epistemic point of view. (d) Unlike the concept of a reliable mode of belief-formation, it represents this "truth-conductivity" as a matter of the belief's being based on an adequate ground within the subject's own cognitive states. Thus it recognizes the "internalist" character of justification; it recognizes that in asking whether a belief is justified we are interested in the prospects for the truth of the belief, given what the subject "has to go on." (e) Thus the concept provides broad guidelines for the specification of conditions of justification, but within those guidelines there is ample room for disagreement over the precise conditions for one or another type of belief. The concept does not leave us totally at a loss as to what to look for. But in adopting J_{eg} we are not building substantive epistemological questions into the concept. As the only candidate to exhibit all these desiderata, J_{eg} is clearly the winner.

VII

It may be useful to bring together the lessons we have learned from this conceptual exploration.

1. Justifying, an activity of showing or establishing something, is much less central for epistemology than is "being justified," as a state or condition.
2. It is central to epistemic justification that *what justifies* is restricted to the subject's "perspective," to the subject's knowledge, justified belief, or experience.

3. Deontological concepts of justification are either saddled with an indefensible assumption of the voluntariness of belief (J_{dv}) or allow for cases in which one believes that p without having any adequate ground for the belief (J_{di}).

4. The notion of one's belief being based on adequate grounds incorporates more of what we are looking for in a concept of epistemic justification than the weaker notion of having adequate grounds for belief.

5. Justification is closely related to reliability, but because of the perspectival character noted in 2, they do not completely coincide; much less can they be identified.

6. The notion of believing that p in a way that is good from an epistemic point of view in that the belief is based on adequate grounds (J_{eg}) satisfies the chief desiderata for a concept of epistemic justification.

VIII

The ultimate payoff of this conceptual exploration is the increased sophistication it gives us in dealing with substantive epistemological issues. Putting our scheme to work is a very large enterprise, spanning a large part of epistemology. In conclusion I will give one illustration of the ways in which our distinctions can be of help in the trenches. For this purpose I will restrict myself to the broad contrast between J_{dv} and J_{eg}.

First, consider what we might term "higher-level requirements" for S's being justified in believing that p. I include under that heading all requirements that S know or justifiably believe something *about* the epistemic status of p, or about the strength of S's grounds for p. This would include requirements that S be justified in believing that:

1. R is an adequate reason for p (where R is alleged to justify S's belief that p).[40]

2. Experience e is an adequate indication that p (where e is alleged to justify S's belief that p).[41]

On J_{eg} there is no temptation to impose such requirements. If R *is* an adequate reason (e is an adequate indication), then if one believes that p on that basis, one is *thereby* in a strong position, epistemically; and the further knowledge, or justified belief, that the reason is adequate (the experience is an adequate indication), though no doubt quite important and valuable for other purposes, will do nothing to improve the truth-conduciveness of one's believing that p. But on J_{dv} we get a different story. If it's a question of being blameless in believing that p, it can be persuasively argued that this requires not only forming the belief on what is in fact an adequate ground, but doing so in the light of the realization that the ground is an adequate one. If I decide to believe that p without knowing whether the ground is adequate, am I not subject to blame for proceeding irresponsibly in my doxastic behavior, whatever the actual strength of the ground? If the higher-level requirements are plausible only if we are using J_{dv}, then the dubiousness of that concept will extend to those requirements.[42]

In the above paragraph we were considering whether S's being justified in believing that his ground is adequate is a *necessary* condition of justification. We can also consider whether it is sufficient. Provided that S is justified in believing that his belief that p is based on an adequate ground, G, does it make any difference, for his being justified in believing that p, whether the ground *is* adequate? Our two contenders will line up here as they did on the previous issue. For J_{eg} the mere fact that S is justified in supposing that G is adequate will cut no ice. What J_{eg} requires is that S *actually be* in an epistemically favorable position; and although S's being justified in supposing G to be adequate is certainly good evidence for that, it doesn't *constitute* being in such a position. Hence J_{eg} requires that the ground of the belief actually be an adequate one. As for J_{dv}, where it is a question of whether S is blameworthy in believing that p, what is decisive is how S's epistemic position appears within S's perspective on the world. If, so far as S could tell, G is an adequate ground, then S is blameless, i.e., J_{dv}, in believing that p on G. Nothing else could be required for justification in that sense. If S has chosen his doxastic state by applying the appropriate principles in the light of all his relevant knowledge and justified belief, then he is totally in the clear. Again the superior viability of J_{eg}, as over against J_{dv}, should tip the scales in favor of the more objective requirement of adequacy.[43]

Notes

1. Of late a number of theorists have been driving a wedge between what it is to *be* P or what *property* P is, on the one hand, and what belongs to the *concept* of P or what is the meaning of "P" on the other. Thus it has been claimed (Kripke, 1972) that *what heat is* is determined by the physical

investigation into the nature of heat, whether or not the results of that investigation are embodied in our *concept* of heat or in the meaning of "heat." I shall take it that no such distinction is applicable to epistemic justification, that here the only reasonable interpretation to be given to "what it is" is "what is involved in the concept" or "what the term means." If someone disagrees with this, that need not be a problem. Such a person can simply read "what concept of justification is being employed" for "what justification is taken to be."

2. I think especially of Chisholm (1977), chap. 1; Ginet (1975), chap. 3; Goldman (1969, 1980); Wolterstorff (1983).

3. It may be claimed that the activity concept is fundamental in another way, viz., by virtue of the fact that one is justified in believing that *p* only if one is *capable* of carrying out a justification of the belief. But if that were so we would be justified in far fewer beliefs than we suppose. Most human subjects are quite incapable of carrying out a justification of any perceptual or introspective beliefs.

4. I am indebted to Alvin Plantinga for helping me to see that this term is more suitable than the term "normative" that I had been using in earlier versions of this paper. The reader should be cautioned that "deontological" as used here does not carry the contrast with "teological" that is common in ethical theory. According to that distinction, a deontological ethical theory, like that of Kant's, does not regard principles of duty or obligation as owing their status to the fact that acting in the way they prescribe tends to realize certain desirable states of affairs. Whereas a teleological theory, like Utilitarianism, holds that this is what renders a principle of obligation acceptable. The fact that we are not using "deontological" with this force is shown by the fact that we are thinking of epistemic obligations as owing their validity to the fact that fulfilling them would tend to lead to the realization of a desirable state of affairs; viz., a large body of beliefs with a favorable truth-falsity ratio.

5. See Chisholm (1977), chap. 1; Ginet (1975), chap. 3; Wolterstorff (1983). An extended development of a deontological concept of epistemic justification is to be found in Naylor (1978). In my development of deontological concepts in this essay I have profited from the writing of all these people and from discussions with them.

6. (1975), p. 28. See also Ayer (1956), pp. 31–34; Chisholm (1977), p. 14; Naylor (1978), p. 8.

7. These examples are meant to be illustrative only; they do not necessarily carry the endorsement of the management.

8. Here I am indebted to Alvin Plantinga.

9. A weaker objective conception would be this. My obligation is to do what in fact is *likely* to bring out A. On this weaker conception I could be said to have fulfilled my obligation in (some) cases in which A is not forthcoming.

10. We could also subjectivize the aimed-at result, instead of or in addition to subjectivizing what it takes to arrive at that result. In this way one would

have subjectively fulfilled one's obligation if one had done what one believed to be one's obligation. Or, to combine the two moves to the subjective, one would have subjectively fulfilled one's obligation if one had done what one believed would lead to the fulfillment of what one believed to be one's obligation. But sufficient unto the day is the distinction thereof.

11. I would call this "epistemic obligation," except that I want to make these same distinctions with respect to epistemic justification, and so I don't want to repeat the generic term for one of the species.

12. Since we are tacitly restricting this to epistemic justification, we will also be, tacitly, restricting ourselves to intellectual obligations.

13. Since this is all on the assumption that S does believe that p, we need not add that to the right-hand side in order to get a sufficient condition.

14. Note that XI, XII, and some forms of XIII are in terms of higher-level beliefs about one's epistemic status vis à vis p. There are less sophisticated sorts of subjectivization. For example: S is subjectively justified in believing that p *iff* S believes that q, and Q is evidence for p. (For the reason this does not count as having adequate evidence see the next paragraph in the text.) Or even more subjectively: S is subjectively justified in believing that p *iff* S believes that q and basis his belief that p on his belief that q. The definitions presented in the text do not dictate what we should say in the case in which S does not have the higher level belief specified in XI and XII, but satisfies either of the above conditions. A thorough treatment of modes of normative status would have to go into all of this.

15. We have been taking it that to be, e.g., subjectively or cognitively justified in believing that p is to not be violating any subjective or cognitive obligations in believing that p. That means that if we opt for cognitive justification we are committed to giving a correspondingly cognitive formulation of what intellectual obligations one has. But that isn't the only way to do it. We could leave all the obligations in a purely objective form, and vary the function that goes from obligation to justification. That is, we could say that one is subjectively justified if one believes that one has not violated an (objective) obligation (or perhaps believes something that is such that, given one's objective obligations, it implies that none of those obligations have been violated). And a similar move could be made for the other modes.

16. Here are a couple of examples of the attraction of XII for J_d. Chisholm (1977) presents an informal explanation of his basic term of epistemic evaluation, "more reasonable than" in terms of an "intellectual requirement." The explanation runs as follows:

> One way, then, of re-expressing the locution "p is more reasonable than q for S at t" is to say this: S is so situated at t that his intellectual requirement, his responsibility as an intellectual being, is better fulfilled by p than by q [p. 14].

The point that is relevant to our present discussion is that Chisholm states our basic intellectual requirement in what I have called "cognitive" rather than "objective" terms; and with a motivational rider.

> We may assume that every person is subject to a purely intellectual requirement—that of trying his best to bring it about that, for every proposition *h* that he considers, he accepts *h* if and only if *h* is true [p. 14].

The "requirement" is that one *try one's best* to bring this about, rather than that one does bring it about. I take it that to try my best to bring about a result, R, is to do what, so far as I can tell, will bring about R, insofar as that is within my power. (It might be claimed that so long as I do what I believe will bring about R I am trying my best, however irresponsible the belief. But it seems to me that so long as I am not acting on the best of the indications available to me I am not "trying my best.") The motivational rider comes in too, since unless I do what I do *because* I am taking it to (have a good chance to) lead to R, I am not trying at all to bring about R.

Of course, Chisholm is speaking in terms of fulfilling an intellectual obligation rather than, as we have been doing, in terms of not violating intellectual obligations. But we are faced with the same choice between our "modes" in either case.

For a second example I turn to Wolterstorff (1983). Wolterstorff's initial formulation of a necessary and sufficient condition of justification (or, as he says, "rationality") for an "eluctable" belief of S that *P* is: S *lacks adequate reasons for ceasing from believing that p* (p. 164). But then by considerations similar to those we have just adduced, he recognizes that even if S does not in fact have adequate reason for ceasing to believe that *p,* he would still be unjustified in continuing to hold the belief if he were "rationally obliged" to believe that he does have adequate reason to cease to believe that *p.* Moreover Wolterstorff recognizes that S would be justified in believing that *p* if, even though he does have adequate reason to cease from believing that *p,* he is rationally justified in supposing that he doesn't. Both these qualifications amount to recognizing that what is crucial is not what reasons S has in fact, but what reasons S is justified in supposing himself to have. The final formulation, embodying these and other qualifications, runs as follows:

> A person is rational in his eluctable and innocently produced belief *Bp* if and only if S does believe *p* and either:
> (i) S neither has nor ought to have adequate reason to cease from believing *p,* and is not rationally obliged to believe that he *does* have adequate reason to cease; or
> (ii) S does have adequate reason to cease from believing *p* but does not realize that he does, and is rationally justified in that [p. 168].

17. See also Ginet (1975), p. 36.

18. Note that I am not restricting the category of what is within my immediate voluntary control to "basic actions." Neither of the actions just mentioned would qualify for that title. The category includes both basic actions and actions that involve other conditions, where I can satisfy those other conditions, when I choose, just at the moment of choice. Thus my point

about believing is not just that it is not a basic action, but that it is not even a non-basic action that is under my effective immediate control. Whatever is required for my believing that there will never be a nuclear war, it is not something that I can bring about immediately by choosing to do so; though, as I am about to point out, I can affect my believing and abstentions in a more long-range fashion.

19. For other accounts of the indirect voluntary control of beliefs see Naylor (1978), pp. 19–20; Wolterstorff (1983), pp. 153–55.

20. Our four "modes" can also be applied to J_{di}. Indeed, the possibilities for variation are even more numerous. For example, with respect to the *subjective* mode we can switch from the objective fact to the subject's belief with respect to (a) the circumstances of a putative violation, (b) whether there was a violation, and (c) whether the violation was causally related to the belief-formation in question. We will leave all this as an exercise for the reader.

21. I must confess that I do not find "justified" an apt term for a favorable or desirable state or condition, when what makes it desirable is cut loose from considerations of obligation and blame. Nevertheless, since the term is firmly ensconced in the literature as the term to use for any concept that satisfies the four conditions set out in section II, I will stifle my linguistic scruples and employ it for a non-deontological concept.

22. There are also problems as to where to draw the line. What about the unconscious "use" of perceptual cues for the depth of an object in the visual field or for "size constancy"? And however we answer that particular question, just where do we draw the line as we move farther and farther from our initial paradigms?

23. For some recent discussion of "based on" see Swain (1981), chap. 3, and Pappas (1979). One additional point I do need to make explicit is this. I mean "based on" to range over both what initially gave rise to the belief, and what sustains it while it continues to be held. To be precise one should speak of *what the belief is based on at time t*. If t is the time of acquisition, one is speaking of what gave rise to the belief; if t is later than that, one is speaking of what sustains it.

24. For an interesting discussion of this point see Quinton (1973), chap. 7. My opponent will be even more hard pressed to make out that beliefs about one's own conscious experience are based on other beliefs. His best move here would be either to deny that there are such beliefs or to deny that they are based on anything.

25. No such restriction would be required just for having grounds (of some sort). Though even here the word "ground" by itself carries a strong suggestion that what is ground is, to some extent, supported. We need a term for anything a belief might be based on, however vainly. "Ground" carries too much positive evaluative force to be ideally suitable for this role.

26. For some recent discussion of this issue see Harman (1973), chap. 2; Lehrer (1974), chap. 6; Firth (1978); Swain (1981), chap. 3; Foley (1984).

27. Even though we have opted for the "based on" formulation as giving us the more fundamental concept of epistemic justification, we have also recognized the "has adequate grounds" formulation as giving us a concept of epistemic justification. Either of these will introduce a "basis of evaluative status" component into the concept.

28. This latter qualification is needed, because I might accept a self-evident proposition on authority. In that case I was not, so to say, taking advantage of its self-evidence.

29. We are not speaking here of a belief that one *is* sleepy. There a ground is readily identifiable—one's feeling of sleepiness.

30. See Bonjour (1980); Kornblith (1984); Bach (1985).

31. See Goldman (1980); Chisholm (1977), chap. 4, pp. 63–64; Ginet (1975), pp. 34–37.

32. Swain (1981), chap. 4.

33. Goldman (1979).

34. For simplicity I shall couch the ensuing formulations solely in terms of belief-formation, but the qualification "or sustenance" is to be understood throughout.

35. These two qualifications testify to the difficulty of getting the concept of reliability in satisfactory shape; and there are other problems to be dealt with, e.g., how to identify the general procedure of which the present belief-formation is an instance.

36. An alternative to explicating "adequate" in terms of reliability would be to use the notion of conditional probability. G is an adequate ground for a belief that p just in case the probability of p on G is high. And since adequacy is closely related both to reliability and to conditional probability, they are presumably closely related to each other. Swain (1981), chap. 4, exploits this connection to explicate reliability in terms of conditional probability, though in a more complex fashion than is indicated by these brief remarks.

37. (1979), p. 18. (See Chapter 7 below.)

38. If you have trouble envisaging his trusting his memory in the face of his parents' story, you may imagine that he is not thinking of that story at the moment he forms the memory belief.

39. In the article in which he introduces this example, Goldman modifies the "reliability is a criterion of justification" view so that it will accommodate the example. The modified formulation runs as follows:

> If S's belief in p at t results from a reliable cognitive process, and there is no reliable or conditionally reliable process available to S which had it been used by S in addition to the process actually used, would have resulted in S's not believing p at t, then S's belief in p at t is justified [p. 20].

On this revised formulation, being formed by a reliable process is sufficient for justification only if there is no other reliable process that the subject could have used and such that if he had used it he would not have come to believe that *p*. In the case cited there is such a reliable process, viz., taking account of the strong reasons for believing one's memory of pre-seven-years-old events to be unreliable. The revised reliability criterion yields the correct result in this case. However, this move leaves unshaken the point that in this case Jones's belief *is* reliably formed but unjustified. That remains true, whatever is to be said about the revised criterion.

40. See, e.g., Armstrong (1973), p. 151; Skyrms (1967), p. 374.

41. See, e.g., Sellars (1963), pp. 168–69; BonJour (1978), pp. 5–6; Lehrer (1974), pp. 103–5.

42. In my paper "What's Wrong with Immediate Knowledge?" *Synthese* 55, no. 2 (May 1983), pp. 73–95, I develop at much greater length this kind of diagnosis of BonJour's deployment of a higher-level requirement in his argument against immediate knowledge (BonJour, 1978).

43. Ancestors of this essay were presented at SUNY at Albany, SUNY at Buffalo, Calvin College, Cornell University, University of California at Irvin, Lehigh University, University of Michigan, University of Nebraska, Syracuse University, and the University of Western Ontario. I wish to thank members of the audience in all these institutions for their helpful comments. I would like to express special appreciation to Robert Audi, Carl Ginet, George Mavrodes, Alvin Plantinga, Fred Schmitt, and Nicholas Wolterstorff for their penetrating comments on earlier versions.

References

Armstrong, D. M. 1973. *Belief, Truth, and Knowledge*. Cambridge: Cambridge University Press.

Ayer, A. J. 1956. *The Problem of Knowledge*. London: Macmillian.

Bach, K. 1985. "A Rationale for Reliabilism." *The Monist* 68.

BonJour, L. 1978. "Can Empirical Knowledge Have a Foundation?" *American Philosophical Quarterly* 15: 1–13. (Chapter 3 below.)

Chisholm, R. M. 1977. *Theory of Knowledge,* 2nd ed. Englewood Cliffs, N.J.: Prentice-Hall.

Dretske, F. 1981. *Knowledge and the Flow of Information*. Cambridge: MIT Press.

Firth, R. 1978. "Are Epistemic Concepts Reducible to Ethical Concepts?" In A. I. Goldman and J. Kim, eds., *Values and Morals*. Dordrecht: D. Reidel.

Foley, R. 1984. "Epistemic Luck and the Purely Epistemic." *American Philosophical Quarterly*.

Ginet, C. 1975. *Knowledge, Perception, and Memory*. Dordrecht.: D. Reidel.

Goldman, A. I. 1979. "What Is Justified Belief?" In G. S. Pappas, ed., *Justification and Knowledge*. Dordrecht: D. Reidel.

————. 1980. "The Internalist Conception of Justification." *Midwest Studies in Philosophy*, vol. 5. Pp. 27–51.

Harman, G. 1973. *Thought*. Princeton: Princeton University Press.

Kornblith, H. 1985. "Ever Since Descartes." *The Monist* 68, no. 2.

Kripke, S. A. 1972. "Naming and Necessity." In D. Davidson and G. Harman, eds., *Semantics of Natural Language*. Dordrecht: D. Reidel.

Lehrer, K. 1974. *Knowledge*. New York: Oxford University Press.

Naylor, M. B. 1978. "Epistemic Justification." Unpublished.

Pappas, G. S. 1979. "Basing Relations." In G. S. Pappas, ed., *Justification and Knowledge*. Dordrecht: D. Reidel.

Pollock, J. 1975. *Knowledge and Justification*. Princeton: Princeton University Press.

Quinton, A. 1973. *The Nature of Things*. London: Routledge & Kegan Paul.

Sellars, W. 1963. *Science, Perception, and Reality*. London: Routledge & Kegan Paul.

Skyrms, B. 1967. "The Explication of 'X Knows that P.' " *Journal of Philosophy* 64: 373–89.

Swain, M. 1981. *Reasons and Knowledge*. Ithaca: Cornell University Press.

Wolterstorff, N. 1983. "Can Belief in God Be Rational if It Has No Foundations?" In A. Plantinga and N. Wolterstorff, eds. *Faith and Rationality*. Notre Dame: University of Notre Dame Press.

2

The Myth of the Given

Roderick Chisholm

1. The doctrine of "the given" involved two theses about our knowledge. We may introduce them by means of a traditional metaphor:

(A) The knowledge which a person has at any time is a structure or edifice, many parts and stages of which help to support each other, but which as a whole is supported by its own foundation.

The second thesis is a specification of the first:

(B) The foundation of one's knowledge consists (at least in part) of the apprehension of what have been called, variously, "sensations," "sense-impressions," "appearances," "sensa," "sense-qualia," and "phenomena."

These phenomenal entities, said to be at the base of the structure of knowledge, are what was called "the given." A third thesis is sometimes associated with the doctrine of the given, but the first two theses do not imply it. We may formulate it in the terms of the same metaphor:

(C) The *only* apprehension which is thus basic to the structure of knowledge is our apprehension of "appearances" (etc.)—our apprehension of the given.

Theses (A) and (B) constitute the "doctrine of the given"; thesis (C), if a label were necessary, might be called "the phenomenalistic version" of the doctrine. The first two theses are essential to the empirical

Reprinted from *Philosophy* (Englewood Cliffs: Prentice-Hall, 1964), pp. 261–86, by permission of the author and the copyright holder. Copyright 1964 by the Trustees of Princeton University.

tradition in Western philosophy. The third is problematic for traditional empiricism and depends in part, but only in part, upon the way in which the metaphor of the edifice and its foundation is spelled out.

I believe it is accurate to say that, at the time at which our study begins, most American epistemologists accepted the first two theses and thus accepted the doctrine of the given. The expression "the given" became a term of contemporary philosophical vocabulary partly because of its use by C. I. Lewis in his *Mind and the World-Order* (1929). Many of the philosophers who accepted the doctrine avoided the expression because of its association with other more controversial parts of Lewis's book—a book which might be taken (though mistakenly, I think) also to endorse thesis (C), the "phenomenalistic version" of the doctrine. The doctrine itself—theses (A) and (B)—became a matter of general controversy during the period of our survey (1930–1960).

Thesis (A) was criticized as being "absolute" and thesis (B) as being overly "subjective." Both criticisms may be found in some of the "instrumentalistic" writings of John Dewey and philosophers associated with him. They may also be found in the writings of those philosophers of science ("logical empiricists") writing in the tradition of the Vienna Circle. (At an early stage of this tradition, however, some of these same philosophers seem to have accepted all three theses.) Discussion became entangled in verbal confusions—especially in connection with the uses of such terms as "doubt," "certainty," "appearance," and "immediate experience." Philosophers, influenced by the work that Ludwig Wittgenstein had been doing in the 1930s, noted such confusions in detail, and some of them seem to have taken the existence of such confusions to indicate that (A) and (B) are false.[1] Many have rejected both theses as being inconsistent with a certain theory of thought and reference; among them, in addition to some of the critics just referred to, we find philosophers in the tradition of nineteenth-century "idealism."

Philosophers of widely diverging schools now believe that "the myth of the given" has finally been dispelled.[2] I suggest, however, that, although thesis (C), "the phenomenalistic version," is false, the two theses (A) and (B), which constitute the doctrine of the given, are true.

The doctrine is not merely the consequence of a metaphor. We are led to it when we attempt to answer certain questions about *justification*—our justification for supposing, in connection with any one of the things that we know to be true, that it is something that we know to be true.

2. To the question "What justification do I have for thinking I know that *a* is true?" one may reply: "I know that *b* is true, and if I know that *b* is true then I also know that *a* is true. And to the question "What justification do I have for thinking I know that *b* is true?" one may reply: "I know that *c* is true, and if I know that *c* is true then I also know that *b* is true." Are we thus led, sooner or later, to something *n* of which one may say: "What justifies me in thinking I know that *n* is true is simply the fact that *n* is true"? If there is such an *n*, then the belief or statement that *n* is true may be thought of either as a belief or statement which "justifies itself" or as a belief or statement which is itself "neither justified nor unjustified." The distinction—unlike that between a Prime Mover which moves itself and a Prime Mover which is neither in motion nor at rest—is largely a verbal one; the essential thing, if there is such an *n*, is that it provides a stopping place in the process, or dialectic, of justification.

We may now re-express, somewhat less metaphorically, the two theses which I have called the "doctrine of the given." The first thesis, that our knowledge is an edifice or structure having its own foundation, becomes (A) "every statement, which we are justified in thinking that we know, is justified in part by some statement which justifies itself." The second thesis, that there are appearances ("the given") at the foundation of our knowledge, becomes (B) "there are statements about appearances which thus justify themselves." (The third thesis—the "phenomenalistic version" of the doctrine of the given—becomes (C) "there are no self-justifying statements which are not statements about appearances.")

Let us now turn to the first of the two theses constituting the doctrine of the given.

3. "Every justified statement is justified in part by some statement which justifies itself." Could it be that the question which this thesis is supposed to answer is a question which arises only because of some mistaken assumption? If not, what are the alternative ways of answering it? And did any of the philosophers with whom we are concerned actually accept any of these alternatives? The first two questions are less difficult to answer than the third.

There are the following points of view to be considered, each of which *seems* to have been taken by some of the philosophers in the period of our survey.

   (1) One may believe that the questions about justification which give rise to our problem are based upon false assumptions and hence that they *should not be asked* at all.

(2) One may believe that no statement or claim is justified unless it is justified, at least in part, by some other justified statement or claim which it does not justify; this belief may suggest that one should continue the process of justifying *ad indefinitium,* justifying each claim by reference to some additional claim.

(3) One may believe that no statement or claim *a* is justified unless it is justified by some other justified statement or claim *b,* and that *b* is not justified unless it in turn is justified by *a;* this would suggest that the process of justifying is, or should be, *circular.*

(4) One may believe that there are some particular claims *n* at which the process of justifying should stop, and one may then hold of any such claim *n* either: (a) *n* is justified by something—viz., *experience* or *observation*—which is not itself a claim and which therefore cannot be said itself either to be justified or unjustified; (b) *n* is itself *unjustified;* (c) *n justifies itself;* or (d) *n* is *neither justified nor unjustified.*

These possibilities, I think, exhaust the significant points of view; let us now consider them in turn.

4. "The questions about justification which give rise to the problem are based upon false assumptions and therefore should not be asked at all."

The questions are *not* based upon false assumptions, but most of the philosophers who discussed the questions put them in such a misleading way that one is very easily misled into supposing that they *are* based upon false assumptions.

Many philosophers, following Descartes, Russell, and Husserl, formulated the questions about justification by means of such terms as "doubt," "certainty," and "incorrigibility," and they used, or misused, these terms in such a way that, when their questions were taken in the way in which one would ordinarily take them, they could be shown to be based upon false assumptions. One may note, for example, that the statement "There is a clock on the mantelpiece" is not self-justifying—for to the question "What is your justification for thinking you know that there is a clock on the mantelpiece?" the proper reply would be to make some other statement (e.g., "I saw it there this morning and no one would have taken it away")—and one may then go on to ask "But are there any statements which can be said to justify themselves?" If we express these facts, as many philosophers did, by saying that the statement "There is a clock on the mantelpiece" is one which is not "certain," or one which may be "doubted," and if we then go on to ask "Does this doubtful statement rest upon other statements which are certain and incorrigible?" then we are using terms in an extraordinarily misleading way. The question "Does this doubtful statement rest upon statements which are certain and incorrigible?"—if taken as one would ordinarily take it—does rest upon a false assumption, for (we may assume) the statement that there is a clock on the mantelpiece is one which is not doubtful at all.

John Dewey, and some of the philosophers whose views were very similar to his, tended to suppose, mistakenly, that the philosophers who asked themselves "What justification do I have for thinking I know this?" were asking the quite different question "What more can I do to verify or confirm that this is so?" and they rejected answers to the first question on the ground that they were unsatisfactory answers to the second.[3] Philosophers influenced by Wittgenstein tended to suppose, also mistakenly, but quite understandably, that the question "What justification do I have for thinking I know this?" contains an implicit challenge and presupposes that one does not have the knowledge concerned. They then pointed out, correctly, that in most of the cases where the question was raised (e.g., "What justifies me in thinking I know that this is a table?") there is no ground for challenging the claim to knowledge and that questions presupposing that the claim is false should not arise. But the question "What justifies me in thinking I know that this is a table?" does not challenge the claim to know that this is a table, much less presuppose that the claim is false.

The "critique of cogency," as Lewis described this concern of epistemology, presupposes that we *are* justified in thinking we know most of the things that we do think we know, and what it seeks to elicit is the nature of this justification. The enterprise is like that of ethics, logic, and aesthetics:

> The nature of the good can be learned from experience only if the content of experience be first classified into good and bad, or grades of better and worse. Such classification or grading already involves the legislative application of the same principle which is sought. In logic, principles can be elicited by generalization from examples only if cases of valid reasoning have first been segregated by some criterion. In esthetics, the laws of the beautiful may be derived from experience only if the criteria of beauty have first been correctly applied.[4]

When Aristotle considered an invalid mood of the syllogism and asked himself "What is wrong with this?" he was not suggesting to himself that perhaps nothing was wrong; he presupposed that the mood *was* invalid, just as he presupposed that others were not, and he attempted, successfully, to formulate criteria which would enable us to distinguish the two types of mood.

When we have answered the question "What justification do I have for thinking I know this?" what we learn, as Socrates taught, is something about ourselves. We learn, of course, what the justification happens to be for the particular claim with which the question is concerned. But we also learn, more generally, what the criteria are, if any, in terms of which we believe ourselves justified in counting one thing as an instance of knowing and another thing not. The truth which the philosopher seeks, when he asks about justification, is "already implicit in the mind which seeks it, and needs only to be elicited and brought to clear expression."[5]

Let us turn, then, to the other approaches to the problem of "the given."

5. "No statement or claim would be justified unless it were justified, at least in part, by some other justified claim or statement which it does not justify."

This regress principle might be suggested by the figure of the building and its support: no stage supports another unless it is itself supported by some other stage beneath it—a truth which holds not only of the upper portions of the building but also of what we call its foundation. And the principle follows if, as some of the philosophers in the tradition of logical empiricism seemed to believe, we should combine a frequency theory of probability with a probability theory of justification.

In *Experience and Prediction* (1938) and in other writings, Hans Reichenbach defended a "probability theory of knowledge" which seemed to involve the following contentions:

(1) To justify accepting a statement, it is necessary to show that the statement is probable.

(2) To say of a statement that it is probable is to say something about statistical frequencies. Somewhat more accurately, a statement of the form "It is *probable* that any particular *a* is a *b*" may be explicated as saying "Most *a's* are *b's*." Or, still more accurately, to say "The probability is *n* that a particular *a* is a *b*" is to say "The limit of the relative frequency with which

the property of being a *b* occurs in the class of things having the property *a* is *n*."

(3) Hence, by (2), to show that a proposition is probable it is necessary to show that a certain statistical frequency obtains; and, by (1), to show that a certain statistical frequency obtains it is necessary to show that it is probable that the statistical frequency obtains; and therefore, by (2), to show that it is probable that a certain statistical frequency obtains, it is necessary to show that a certain frequency of frequencies obtains . . .

(4) And therefore "there is no Archimedean point of absolute certainty left to which to attach our knowledge of the world; all we have is an elastic net of probability connections floating in open space (Reichenbach, p. 192).

This reasoning suggests that an infinite number of steps must be taken in order to justify acceptance of any statement. For, according to the reasoning, we cannot determine the probability of one statement until we have determined that of a second, and we cannot determine that of the second until we have determined that of a third, and so on. Reichenbach does not leave the matter here, however. He suggests that there is a way of "descending" from this "open space" of probability connections, but if I am not mistaken, we can make the descent only by letting go of the concept of justification.

He says that, if we are to avoid the regress of probabilities of probabilities of probabilities . . . , we must be willing at some point merely to make a guess; "there will always be some blind posits on which the whole concatenation is based" (p. 367). The view that knowledge is to be identified with certainty and that probable knowledge must be "imbedded in a framework of certainty" is "a remnant of rationalism. An empiricist theory of probability can be constructed only if we are willing to regard knowledge as a system of posits."[6]

But if we begin by assuming, as we do, that there is a distinction between knowledge, on the one hand, and a lucky guess, on the other, then we must reject at least one of the premises of an argument purporting to demonstrate that knowledge is a system of "blind posits." The unacceptable conclusion of Reichenbach's argument may be so construed as to follow from premises (1) and (2); and premise (2) may be accepted as a kind of definition (though there are many who believe that this definition is not adequate to all of the uses of the term "probable" in science and everyday life). Premise (1), therefore, is the one we should reject, and there are good reasons, I think, for rejecting

(1), the thesis that "to justify accepting a proposition it is necessary to show that the proposition is probable." In fairness to Reichenbach, it should be added that he never explicitly affirms premise (1); but some such premise is essential to his argument.

6. "No statement or claim *a* would be justified unless it were justified by some other justified statement or claim *b* which would not be justified unless it wee justified in turn by *a*."

The "coherence theory of truth," to which some philosophers committed themselves, is sometimes taken to imply that justification may thus be circular; I believe, however, that the theory does not have this implication. It does define "truth" as a kind of systematic consistency of beliefs or propositions. The truth of a proposition is said to consist not in the fact that the proposition "corresponds" with something which is not itself a proposition, but in the fact that it fits consistently into a certain more general system of propositions. This view may even be suggested by the figure of the building and its foundations. There is no difference in principle between the way in which the upper stories are supported by the lower, and that in which the cellar is supported by the earth just below it, or the way in which that stratum of earth is supported by various substrata farther below; a good building appears to be a part of the terrain on which it stands, and a good system of propositions is a part of the wider system which gives it its truth. But these metaphors do not solve philosophical problems.

The coherence theory did in fact appeal to something other than logical consistency; its proponents conceded that a system of false propositions may be internally consistent and hence that logical consistency alone is no guarantee of truth. Brand Blanshard, who defended the coherence theory in *The Nature of Thought,* said that a proposition is true provided it is a member of an internally consistent system of propositions and *provided further* this system is "the system in which everything real and possible is coherently included."[7] In one phase of the development of "logical empiricism" its proponents seem to have held a similar view: a proposition—or, in this case, a statement—is true provided it is a member of an internally consistent system of statements and *provided further* this system is "the system which is actually adopted by mankind, and especially by the scientists in our culture circle."[8]

A theory of truth is not, as such, a theory of justification. To say that a proposition is true is not to say that we are justified in accepting it as true, and to say that we are justified in accepting it as true is not to say that it is true. Whatever merits the coherence theory may have as an answer to certain questions about truth, it throws no light upon our present epistemological question. If we accept the coherence theory, we may still ask, concerning any proposition *a* which we think we know to be true, "What is my justification for thinking I know that *a* is a member of the system of propositions in which everything real and possible is coherently included, or that *a* is a member of the system of propositions which is actually adopted by mankind and by the scientists of our culture circle?" And when we ask such a question, we are confronted, once again, with our original alternatives.

7. If our questions about justification do have a proper stopping place, then, as I have said, there are still four significant possibilities to consider. We may stop with some particular claim and say of it that either:

    (a) it is justified by something—by experience, or by observation—which is not itself a claim and which, therefore, cannot be said either to be justified or to be unjustified;

    (b) it is justified by some claim which refers to our experience or observation, and the claim referring to our experience or observation has *no* justification;

    (c) it justifies itself; or

    (d) it is itself neither justified nor unjustified.

The first of these alternatives leads readily to the second, and the second to the third or to the fourth. The third and the fourth—which differ only verbally, I think—involve the doctrine of "the given."

Carnap wrote, in 1936, that the procedure of scientific testing involves two operations: the "confrontation of a statement with observation" and the "confrontation of a statement with previously accepted statements." He suggested that those logical empiricists who were attracted to the coherence theory of truth tended to lose sight of the first of these operations—the confrontation of a statement with observation. He proposed a way of formulating simple "acceptance rules" for such confrontation, and he seemed to believe that, merely by applying such rules, we could avoid the epistemological questions with which the adherents of "the given" had become involved.

Carnap said this about his acceptance rules: "If no foreign language or introduction of new terms is involved, the rules are trivial. For example: 'If one is hungry, the statement "I am hungry" may be accepted'; or: 'If one sees a key one may accept the statement "there lies a key." ' "[9] As we shall note later, the first of these rules differs in an important way from the second. Confining ourselves for the moment to rules of the second sort—"If one sees a key one may accept the statement 'there lies a key' "—let us ask ourselves whether the appeal to such rules enables us to solve our problem of the stopping place.

When we have made the statement "There lies a key," we can, of course, raise the question "What is my justification for thinking I know, or for believing, that there lies a key?" The answer would be "I see the key." We cannot ask "What is my justification for seeing a key?" But we *can* ask "What is my justification for thinking that it is a *key* that I see?" and, if we *do* see that the thing is a key, the question will have an answer. The answer might be "I see that it's shaped like a key and that it's in the lock, and I remember that a key is usually here." The possibility of this question, and its answer, indicates that we cannot stop our questions about justification merely by appealing to observation or experience. For, of the statement "I observe that this is an A," we can ask, and answer, the question "What is my justification for thinking that I observe that there is an A?"

It is relevant to note, moreover, that there may be conditions under which seeing a key does *not* justify one in accepting the statement "There is a key" or in believing that one sees a key. If the key were so disguised or concealed that the man who saw it did not recognize it to be a key, then he might not be justified in accepting the statement "There is a key." Just as, if Mr. Jones unknown to anyone but himself is a thief, then the people who see him may be said to see a thief—but none of those who thus sees a thief is justified in accepting the statement "There is a thief."[10]

Some of the writings of logical empiricists suggest that, although some statements may be justified by reference to other statements, those statements which involve "confrontation with observation" are not justified at all. C. G. Hempel, for example, wrote that "the acknowledgement of an experiential statement as true is psychologically motivated by certain experiences; but within the system of statements which express scientific knowledge or one's beliefs at a given time, they function in the manner of postulates for which no grounds are offered."[11] Hempel conceded, however, that this use of the term "postulate" is misleading, and he added the following note of clarification: "When an experiential sentence is accepted 'on the basis of direct experiential evidence,' it is indeed not asserted arbitrarily; but to describe the evidence in question would simply mean to repeat the experiential statement itself. Hence, in the context of cognitive justification, the statement functions in the manner of a primitive sentence."[12]

When we reach a statement having the property just referred to—an experiential statement such that to describe its evidence "would simply mean to repeat the experiential statement itself"—we have reached a proper stopping place in the process of justification.

8. We are thus led to the concept of a belief, statement, claim, proposition, or hypothesis, which justifies itself. To be clear about the concept, let us note the way in which we would justify the statement that we have a certain belief. It is essential, of course, that we distinguish justifying the statement *that* we have a certain belief from justifying the belief itself.

Suppose, then, a man is led to say "I believe that Socrates is mortal" and we ask him "What is your justification for thinking that you believe, or for thinking that you know that you believe, that Socrates is mortal?" To this strange question, the only appropriate reply would be "My justification for thinking I believe, or for thinking that I know that I believe, that Socrates is mortal is simply the fact that I *do* believe that Socrates is mortal." One justifies the statement simply by reiterating it; the statement's justification is what the statement says. Here, then, we have a case which satisfies Hempel's remark quoted above; we describe the evidence for a statement merely by repeating the statement. We could say, as C. J. Ducasse did, that "the occurrence of belief is its own evidence."[13]

Normally, as I have suggested, one cannot justify a statement merely by reiterating it. To the question "What justification do you have for thinking you know that there can be no life on the moon?" it would be inappropriate, and impertinent, to reply by saying simply "There *can* be life on the moon," thus reiterating the fact at issue. An appropriate answer would be one referring to certain *other* facts—for example, the fact that we know there is insufficient oxygen on the moon to support any kind of life. But to the question "What is your justification for thinking you know that you believe so and so?" there is nothing to say other than "I *do* believe so and so."

We may say, then, that there are some statements which are self-justifying, or which justify themselves. And we may say, analogously, that there are certain beliefs, claims, propositions, or hypotheses which are self-justifying, or which justify themselves. A statement, belief, claim, proposition, or hypothesis may be said to be self-justifying for a person, if the person's justification for thinking he knows it to be true is simply the fact that it *is* true.

Paradoxically, these things I have described by saying that they "justify themselves" may *also* be described by saying they are "neither justified nor unjustified." The two modes of description are two different ways of saying the same thing.

If we are sensitive to ordinary usage, we may note that the expression "I believe that I believe" is ordinarily used not to refer to a second-order belief about the speaker's own beliefs, but to indicate that the speaker has not yet made up his mind. "*I believe that I believe* that Johnson is a good president" might properly be taken to indicate that, if the speaker *does* believe that Johnson is a good president, he is not yet firm in that belief. Hence there is a temptation to infer that, if we say of a man who is firm in his belief that Socrates is mortal, that he is "justified in believing that he believes that Socrates is mortal," our statement "makes no sense." And there is also a temptation to go on and say that it "makes no sense" even to say of such a man, that his *statement* "I believe that Socrates is mortal" is one which is "justified" for him.[14] After all, what would it mean to say of a man's statement about his own belief, that he is *not* justified in accepting it?[15]

The questions about what does or does not "make any sense" need not, however, be argued. We *may* say, if we prefer, that the statements about the beliefs in question are "neither justified nor unjustified." Whatever mode of description we use, the essential points are two. First, we may appeal to such statements in the process of justifying some *other* statement or belief. If they *have* no justification they may yet *be* a justification—for something other than themselves. ("What justifies me in thinking that he and I are not likely to agree? The fact that I believe that Socrates is mortal and he does not.") Second, the making of such a statement does provide what I have been calling a "stopping place" in the dialectic of justification; but now, instead of signalizing the stopping place by reiterating the questioned statement, we do it by saying that the question of its justification is one which "should not arise."

It does not matter, then, whether we speak of certain statements which "justify themselves" or of certain statements which are "neither justified nor unjustified," for in either case we will be referring to the same set of statements. I shall continue to use the former phrase.

There are, then, statements about one's own beliefs ("I believe that

Socrates is mortal")—and statements about many other psychological attitudes—which are self-justifying "What justifies me in believing, or in hope to come tomorrow." Thinking, desiring, wondering, loving, hating, and other such attitudes are similar. Some, but by no means all, of the statements we can make about such attitudes, when the attitudes are our own, are self-justifying—as are statements containing such phrases as "I think I remember" or "I seem to remember" (as distinguished from "I remember"), and "I think that I see" and "I think that I perceive" (as distinguished from "I see" and "I perceive"). Thus, of the two examples which Carnap introduced in connection with his "acceptance rules" discussed above, viz., "I am hungry" and "I see a key," we may say that the first is self-justifying and the second not.

The "doctrine of the given," it will be recalled, tells us (A) that every justified statement, about what we think we know, is justified in part by some statement which justifies itself and (B) that there are statements about appearances which thus justify themselves. The "phenomenalistic version" of the theory adds (C) that statements about appearances are the *only* statements which justify themselves. What we have been saying is that the first thesis, (A), of the doctrine of the given is true and that the "phenomenalistic version," (C), is false; let us turn now to thesis (B).

9. In addition to the self-justifying statements about psychological attitudes, are there self-justifying statements about "appearances"? Now we encounter difficulties involving the word "appearance" and its cognates.

Sometimes such words as "appears," "looks," and "seems" are used to convey what one might also convey by such terms as "believe." For example, if I say "It appears to me that General de Gaulle was successful," or "General de Gaulle seems to have been successful," I am likely to mean only that I believe, or incline to believe, that he has been successful; the words "appears" and "seems" serve as useful hedges, giving me an out, should I find out later that de Gaulle was not successful. When "appear"-words are used in this way, the statements in which they occur add nothing significant to the class of "self-justifying" statements we have just provided. Philosophers have traditionally assumed, however, that such terms as "appear" may also be used in a quite different way. If this assumption is correct, as I believe it is, then this additional use does lead us to another type of self-justifying statement.

Later we shall have occasion to note some of the confusion to which the substantival expression "appearance" gave rise. The philosophers who exposed these confusions were sometimes inclined to forget, I think, that things do appear to us in various ways.[16] We can alter the appearance of anything we like merely by doing something which will affect our sense organs or the conditions of observation. One of the important epistemological questions about appearances is "Are there self-justifying statements about the ways in which things appear?"

Augustine, refuting the skeptics of the late Platonic Academy, wrote: "I do not see how the Academician can refute him who says: 'I know that this appears white to me, I know that my hearing is delighted with this, I know this has an agreeable odor, I know this tastes sweet to me, I know that this feels cold to me.' . . . When a person tastes something, he can honestly swear that he knows it is sweet to his palate or the contrary, and that no trickery of the Greeks can dispossess him of that knowledge."[17] Suppose, now, one were to ask "What justification do you have for believing, or thinking you know, that this appears white to you, or that that tastes bitter to you?" Here, too, we can only reiterate the statement: "What justifies me in believing, or in thinking I know, that this appears white to me and that that tastes bitter to me is the fact that this *does* appear white to me and that *does* taste bitter."

An advantage of the misleading substantive "appearance," as distinguished from the verb "appears," is that the former may be applied to those sensuous experiences which, though capable of being appearances of things, are actually not appearances of anything. Feelings, imagery, and the sensuous content of dreams and hallucination are very much like the appearances of things and they are such that, under some circumstances, they could be appearances of things. But if we do not wish to say that they are experiences wherein some external physical thing *appears* to us, we must use some expression other than "appear." For "appear," in its active voice, requires a grammatical subject and thus requires a term which refers, not merely to a way of appearing, but also to something *which* appears.

But we may avoid *both* the objective "*Something* appears blue to me," and the substantival "I sense a blue *appearance*." We may use another verb, say "sense," in a technical way, as many philosophers did, and equate it in meaning with the passive voice of "appear," thus saying simply "I *sense* blue," or the like. Or better still, it seems to me, and at the expense only of a little awkwardness, we can use "appear" in its passive voice and say "I am *appeared to* blue."

Summing up, in our new vocabulary, we may say that the philosophers who talked of the "empirically given" were referring, not to "self-justifying" statements and beliefs generally, but only to those pertaining to certain "ways of being appeared to." And the philosophers who objected to the doctrine of the given, or some of them, argued that no statement about "a way of being appeared to" can be "self-justifying."

10. Why would one suppose that "This appears white" (or, more exactly, "I am now appeared white to") is not self-justifying? The most convincing argument was this: If I say "This appears white," then, as Reichenbach put it, I am making a "comparison between a present object and a formerly seen object."[18] What I am saying *could* have been expressed by "The present way of appearing is the way in which white objects, or objects which I believe to be white, ordinarily appear." And this new statement, clearly, is not self-justifying; to justify it, as Reichenbach intimated, I must go and say something further—something about the way in which I remember white objects to have appeared.

"Appears white" *may* thus be used to abbreviate "appears the way in which white things normally appear." Or "white thing," on the other hand, *may* be used to abbreviate "thing having the color of things which ordinarily appear white." The phrase "appear white" as it is used in the second quoted expression cannot be spelled out in the manner of the first; for the point of the second can hardly be put by saying that "white thing" may be used to abbreviate "thing having the color of things which ordinarily appear the way in which *white things* normally appear." In the second expression, the point of "appears white" is not to *compare* a way of appearing with something else; the point is to say something about the way of appearing itself. It is in terms of this second sense of "appears white"—that in which one may say significantly and without redundancy "Things that are white may normally be expected to appear white"—that we are to interpret the quotation from Augustine above. And, more generally, when it was said that "appear"-statements constitute the foundation of the edifice of knowledge, it was not intended that the "appear"-statements be interpreted as statements asserting a comparison between a present object and any other object or set of objects.

The question now becomes "Can we formulate any significant "appear"-statements *without* thus comparing the way in which some object appears with the way in which some other object appears, or with the way in which the object in question has appeared at some other time? Can we interpret "This appears white" in such a way that it may be understood to refer to a present way of appearing *without* relating that way of appearing to any other object?" In *Experience and Prediction,* Reichenbach defended his own view (and that of a good many others) in this way:

> The objection may be raised that a comparison with formerly seen physical objects should be avoided, and that a basic statement is to concern the present fact only, as it is. But such a reduction would make the basic statement empty. Its content is just that there is a similarity between the present object and one formerly seen; it is by means of this relation that the present object is described. Otherwise the basic statement would consist in attaching an individual symbol, say a number, to the present object; but the introduction of such a symbol would help us in no way, since we could not make use of it to construct a comparison with other things. Only in attaching the same symbols to different objects, do we arrive at the possibility of constructing relations between the objects [pp. 176–77].

It is true that, if an "appear"-statement is to be used successfully in communication, it must assert some comparison of objects. Clearly, if I wish *you* to know the way things are now appearing to me, I must relate these ways of appearing to something that is familiar to you. But our present question is not "Can you understand me if I predicate something of the way in which something now appears to me without relating that way of appearing to something that is familiar to you?" The question is, more simply, "Can I predicate anything of the way in which something now appears to me without thereby comparing that way of appearing with something else?" From the fact that the first of these two questions must be answered in the negative it does not follow that the second must also be answered in the negative.[19]

The issue is not one about communication, nor is it, strictly speaking, an issue about language; it concerns, rather, the nature of thought itself. Common to both "pragmatism" and "idealism," as traditions in American philosophy, is the view that to *think* about a thing, or to *interpret* or *conceptualize* it, and hence to have a *belief* about it, is essentially to relate the thing to *other* things, actual or possible, and therefore to "refer beyond it." It is this view—and not any view about language or communication—that we must oppose if we are to say of some statements about appearing, or of any other statements, that they "justify themselves."

To think about the way in which something is now appearing, according to the view in question, is to relate that way of appearing to something else, possibly to certain future experiences, possibly to the way in which things of a certain sort may be commonly expected to appear. According to the "conceptualistic pragmatism" of C. I. Lewis's *Mind and the World-Order* (1929), we grasp the present experience, any present way of appearing only to the extent to which we relate it to some future experience.[20] According to one interpretation of John Dewey's "instrumentalistic" version of pragmatism, the present experience may be used to present or disclose something else, but it does not present or disclose itself. And according to the idealistic view defended in Brand Blanshard's *The Nature of Thought,* we grasp our present experience only to the extent that we are able to include it in the one "intelligible system of universals" (vol. 1, p. 632).

This theory of reference, it should be noted, applies not only to statements and beliefs about "ways of being appeared to" but also to those other statements and beliefs which I have called "self-justifying." If "This appears white," or "I am appeared white to," compares the present experience with something else, and thus depends for its justification upon what we are justified in believing about the something else, then so, too, does "I believe that Socrates is mortal" and "I hope that the peace will continue." This general conception of thought, therefore, would seem to imply that no belief or statement can be said to justify itself. But according to what we have been saying, if there is no belief or statement which justifies itself, then it is problematic whether any belief or statement is justified at all. And therefore, as we might expect, this conception of thought and reference has been associated with skepticism.

Blanshard conceded that his theory of thought "does involve a degree of scepticism regarding our present knowledge and probably all future knowledge. In all likelihood there will never be a proposition of which we can say, 'This that I am asserting, with precisely the meaning I now attach to it, is absolutely true.'"[21] On Dewey's theory, or on one common interpretation of Dewey's theory, it is problematic whether anyone can now be said to *know* that Mr. Jones is working in his garden. A. O. Lovejoy is reported to have said that, for Dewey, "I am about to have known" is as close as we ever get to "I know."[22] C. I. Lewis, in his *An Analysis of Knowledge and Valuation* (1946), conceded in effect that the conception of thought suggested by his earlier *Mind and the World-Order* does lead to a kind of skepticism; according to the later work there *are* "apprehensions of the given" (cf. pp. 182–83)—and thus beliefs which justify themselves.

What is the plausibility of a theory of thought and reference which seems to imply that no one knows anything?

Perhaps it is correct to say that when we think about a thing we think about it as having certain properties. But why should one go on to say that to think about a thing must always involve thinking about some *other* thing as well? Does thinking about the other thing then involve thinking about some third thing? Or can we think about one thing in relation to a second thing without thereby thinking of a third thing? And if we can, then why can we not think of one thing—of one thing as having certain properties—without thereby relating it to another thing?

The linguistic analogue of this view of thought is similar. Why should one suppose—as Reichenbach supposed in the passage cited above and as many others have also supposed—that to *refer* to a thing, in this instance to refer to a way of appearing, is necessarily to relate the thing to some *other* thing?

Some philosophers seem to have been led to such a view of reference as a result of such considerations as the following: We have imagined a man saying, in agreement with Augustine, "It just does appear white—and that is the end of the matter." Let us consider now the possible reply "That it is not the end of the matter. You are making certain assumptions about the language you are using; you are assuming, for example, that you are using the word 'white,' or the phrase 'appears white,' in the way in which you have formerly used it, or in the way in which it is ordinarily used, or in the way in which it would ordinarily be understood. And if you state your justification for this assumption, you *will* refer to certain other things—to yourself and to other people, to the word "white," or to the phrase 'appears white,' and to what the word or phrase has referred to or might refer to on other occasions. And therefore, when you say 'This appears white' you are saying something, not only about your present experience, but also about all of these other things as well."

The conclusion of this argument—the part that follows the "therefore"—does not follow from the premises. In supposing that the argument is valid, one fails to distinguish between (a) *what* it is that a man means to say when he uses certain words and (b) his assumptions concerning the adequacy of these words for *expressing* what it is that he means to say; one supposes, mistakenly, that what justifies (b) must be included in what justifies (a). A Frenchman, not yet sure of his

English, may utter the words "There are apples in the basket," intending thereby to express his belief that there are potatoes in the basket. If we show him that he has used the word "apples" incorrectly, and hence that he is mistaken in his assumptions about the ways in which English-speaking people use and understand the word "apples," we have not shown him anything relevant to his *belief* that there are apples in the basket.

Logicians now take care to distinguish between the *use* and *mention* of language (e.g., the English word "Socrates" is mentioned in the sentence " 'Socrates' has eight letters" and is used but not mentioned, in "Socrates is a Greek").[23] But the distinction has not always been observed in writings on epistemology.

11. If we decide, then, that there is a class of beliefs or statements which are "self-justifying," and that this class is limited to certain beliefs or statements about our own psychological states and about the ways in which we are "appeared to," we may be tempted to return to the figure of the edifice: our knowledge of the world is a structure supported entirely by a foundation of such self-justifying statements or beliefs. We should recall, however, that the answers to our original Socratic questions had *two* parts. When asked "What is your justification for thinking that you know *a?*" one may reply "I am justified in thinking I know *a*, because (1) I know *b* and (2) if I know *b* then I know *a*." We considered our justification for the *first* part of this answer, saying "I am justified in thinking I know *b*, because (1) I know *c* and (2) if I know *c* then I know *b*." And then we considered our justification for the first part of the second answer, and continued in this fashion until we reached the point of self-justification. In thus moving toward "the given," we accumulated, step by step, a backlog of claims that we did not attempt to justify—those claims constituting the *second* part of each of our answers. Hence our original claim—"I know that *a* is true"—does not rest upon "the given" alone; it also rests upon all of those other claims that we made en route. And it is not justified unless these other claims are justified.

A consideration of these other claims will lead us, I think, to at least three additional types of "stopping place," which are concerned, respectively, with memory, perception, and what Kant called the a priori. I shall comment briefly on the first two.

It is difficult to think of any claim to empirical knowledge, other than the self-justifying statements we have just considered, which does not to some extent rest upon an appeal to memory. But the appeal to memory—"I remember that A occurred"—is not self-justifying. One may ask "And what is your justification for thinking that you remember that A occurred?" and the question will have an answer—even if the answer is only the self-justifying "I think that I remember that A occurred." The statement "I remember that A occurred" does, of course, imply "A occurred"; but "I think that I remember that A occurred" does not imply "A occurred" and hence does not imply "I remember that A occurred." For we can remember occasions—at least we think we can remember them—when we learned, concerning some event we had thought we remembered, that the event had not occurred at all, and consequently that we had not really remembered it. When we thus find that one memory conflicts with another, or, more accurately, when we thus find that one thing that we think we remember conflicts with another thing that we think we remember, we may correct one or the other by making further inquiry; but the results of any such inquiry will always be justified in part by other memories, or by other things that we think that we remember. How then are we to choose between what seem to be conflicting memories? Under what conditions does "I think that I remember that A occurred" serve to justify "I remember that A occurred"?

The problem is one of formulating a rule of evidence—a rule specifying the conditions under which statements about what we think we remember can justify statements about what we do remember. A possible solution, in very general terms, is "When we think that we remember, then we are justified in believing that we do remember, provided that what we think we remember does not conflict with anything else that we think we remember; when what we think we remember does conflict with something else we think we remember, then, of the two conflicting memories (more accurately, ostensible memories) the one that is justified is the one that fits in better with the other things that we think we remember." Ledger Wood made the latter point by saying that the justified memory is the one which "coheres with the system of related memories"; C. I. Lewis used "congruence" instead of "coherence."[24] But we cannot say precisely what is meant by "fitting in," "coherence," or "congruence" until certain controversial questions of confirmation theory and the logic of probability have been answered. And it may be that the rule of evidence is too liberal; perhaps we should say, for example, that when two ostensible memories conflict neither one of them is justified. But these are questions which have not yet been satisfactorily answered.

If we substitute "perceive" for "remember" in the foregoing, we can formulate a similar set of problems about perception; these problems, too, must await solution.[25]

The problems involved in formulating such rules of evidence, and in determining the validity of these rules, do not differ in any significant way from those which arise in connection with the formulation, and validity, of the rules of logic. Nor do they differ from the problems posed by the moral and religious "cognitivists" (the "nonintuitionistic cognitivists"). The status of ostensible memories and perceptions, with respect to that experience which is their "source," is essentially like that which such "cognitivists" claim for judgments having an ethical or theological subject matter. Unfortunately, it is also like that which other "enthusiasts" claim for still other types of subject matter.

12. What, then, is the status of the doctrine of "the given"—of the "myth of the given"? In my opinion, the doctrine is correct in saying that there are some beliefs or statements which are "self-justifying" and that among such beliefs and statements are some which concern appearances or "ways of being appeared to"; but the "phenomenalistic version" of the doctrine is mistaken in implying that our knowledge may be thought of as an edifice which is supported by appearances alone.[26] The cognitive significance of "the empirically given" was correctly described—in a vocabulary rather different from that which I have been using—by John Dewey:

> The alleged primacy of sensory meanings is mythical. They are primary only in logical status; they are primary as tests and confirmation of inferences concerning matters of fact, not as historic originals. For, while it is not usually needful to carry the check or test of theoretical calculations to the point of irreducible sensa, colors, sounds, etc., these sensa form a limit approached in careful analytic certifications, and upon critical occasions it is necessary to touch the limit. . . . Sensa are the class of irreducible meanings which are employed in verifying and correcting other meanings. We actually set out with much coarser and more inclusive meanings and not till we have met with failure from their use do we even set out to discover those ultimate and harder meanings which are sensory in character.[27]

The Socratic questions leading to the concept of "the given" also lead to the concept of "rules of evidence." Unfortunately some of the philosophers who stressed the importance of the former concept tended to overlook that of the latter.

Notes

1. Philosophers in other traditions also noted these confusions. See, for example, John Wild, "The Concept of the Given in Contemporary Philosophy," *Philosophy and Phenomenological Research* 1 (1940): 70–82.

2. The expression "myth of the given" was used by Wilfrid Sellars in "Empiricism and the Philosophy of Mind," in *Science, Perception and Reality*.

3. Dewey also said that, instead of trying to provide "Foundations for Knowledge," the philosopher should apply "what is known to intelligent conduct of the affairs of human life" to "the problems of men." John Dewey, *Problems of Men* (Philosophical Library, 1946), pp. 6–7.

4. C. I. Lewis, *Mind and the World-Order*, p. 29.

5. Ibid., p. 19. Cf. Hans Reichenbach, *Experience and Prediction* (U. of Chicago Press, 1938), p. 6; C. J. Ducasse, "Some Observations Concerning the Nature of Probability," *Journal of Philosophy* 38 (1941), esp. 400–401.

6. Hans Reichenbach, "Are Phenomenal Reports Absolutely Certain?," *Philosophical Review* 61 (1952): 147–59; the quotation is from p. 150.

7. Brand Blanshard, *The Nature of Thought*, vol. 2, p. 276.

8. C. G. Hempel, "On the Logical Positivists' Theory of Truth," *Analysis* (1935): 49–59; the quotation is from p. 57.

9. Rudolf Carnap, "Truth and Confirmation," in *Readings in Philosophical Analysis,* ed. Herbert Feigl and W. S. Sellars (Appleton, 1949), p. 125. The portions of the article quoted above first appeared in "Wahrheit und Bewährung," *Actes du congrès internationale de philosophie scientifique,* Vol. 4 (Paris, 1936), 18–23.

10. Cf. Nelson Goodman, *The Structure of Appearance* (Harvard, 1951), p. 104.

11. C. G. Hempel, "Some Theses on Empirical Certainty," *Review of Metaphysics* 5 (1952): 621–29; the quotation is from p. 621.

12. Ibid., p. 628. Hempel's remarks were made in an "Exploration" in which he set forth several theses about 'empirical certainty' and then replied to objections by Paul Weiss, Roderick Firth, Wilfrid Sellars and myself.

13. C. J. Ducasse, "Propositions, Truth, and the Ultimate Criterion of Truth," *Philosophy and Phenomenological Research* 4 (1939): 317–40; the quotation is from p. 339.

14. Cf. Norman Malcolm, "Knowledge of Other Minds," *Journal of Philosophy* 55 (1958): 969–78. Reprinted in Malcolm, *Knowledge and Certainty*.

15. The principle behind this way of looking at the matter is defended in detail by Max Black in *Language and Philosophy* (Cornell, 1949), p. 16ff.

16. One of the best criticisms of the "appearance" (or "sense-datum") terminology was O. K. Bouwsma's "Moore's Theory of Sense-Data," in *The Philosophy of G. E. Moore,* ed. Schilpp, pp. 201–21. In *Perceiving: A Philosophical Study,* I tried to call attention to certain facts about appearing which, I believe, Bouwsma may have overlooked.

17. Augustine, *Contra academicos,* xi, 26; translated by Sister Mary Patricia Garvey as *Saint Augustine Against the Academicians* (Marquette, 1942); the quotations are from pp. 68–69.

18. Reichenbach, *Experience and Prediction,* p. 176.

19. It may follow, however, that "the vaunted incorrigibility of the sense-datum language can be achieved only at the cost of its perfect utility as a means of communication" (Max Black, *Problems of Analysis* [Cornell, 1954], p. 66), and doubtless, as Black added, it would be "misleading, to say the least" to speak of a "language that cannot be communicated"—cf. Wilfrid Sellars, "Empiricism and the Philosophy of Mind"—but these points do affect the epistemological question at issue.

20. This doctrine was modified in Lewis's later *An Analysis of Knowledge and Valuation* in a way which enabled him to preserve the theory of the given.

21. *The Nature of Thought,* pp. 269–70. Blanshard added, however, that "for all the ordinary purposes of life" we *can* justify some beliefs by showing that they cohere "with the system of present knowledge"; and therefore, he said, his theory should not be described as being "simply sceptical" (Vol. II, p. 271). Cf. W. H. Werkmeister, *The Basis and Structure of Knowledge* (Harper, 1948), part 2.

22. Quoted by A. E. Murphy in "Dewey's Epistemology and Metaphysics," in *The Philosophy of John Dewey,* ed. P. A. Schilpp (Northwestern, 1939), p. 203. Dewey's theory of inquiry, however, was not intended to be an epistemology and he did not directly address himself to the question with which we are here concerned.

23. Cf. W. V. Quine, *Mathematical Logic* (Norton, 1940; rev. ed. Harvard, 1951), sec. 4.

24. Ledger Wood, *The Analysis of Knowledge* (Princeton, 1941), p. 81; C. I. Lewis, *An Analysis of Knowledge and Valuation,* p. 334.

25. Important steps toward solving them were taken by Nelson Goodman in "Sense and Certainty," *Philosophical Review* 61 (1952): 160–67; and by Israel Scheffler in "On Justification and Commitment," *Journal of Philosophy* 51 (1954): 180–90. The former paper is reprinted in *Philosophy of Knowledge,* ed. Roland Houde and J. P. Mullally (Lippincott, 1960), 97–103.

26. Alternatives to the general metaphor of the edifice are proposed by W. V. Quine in the introduction to *Methods of Logic* (Holt, 1950; rev. ed., 1959), in *From a Logical Point of View* (Harvard, 1953), and in *Word and Object* (Wiley, 1960).

27. John Dewey, *Experience and Nature,* 2nd ed. (Norton, 1929), p. 327.

3

Can Empirical Knowledge
Have a Foundation?

Laurence BonJour

The idea that empirical knowledge has, and must have, a *foundation* has been a common tenet of most major epistemologists, both past and present. There have been, as we shall see further below, many importantly different variants of this idea. But the common denominator among them, the central thesis of epistemological foundationism as I shall understand it here, is the claim that certain empirical beliefs possess a degree of epistemic justification or warrant which does not depend, inferentially or otherwise, on the justification of other empirical beliefs, but is instead somehow immediate or intrinsic. It is these noninferentially justified beliefs, the unmoved (or self-moved) movers of the epistemic realm, as Chisholm has called them,[1] that constitute the foundation upon which the rest of empirical knowledge is alleged to rest.

In recent years, the most familiar foundationist views have been subjected to severe and continuous attack. But this attack has rarely been aimed directly at the central foundationist thesis itself, and new versions of foundationism have been quick to emerge, often propounded by the erst-while critics themselves. Thus foundationism has become a philosophical hydra, difficult to come to grips with and seemingly impossible to kill. The purposes of this essay are, first, to distinguish and clarify the main dialectical variants of foundationism, by viewing them as responses to one fundamental problem which is both the main motivation and the primary obstacle for foundationism;

Reprinted from the *American Philosophical Quarterly* 15 (1978): 1–13, by permission of the editor and the author. Copyright 1978, *American Philosophical Quarterly*.

and second, as a result of this discussion to offer schematic reasons for doubting whether any version of foundationism is finally acceptable.

The main reason for the impressive durability of foundationism is not any overwhelming plausibility attaching to the main foundationist thesis in itself, but rather the existence of one apparently decisive argument which seems to rule out all nonskeptical alternatives to foundationism, thereby showing that *some* version of foundationism must be true (on the assumption that skepticism is false). In a recent statement by Quinton, this argument runs as follows:

> If any beliefs are to be justified at all, . . . there must be some terminal beliefs that do not owe their . . . credibility to others. For a belief to be justified it is not enough for it to be accepted, let alone merely entertained: there must also be good reason for accepting it. Furthermore, for an inferential belief to be justified the beliefs that support it must be justified themselves. There must, therefore, be a kind of belief that does not owe its justification to the support provided by others. Unless this were so no belief would be justified at all, for to justify any belief would require the antecedent justification of an infinite series of beliefs. The terminal . . . beliefs that are needed to bring the regress of justification to a stop need not be strictly self-evident in the sense that they somehow justify themselves. All that is required is that they should not owe their justification to any other beliefs.[2]

I shall call this argument *the epistemic regress argument,* and the problem which generates it, the *epistemic regress problem.* Since it is this argument which provides the primary rationale and argumentative support for foundationism, a careful examination of it will also constitute an exploration of the foundationist position itself. The main dialectical variants of foundationism can best be understood as differing attempts to solve the regress problem, and the most basic objection to the foundationist approach is that it is doubtful that any of these attempts can succeed. (In this essay, I shall be concerned with the epistemic regress argument and the epistemic regress problem only as they apply to empirical knowledge. It is obvious that an analogous problem arises also for *a priori* knowledge, but there it seems likely that the argument would take a different course. In particular, a foundationist approach might be inescapable in an account of *a priori* knowledge.)

I

This epistemic regress problem arises directly out of the traditional conception of knowledge as *adequately justified true belief*[3]—whether this be taken as a fully adequate definition of knowledge, or in light of the apparent counterexamples discovered by Gettier,[4] as merely a necessary but not sufficient condition. (I shall assume throughout that the elements of the traditional conception are at least necessary for knowledge.) Now the most natural way to justify a belief is by producing a justificatory argument: belief A is justified by citing some other (perhaps conjunctive) belief B, from which A is inferable in some acceptable way and which is thus offered as a reason for accepting A.[5] Call this *inferential justification*. It is clear, as Quinton points out in the passage quoted above, that for A to be genuinely justified by virtue of such a justificatory argument, B must itself be justified in some fashion; merely being inferable from an unsupported guess or hunch, e.g., would confer no genuine justification upon A.

Two further points about inferential justification, as understood here, must be briefly noted. First, the belief in question need not have been *arrived at* as the result of an inference in order to be inferentially justified. This is obvious, since a belief arrived at in some other way (e.g., as a result of wishful thinking) may later come to be maintained solely because it is now seen to be inferentially justifiable. Second, less obviously, a person for whom a belief is inferentially justified need not have explicitly rehearsed the justificatory argument in question to others or even to himself. It is enough that the inference be available to him if the belief is called into question by others or by himself (where such availability may itself be less than fully explicit) and that the availability of the inference be, in the final analysis, his reason for holding the belief.[6] It seems clear that many beliefs which are quite sufficiently justified to satisfy the justification criterion for knowledge depend for their justification on inferences which have not been explicitly formulated and indeed which could not be explicitly formulated without considerable reflective effort (e.g., my current belief that this is the same piece of paper upon which I was typing yesterday).[7]

Suppose then that belief A is (putatively) justified via inference, thus raising the question of how the justifying premise-belief B is justified. Here again the answer may be in inferential terms: B may be (putatively) justified in virtue of being inferable from some further belief C. But then the same question arises about the justification of C, and so on, threatening an infinite and apparently vicious regress of epistemic justification. Each belief is justified only if an epistemically prior belief is justified, and that epistemically prior belief is justified only if a still prior belief is justified, etc., with the apparent result that justification can never get started—and hence that there is no justification and no knowledge. The foundationist claim is that only through the adoption of some version of foundationism can this skeptical consequence be avoided.

Prima facie, there seem to be only four basic possibilities with regard to the eventual outcome of this potential regress of epistemic justification: (i) the regress might terminate with beliefs for which no justification of any kind is available, even though they were earlier offered as justifying premises; (ii) the regress might proceed infinitely backwards with ever more new premise-beliefs being introduced and then themselves requiring justification; (iii) the regress might circle back upon itself, so that at some point beliefs which appeared earlier in the sequence of justifying arguments are appealed to again as premises; (iv) the regress might terminate because beliefs are reached which are justified—unlike those in alternative (i)—but whose justification does not depend inferentially on other empirical beliefs and thus does not raise any further issue of justification with respect to such beliefs.[8] The foundationist opts for the last alternative. His argument is that the other three lead inexorably to the skeptical result, and that the second and third have additional fatal defects as well, so that some version of the fourth, foundationist alternative must be correct (assuming that skepticism is false).

With respect to alternative (i), it seems apparent that the foundationist is correct. If this alternative were correct, empirical knowledge would rest ultimately on beliefs which were, from an epistemic standpoint at least, entirely arbitrary and hence incapable of conferring any genuine justification. What about the other two alternatives?

The argument that alternative (ii) leads to a skeptical outcome has in effect already been sketched in the original formulation of the problem. One who opted for this alternative could hope to avoid skepticism only by claiming that the regress, though infinite, is not vicious; but there seems to be no plausible way to defend such a claim. Moreover, a defense of an infinite regress view as an account of how empirical knowledge is actually justified—as opposed to how it might in principle be justified—would have to involve the seemingly dubious thesis that an ordinary knower holds a literally infinite number of distinct beliefs. Thus it is not surprising that no important philosopher, with the rather uncertain exception of Peirce,[9] seems to have advocated such a position.

Alternative (iii), the view that justification ultimately moves in a closed curve, has been historically more prominent, albeit often only as a dialectical foil for foundationism. At first glance, this alternative might seem even less attractive than the second. Although the problem of the knower having to have an infinite number of beliefs is no longer present, the regress itself, still infinite, now seems undeniably vicious. For the justification of each of the beliefs which figure in the circle seems now to presuppose *its own* epistemically prior justification: such a belief must, paradoxically, be justified before it can be justified. Advocates of views resembling alternative (iii) have generally tended to respond to this sort of objective by adopting a holistic conception of justification in which the justification of individual beliefs is subordinated to that of the closed systems of beliefs which such a view implies; the property of such systems usually appealed to as a basis for justification is internal *coherence*. Such coherence theories attempt to evade the regress problem by abandoning the view of justification as essentially involving a linear order of dependence (though a non-linear view of justification has never been worked out in detail).[10] Moreover, such a coherence theory of empirical knowledge is subject to a number of other familiar and seemingly decisive objections.[11] Thus alternative (iii) seems unacceptable, leaving only alternative (iv), the foundationist alternative, as apparently viable.

As thus formulated, the epistemic regress argument makes an undeniably persuasive case for foundationism. Like any argument by elimination however, it cannot be conclusive until the surviving alternative has itself been carefully examined. The foundationist position may turn out to be subject to equally serious objections, thus forcing a reexamination of the other alternatives, a search for a further non-skeptical alternative, or conceivably the reluctant acceptance of the skeptical conclusion.[12] In particular, it is not clear on the basis of the argument thus far whether and how foundationism can itself solve the regress problem; and thus the possibility exists that the epistemic regress argument will prove to be a two-edged sword, as lethal to the foundationist as it is to his opponents.

II

The most straightforward interpretation of alternative (iv) leads directly to a view which I will here call *strong foundationism*. According to strong foundationism, the foundational beliefs which terminate the regress of justification possesses sufficient epistemic warrant, independently of any appeal to inference from (or coherence with) other empirical beliefs, to satisfy the justification condition of knowledge and qualify as acceptable justifying premises for further beliefs. Since the justification of these *basic beliefs,* as they have come to be called, is thus allegedly not dependent on that of any other empirical belief, they are uniquely able to provide secure starting-points for the justification of empirical knowledge and stopping-points for the regress of justification.

The position just outlined is in fact a fairly modest version of strong foundationism. Strong foundationists have typically made considerably stronger claims on behalf of basic beliefs. Basic beliefs have been claimed not only to have sufficient non-inferential justification to qualify as knowledge, but also to be *certain, infallible, indubitable,* or *incorrigible* (terms which are usually not very carefully distinguished).[13] And most of the major attacks on foundationism have focused on these stronger claims. Thus it is important to point out that nothing about the basic strong foundationist response to the regress problem demands that basic beliefs be more than adequately justified. There might of course be other reasons for requiring that basic beliefs have some more exalted epistemic status or for thinking that in fact they do. There might even be some sort of indirect argument to show that such a status is a consequence of the sorts of epistemic properties which are directly required to solve the regress problem. But until such an argument is given (and it is doubtful that it can be), the question of whether basic beliefs are or can be certain, infallible, etc., will remain a relatively unimportant side-issue.

Indeed, many recent foundationists have felt that even the relatively modest version of strong foundationism outlined above is still too strong. Their alternative, still within the general aegis of the foundationist position, is a view which may be called *weak foundationism.* Weak foundationism accepts the central idea of foundationism—viz. that certain empirical beliefs possess a degree of independent epistemic justification or warrant which does not derive from inference or coherence relations. But the weak foundationist holds that these foundational beliefs have only a quite low degree of warrant, much lower than that attributed to them by even modest strong foundationism and insufficient by itself to satisfy the justification condition for knowledge or to qualify them as acceptable justifying premises for other beliefs. Thus this independent warrant must somehow be augmented if knowledge is to be achieved, and the usual appeal here is to coherence with other such minimally warranted beliefs. By combining such beliefs into larger and larger coherent systems, it is held, their initial, minimal degree of warrant can gradually be enhanced until knowledge is finally achieved. Thus weak foundationism, like the pure coherence theories mentioned above, abandons the linear conception of justification.[14]

Weak foundationism thus represents a kind of hybrid between strong foundationism and the coherence views discussed earlier, and it is often thought to embody the virtues of both and the vices of neither. Whether or not this is so in other respects, however, relative to the regress problem weak foundationism is finally open to the very same basic objection as strong foundationism, with essentially the same options available for meeting it. As we shall see, the key problem for any version of foundationism is whether it can itself solve the regress problem which motivates its very existence, without resorting to essentially *ad hoc* stipulation. The distinction between the two main ways of meeting this challenge both cuts across and is more basic than that between strong and weak foundationism. This being so, it will suffice to concentrate here on strong foundationism, leaving the application of the discussion to weak foundationism largely implicit.

The fundamental concept of strong foundationism is obviously the concept of a basic belief. It is by appeal to this concept that the threat of an infinite regress is to be avoided and empirical knowledge given a secure foundation. But how can there be any empirical beliefs which are thus basic? In fact, though this has not always been noticed, the very idea of an epistemically basic empirical belief is extremely paradoxical. For on what basis is such a belief to be justified, once appeal to further empirical beliefs is ruled out? Chisholm's theological analogy, cited earlier, is most appropriate: a basic belief is in effect an epistemological unmoved (or self-moved) mover. It is able to confer justification on other beliefs, but apparently has no need to have justification conferred on it. But is such a status any easier to understand in epistemology than it is in theology? How can a belief impart epistemic "motion" to other beliefs unless it is itself in "motion"? And, even more paradoxically, how can a belief epistemically "move" itself?

This intuitive difficulty with the concept of a basic empirical belief may be elaborated and clarified by reflecting a bit on the concept of epistemic justification. The idea of justification is a generic one, admitting in principle of many specific varieties. Thus the acceptance of an empirical belief might be morally justified, i.e., justified as morally obligatory by reference to moral principles and standards; or pragmatically justified, i.e., justified by reference to the desirable practical consequences which will result from such acceptance; or religiously justified, i.e., justified by reference to specified religious texts or theological dogmas, etc. But none of these other varieties of justification can satisfy the justification condition for knowledge. Knowledge requires *epistemic* justification, and the distinguishing characteristic of this particular species of justification is, I submit, its essential or internal relationship to the cognitive goal of truth. Cognitive doings are epistemically justified, on this conception, only if and to the extent that they are aimed at this goal—which means roughly that one accepts all and only beliefs which one has good reason to think are true.[15] To accept a belief in the absence of such a reason, however appealing or even mandatory such acceptance might be from other standpoints, is to neglect the pursuit of truth; such acceptance is, one might say, *epistemically irresponsible*. My contention is that the idea of being epistemically responsible is the core of the concept of epistemic justification.[16]

A corollary of this conception of epistemic justification is that a satisfactory defense of a particular standard of epistemic justification must consist in showing it to be truth-conducive, i.e., in showing that accepting beliefs in accordance with its dictates is likely to lead to truth (and more likely than any proposed alternative). Without such a meta-justification, a proposed standard of epistemic justification lacks any underlying rationale. Why after all should an epistemically responsible inquirer prefer justified beliefs to unjustified ones, if not that the former are more likely to be true? To insist that a certain belief is epistemically justified, while confessing in the same breath that this fact about it provides no good reason to think that it is true, would be to render nugatory the whole concept of epistemic justification.

This general remarks about epistemic justification apply in full measure to any strong foundationist position and to its constituent account of basic beliefs. If basic beliefs are to provide a secure foundation for empirical knowledge, if inference from them is to be the sole basis for the justification of other empirical beliefs, then that feature, whatever it may be, in virtue of which a belief qualifies as a basic must also constitute a good reason for thinking that the belief is true. If we let 'ф' represent this feature, then for a belief B to qualify as basic in an acceptance foundationist account, the premises of the following justificatory argument must themselves be at least justified:[17]

(i) Belief B has feature ф.
(ii) Beliefs having feature ф are highly likely to be true.

Therefore, B is highly likely to be true.

Notice further that while either premise taken separately might turn out to be justifiable on an *a priori* basis (depending on the particular choice of φ), it seems clear that they could not both be thus justifiable. For B is *ex hypothesi* an empirical belief, and it is hard to see how a particular empirical belief could be justified on a purely *a priori* basis.[18] And if we now assume, reasonably enough, that for B to be justified for a particular person (at a particular time) it is necessary, not merely that a justification for B exist in the abstract, but that the person in question be in cognitive possession of that justification, we get the result that B is not basic after all since its justification depends on that of at least one other empirical belief. If this is correct, strong foundationism is untenable as a solution to the regress problem (and an analogous argument will show weak foundationism to be similarly untenable).

The foregoing argument is, no doubt, exceedingly obvious. But how is the strong foundationist to answer it? *Prima facie,* there seem to be only two general sorts of answer which are even remotely plausible, so long as the strong foundationist remains within the confines of the traditional conception of knowledge, avoids tacitly embracing skepticism, and does not attempt the heroic task of arguing that an empirical belief could be justified on a purely *a priori* basis. First, he might argue that although it is indeed necessary for a belief to be justified and *a fortiori* for it to be basic that a justifying argument of the sort schematized above be in principle available in the situation, it is *not* always necessary that the person for whom the belief is basic (or anyone else) know or even justifiably believe that it is available; instead, in the case of basic beliefs at least, it is sufficient that the premises for an argument of that general sort (or for some favored particular variety of such argument) merely be *true,* whether or not that person (or anyone else) justifiably believes that they are true. Second, he might grant that it is necessary both that such justification exist and that the person for whom the belief is basic be in cognitive possession of it, but insist that his cognitive grasp of the premises required for that justification does not involve further empirical beliefs which would then require justification, but instead involves cognitive states of a more rudimentary sort which do not themselves require justification: *intuitions or immediate apprehensions.* I will consider each of these alternatives in turn.

III

The philosopher who has come the closest to an explicit advocacy of the view that basic beliefs may be justified even though the person for whom they are basic is not in any way in cognitive possession of the appropriate justifying argument is D. M. Armstrong. In his recent book, *Belief, Truth, and Knowledge*,[19] Armstrong presents a version of the epistemic regress problem (though one couched in terms of knowledge rather than justification) and defends what he calls an "Externalist" solution:

> According to 'Externalist' accounts of non-inferential knowledge, what makes a true non-inferential belief a case of *knowledge* is some natural relation which holds between the belief-state . . . and the situation which makes the belief true. It is a matter of a certain relation holding between the believer and the world [157].

Armstrong's own candidate for this "natural relation" is "that there must be a *law-like connection* between the state of affairs *Bap* [i.e. *a*'s believing that *p*] and the state of affairs makes '*p*' true such that, given *Bap*, it must be the case that *p*." [166] A similar view seems to be implicit in Dretske's account of perceptual knowledge in *Seeing and Knowing*, with the variation that Dretske requires for knowledge not only that the relation in question obtain, but also that the putative knower *believe* that it obtains—though *not* that this belief be justified.[20] In addition, it seems likely that various views of an ordinary-language stripe which appeal to facts about how language is learned either to justify basic belief or to support the claim that no justification is required would, if pushed, turn out to be positions of this general sort. Here I shall mainly confine myself to Armstrong, who is the only one of these philosophers who is explicitly concerned with the regress problem.

There is, however, some uncertainty as to how views of this sort in general and Armstrong's view in particular are properly to be interpreted. On the one hand, Armstrong might be taken as offering an account of how basic beliefs (and perhaps others as well satisfy the adequate-justification condition for knowledge; while on the other hand, he might be taken as simply repudiating the traditional conception of knowledge and the associated concept of epistemic justification, and offering a surrogate conception in its place—one which better accords with the "naturalistic" world-view which Armstrong prefers.[21] But it is only when understood in the former way that externalism (to adopt Armstrong's useful term) is of any immediate interest here, since it is only on that interpretation that it constitutes a version of foundation-ism and offers a direct response to the anti-foundationist argument set out above. Thus I shall mainly focus on this interpretation of externalism, remarking only briefly at the end of the present section on the alternative one.

Understood in this way, the externalist solution to the regress problem is quite simple: the person who has a basic belief need not be in possession of any justified reason for his belief and indeed, except in Dretske's version, need not even think that there is such a reason; the status of his belief as constituting knowledge (if true) depends solely on the external relation and not at all on his subjective view of the situation. Thus there are no further empirical beliefs in need of justification and no regress.

Now it is clear that such an externalist position succeeds in avoiding the regress problem and the anti-foundationist argument. What may well be doubted, however, is whether this avoidance deserves to be considered a *solution,* rather than an essentially *ad hoc* evasion, of the problem. Plainly the sort of "external" relation which Armstrong has in mind would, if known, provide a basis for a justifying argument along the lines sketched earlier, roughly as follows:

(i) Belief *B* is an instance of kind *K*.
(ii) Beliefs of kind *K* are connected in a law-like way with the sorts of states of affairs which would make them true, and therefore are highly likely to be true.

Therefore, *B* is highly likely to be true.

But precisely what generates the regress problem in the first place is the requirement that for a belief *B* to be epistemically justified for a given person *P,* it is necessary, not just that there be justifiable or even true premises available in the situation which could in principle provide a basis for a justification of *B,* but that *P* himself know or at least justifiably believe some such set of premises and thus be in a position to employ the corresponding argument. The externalist position seems to amount merely to waiving this general requirement in cases where the justification takes a certain form, and the question is why this should be acceptance in these cases when it is not acceptance generally. (If it were acceptance generally, then it would seem that any true belief would be justified for any person, and the distinction between knowledge and true belief would collapse.) Such a move seems rather analogous to solving a regress of causes by simply stipulating that although most events must have a cause, events of a certain kind need not.

Whatever plausibility attaches to externalism seems to derive from the fact that if the external relation in question genuinely obtains, then *P* will not go wrong in accepting the belief, and it is, in a sense, not an accident that this is so. But it remains unclear how these facts are supposed to justify *P*'s acceptance of *B*. It is clear, of course, that an external observer who knew both that that *P* accepted *B* and that there was a law-like connection between such acceptance and the truth of *B* would be in a position to construct an argument to justify *his own* acceptance of *B*. *P* could thus serve as a useful epistemic instrument, a kind of cognitive thermometer, for such an external observer (and in fact the example of a thermometer is exactly the analogy which Armstrong employs to illustrate the relationship which is supposed to obtain between the person who has the belief and the external state of affairs [166ff.]). But *P* himself has no reason at all for thinking that *B* is likely to be true. From his perspective, it *is* an accident that the belief is true.[22] And thus his acceptance of *B* is no more rational or responsible from an epistemic standpoint than would be the acceptance of a subjectively similar belief for which the external relation in question failed to obtain.[23]

Nor does it seem to help matters to move from Armstrong's version of externalism, which requires only that the requisite relationship between the believer and the world obtain, to the superficially less radical version apparently held by Dretske, which requires that *P* also believe that the external relation obtains, but does not require that this latter belief be justified. This view may seem slightly less implausible, since it at least requires that the person have some idea, albeit unjustified, of why *B* is likely to be true. But this change is not enough to save externalism. One way to see this is to suppose that the person believes the requisite relation to obtain on some totally irrational and irrelevant basis, e.g., as a result of reading tea leaves or studying astrological charts. If *B* were an ordinary, non-basic belief, such a situation would surely preclude its being justified, and it is hard to see why the result should be any different for an allegedly basic belief.

Thus it finally seems possible to make sense of externalism only by construing the externalist as simply abandoning the traditional notion of epistemic justification and along with it anything resembling the traditional conception of knowledge. (As already remarked, this may be precisely what the proponents of externalism intend to be doing, though most of them are not very clear on this point.) Thus consider Armstrong's final summation of his conception of knowledge:

> *Knowledge of the truth of particular matters of fact* is a belief which
> must be true, where the 'must' is a matter of law-like necessity. Such
> knowledge is a reliable representation or 'mapping' of reality [220].

Nothing is said here of reasons or justification or evidence or having
the right to be sure. Indeed the whole idea, central to the western
epistemological tradition, of knowledge as essentially the product of
reflective, critical, and rational inquiry has seemingly vanished without
a trace. It is possible of course that such an altered conception of
knowledge may be inescapable or even in some way desirable, but it
constitutes a solution to the regress problem or any problem arising
out of the traditional conception of knowledge only in the radical and
relatively uninteresting sense that to reject that conception is also to
reject the problems arising out of it. In this essay, I shall confine
myself to less radical solutions.

IV

The externalist solution just discussed represents a very recent approach to the justification of basic beliefs. The second view to be
considered is, in contrast, so venerable that it deserves to be called the
standard foundationist solution to the problem in question. I refer of
course to the traditional doctrine of cognitive givenness, which has
played a central role in epistemological discussions at least since
Descartes. In recent years, however, the concept of the given, like
foundationism itself, has come under serious attack. One upshot of the
resulting discussion has been a realization that there are many different
notions of givenness, related to each other in complicated ways, which
almost certainly do not stand or fall together. Thus it will be well to
begin by formulating the precise notion of givenness which is relevant
in the present context and distinguishing it from some related conceptions.

In the context of the epistemic regress problem, givenness amounts
to the idea that basic beliefs are justified by reference not to further
beliefs, but rather to states of affairs in the world which are "immediately apprehended" or "directly presented" or "intuited." This justification by reference to non-cognitive states of affairs thus allegedly
avoids the need for any further justification and thereby stops the
regress. In a way, the basic gambit of givenism (as I shall call positions
of this sort) thus resembles that of the externalist positions considered

above. In both cases the justificatory appeal to further beliefs which generates the regress problem is avoided for basic beliefs by an appeal directly to the non-cognitive world; the crucial difference is that for the givenist, unlike the externalist, the justifying state of affairs in the world is allegedly apprehended *in some way* by the believer.

The givenist position to be considered here is significantly weaker than more familiar versions of the doctrine of givenness in at least two different respects. In the first place, the present version does not claim that the given (or, better, the apprehension thereof) is certain or even incorrigible. As discussed above, these stronger claims are inessential to the strong foundationist solution to the regress problem. If they have any importance at all in this context it is only because, as we shall see, they might be thought to be entailed by the only very obvious intuitive picture of how the view is supposed to work. In the second place, givenism as understood here does not involve the usual stipulation that only one's private mental and sensory states can be given. There may or may not be other reasons for thinking that this is in fact the case, but such a restriction is not part of the position itself. Thus both positions like that of C. I. Lewis, for whom the given is restricted to private states apprehended with certainty, and positions like that of Quinton, for whom ordinary physical states of affairs are given with no claim of certainty or incorrigibility being involved, will count as versions of givenism.

As already noted, the idea of givenness has been roundly criticized in recent philosophical discussion and widely dismissed as a piece of philosophical mythology. But much at least of this criticism has to do with the claim of certainty on behalf of the given or with the restriction to private, subjective states. And some of it at least has been mainly concerned with issues in the philosophy of mind which are only distantly related to our present epistemological concerns. Thus even if the objections offered are cogent against other and stronger versions of givenness, it remains unclear whether and how they apply to the more modest version at issue here. The possibility suggests itself that modest givenness may not be a myth, even if more ambitious varieties are, a result which would give the epistemological foundationist all he really needs, even though he has usually, in a spirit of philosophical greed, sought considerably more. In what follows, however, I shall sketch a line of argument which, if correct, will show that even modest givenism is an untenable position.[24]

The argument to be developed depends on a problem within the givenist position which is surprisingly easy to overlook. I shall therefore proceed in the following way. I shall first state the problem in an initial way, then illustrate it by showing how it arises in one recent version of givenism, and finally consider whether any plausible solution is possible. (It will be useful for the purposes of this discussion to make two simplifying assumptions, without which the argument would be more complicated, but not essentially altered. First, I shall assume that the basic belief which is to be justified by reference to the given or immediately apprehended state of affairs is just the belief that this same state of affairs obtains. Second, I shall assume that the given or immediately apprehended state of affairs is not itself a belief or other cognitive state.)

Consider then an allegedly basic belief that-p which is supposed to be justified by reference to a given or immediately apprehended state of affairs that-p. Clearly what justifies the belief is not the state of affairs simpliciter, for to say that would be to return to a form of externalism. For the givenist, what justifies the belief is the *immediate apprehension* or *intuition* of the state of affairs. Thus we seem to have three items present in the situation: the belief, the state of affairs which is the object of the belief, and the intuition or immediate apprehension of that state of affairs. The problem to be raised revolves around the nature of the last of these items, the intuition or immediate apprehension (hereafter I will use mainly the former term). It *seems* to be a cognitive state, perhaps somehow of a more rudimentary sort than a belief, which involves the thesis or assertion that-p. Now if this is correct, it is easy enough to understand in a rough sort of way how an intuition can serve to justify a belief with this same assertive content. The problem is to understand why the intuition, involving as it does the cognitive thesis that-p, does not *itself* require justification. And if the answer is offered that the intuition is justified by reference to the state of affairs that-p, then the question will be why this would not require a second intuition or other apprehension of the state of affairs to justify the original one. For otherwise one and the same cognitive state must somehow constitute both an apprehension of the state of affairs and a justification of that very apprehension, thus pulling itself up by its own cognitive bootstraps. One is reminded here of Chisholm's claim that certain cognitive states justify themselves,[25] but that extremely paradoxical remark hardly constitutes an explanation of how this is possible.

If, on the other hand, an intuition is not a cognitive state and thus involves no cognitive grasp of the state of affairs in question, then the need for a justification for the intuition is obviated, but at the serious cost of making it difficult to see how the intuition is supposed to justify the belief. If the person in question has no cognitive grasp of that state of affairs (or of any other) by virtue of having such an intuition, then how does the intuition give him a *reason* for thinking that his belief is true or likely to be true? We seem again to be back to an externalist position, which it was the whole point of the category of intuition or givenness to avoid.

As an illustration of this problem, consider Quinton's version of givenism, as outlined in his book *The Nature of Things*.[26] As noted above, basic beliefs may, according to Quinton, concern ordinary perceptible states of affairs and need not be certain or incorrigible. (Quinton uses the phrase "intuitive belief" as I have been using "basic belief" and calls the linguistic expression of an intuitive belief a "basic statement"; he also seems to pay very little attention to the difference between beliefs and statements, shifting freely back and forth between them, and I will generally follow him in this.) Thus "this book is red" might, in an appropriate context, be a basic statement expressing a basic or intuitive belief. But how are such basic statements (or the correlative beliefs) supposed to be justified? Here Quinton's account, beyond the insistence that they are not justified by reference to further beliefs, is seriously unclear. He says rather vaguely that the person is "aware" [129] or "directly aware" [139] of the appropriate state of affairs, or that he has "direct knowledge" [126] of it, but he gives no real account of the nature or epistemological status of this state of "direct awareness" or "direct knowledge," though it seems clear that it is supposed to be a cognitive state of some kind. (In particular, it is not clear what "direct" means, over and above "non-inferential.")[27]

The difficulty with Quinton's account comes out most clearly in his discussion of its relation to the correspondence theory of truth:

> The theory of basic statements is closely connected with the correspondence theory of truth. In its classical form that theory holds that to each true statement, whatever its form may be, a fact of the same form corresponds. The theory of basic statements indicates the point at which correspondence is established, at which the system of beliefs makes its justifying contact with the world [139].

And further on he remarks that the truth of basic statements "is directly determined by their correspondence with fact" [143]. (It is clear that "determined" here means "epistemically determined.") Now it is a familiar but still forceful idealist objection to the correspondence theory of truth that if the theory were correct we could never know whether any of our beliefs were true, since we have no perspective outside our system of beliefs from which to see that they do or do not correspond. Quinton, however, seems to suppose rather blithely that intuition or direct awareness provides just such a perspective, from which we can in some cases apprehend both beliefs and world and judge whether or not they correspond. And he further supposes that the issue of justification somehow does not arise for apprehensions made from this perspective, though without giving any account of how or why this is so.

My suggestion here is that no such account can be given. As indicated above, the givenist is caught in a fundamental dilemma: if his intuitions or immediate apprehensions are construed as cognitive, then they will be both capable of giving justification and in need of it themselves; if they are non-cognitive, then they do not need justification but are also apparently incapable of providing it. This, at bottom, is why epistemological givenness is a myth.[28]

Once the problem is clearly realized, the only possible solution seems to be to split the difference by claiming that an intuition is a semi-cognitive or quasi-cognitive state,[29] which resembles a belief in its capability to confer justification, while differing from a belief in not requiring justification itself. In fact, some such conception seems to be implicit in most if not all givenist positions. But when stated thus baldly, this "solution" to the problem seems hopelessly contrived and *ad hoc*. If such a move is acceptable, one is inclined to expostulate, then once again any sort of regress could be solved in similar fashion. Simply postulate a final term in the regress which is sufficiently similar to the previous terms to satisfy, with respect to the penultimate term, the sort of need or impetus which originally generated the regress; but which is different enough from previous terms so as not itself to require satisfaction by a further term. Thus we would have semi-events, which could cause but need not be caused; semi-explanatia, which could explain but need not be explained; and semi-beliefs, which could justify but need not be justified. The point is not that such a move is always incorrect (though I suspect that it is), but simply that the nature and possibility of such a convenient regress-stopper needs at the very least to be clearly and convincingly established and explained before it can constitute a satisfactory solution to any regress problem.

The main account which has usually been offered by givenists of such semi-cognitive states is well suggested by the terms in which immediate or intuitive apprehensions are described: "immediate," "direct," "presentation," etc. The underlying idea here is that of *confrontation:* in intuition, mind or consciousness is directly confronted with its object, without the intervention of any sort of intermediary. It is in this sense that the object is *given* to the mind. The root metaphor underlying this whole picture is vision: mind or consciousness is likened to an immaterial eye, and the object of intuitive awareness is that which is directly before the mental eye and open to its gaze. If this metaphor were to be taken seriously, it would become relatively simple to explain how there can be a cognitive state which can justify but does not require justification. (If the metaphor is to be taken seriously enough to do the foundationist any real good, it becomes plausible to hold that the intuitive cognitive states which result would after all have to be infallible. For if all need for justification is to be precluded, the envisaged relation of confrontation seemingly must be conceived as too intimate to allow any possibility of error. To the extent that this is so, the various arguments which have been offered against the notion of infallible cognitive states count also against this version of givenism.)

Unfortunately, however, it seems clear that the mental eye metaphor will not stand serious scrutiny. The mind, whatever else it may be, is not an eye or, so far as we know, anything like an eye. Ultimately the metaphor is just far too simple to be even minimally adequate to the complexity of mental phenomena and to the variety of conditions upon which such phenomena depend. This is not to deny that there is considerable intuitive appeal to the confrontational model, especially as applied to perceptual consciousness, but only to insist that this appeal is far too vague in its import to adequately support the very specific sorts of epistemological results which the strong foundationist needs. In particular, even if empirical knowledge at some point involves some sort of confrontation or seeming confrontation, this by itself provides no clear reason for attributing epistemic justification or reliability, let alone certainty, to the cognitive states, whatever they may be called, which result.

Moreover, quite apart from the vicissitudes of the mental eye metaphor, there are powerful independent reasons for thinking that the attempt to defend givenism by appeal to the idea of a semi-cognitive or quasi-cognitive state is fundamentally misguided. The basic idea, after all, is to distinguish two aspects of a cognitive state, its capacity to justify other states and its own need for justification, and then try to find a state which possesses only the former aspect and not the latter. But it seems clear on reflection that these two aspects cannot be separated, that it is one and the same feature of a cognitive state, viz., its assertive content, which both enables it to confer justification on other states and also requires that it be justified itself. If this is right, then it does no good to introduce semi-cognitive states in an attempt to justify basic beliefs, since to whatever extent such a state is capable of conferring justification, it will to that very same extent require justification. Thus even if such states do exist, they are of no help to the givenist in attempting to answer the objection at issue here.[30]

Hence the givenist response to the anti-foundationist argument seems to fail. There seems to be no way to explain how a basic cognitive state, whether called a belief or an intuition, can be directly justified by the world without lapsing back into externalism—and from there into skepticism. I shall conclude with three further comments aimed at warding off certain likely sorts of misunderstanding. First. It is natural in this connection to attempt to justify basic beliefs by appealing to *experience*. But there is a familiar ambiguity in the term "experience," which in fact glosses over the crucial distinction upon which the foregoing argument rests. Thus "experience" may mean either an *experiencing* (i.e., a cognitive state) or something *experienced* (i.e., an object of cognition). And once this ambiguity is resolved, the concept of experience seems to be of no particular help to the givenist. Second. I have concentrated, for the sake of simplicity, on Quinton's version of givenism in which ordinary physical states of affairs are among the things which are given. But the logic of the argument would be essentially the same if it were applied to a more traditional version like Lewis's in which it is private experiences which are given; and I cannot see that the end result would be different—though it might be harder to discern, especially in cases where the allegedly basic belief is a belief about another cognitive state. Third. Notice carefully that the problem raised here with respect to givenism is a logical problem (in a broad sense of "logical"). Thus it would be a mistake to think that it can be solved simply by indicating some sort of state which seems intuitively to have the appropriate sorts of characteristics; the problem is to understand how it is *possible* for any state to have those characteristics. (The mistake would be analogous to one occasionally made in connection with the free-will problem: the mistake of attempting to solve the logical problem of how an action can be not determined but also not merely random by indicating a subjective act of effort or similar state, which seems intuitively to satisfy such a description.)

Thus foundationism appears to be doomed by its own internal

momentum. No account seems to be available of how an empirical belief can be genuinely justified in an epistemic sense, while avoiding all reference to further empirical beliefs or cognitions which themselves would require justification. How then is the epistemic regress problem to be solved? The natural direction to look for an answer is to the coherence theory of empirical knowledge and the associated non-linear conception of justification which were briefly mentioned above.[31] But arguments by elimination are dangerous at best: there may be further alternatives which have not yet been formulated, and the possibility still threatens that the epistemic regress problem may in the end be of aid and comfort only to the skeptic.[32]

Notes

1. Roderick M. Chisholm, *Theory of Knowledge,* 1st. ed., p. 30.

2. Anthony Quinton, *The Nature of Things,* p. 119. This is an extremely venerable argument, which has played a central role in epistemological discussion at least since Aristotle's statement of it in the *Posterior Analytics,* Book I, ch. 2–3. (Some have found an anticipation of the argument in the *Theaetetus* at 209E–210B, but Plato's worry in that passage appears to be that the proposed definition of knowledge is circular, not that it leads to an infinite regress of justification.)

3. "Adequately justified" because a belief could be justified to some degree without being sufficiently justified to qualify as knowledge (if true). But it is far from clear just how much justification is needed for adequacy. Virtually all recent epistemologists agree that certainty is not required. But the lottery paradox shows that adequacy cannot be understood merely in terms of some specified level of probability. (For a useful account of the lottery paradox, see Robert Ackermann, *Belief and Knowledge,* pp. 39–50.) Armstrong, in *Belief, Truth, and Knowledge,* argues that what is required is that one's reasons for the belief be "conclusive," but the precise meaning of this is less than clear. Ultimately, it may be that the concept of knowledge is simply too crude for refined epistemological discussion, so that it may be necessary to speak instead of degrees of belief and corresponding degrees of justification. I shall assume (perhaps controversially) that the proper solution to this problem will not affect the issues to be discussed here, and speak merely of the reasons or justification making the belief *highly likely* to be true, without trying to say exactly what this means.

4. See Edmund Gettier, "Is Justified True Belief Knowledge?" Chap. 9 below. Also Ackermann, *Belief and Knowledge,* chap. 5, and the corresponding references.

5. For simplicity, I will speak of inference relations as obtaining between

beliefs rather than, more accurately, between the propositions which are believed. "Inference" is to be understood here in a very broad sense; any relation between two beliefs which allows one, if accepted, to serve as a good reason for accepting the other will count as inferential.

6. It is difficult to give precise criteria for when a given reason is *the* reason for a person's holding a belief. G. Harman, in *Thought,* argues that for a person to believe for a given reason is for that reason to *explain* why he holds that belief. But this suggestion, though heuristically useful, hardly yields a usable criterion.

7. Thus it is a mistake to conceive the regress as a *temporal* regress, as it would be if each justifying argument had to be explicitly given before the belief in question was justified.

8. Obviously these views could be combined, with different instances of the regress being handled in different ways. I will not consider such combined views here. In general, they would simply inherit all of the objections pertaining to the simpler views.

9. Peirce seems to suggest a virtuous regress view in "Questions Concerning Certain Faculties Claimed for Man," *Collected Papers,* 5, pp. 135–55. But the view is presented metaphorically and it is hard to be sure exactly what it comes to or to what extent it bears on the present issue.

10. The original statement of the non-linear view was by Bernard Bosanquet in *Implication and Linear Inference* (London, 1920). For more recent discussions, see Gilbert Harman, *Thought;* and Nicholas Rescher, "Foundationism, Coherentism, and the Idea of Cognitive Systematization," *Journal of Philosophy* 71 (1974): 695–708.

11. I have attempted to show how a coherence view might be defended against the most standard of these objections in "The Coherence Theory of Empirical Knowledge," *Philosophical Studies* 30 (1976): 281–312.

12. The presumption against a skeptical outcome is strong, but I think it is a mistake to treat it as absolute. If no non-skeptical theory can be found which is at least reasonably plausible in its own right, skepticism might become the only rational alternative.

13. For some useful distinctions among these terms, see William Alston, "Varieties of Privileged Access," *American Philosophical Quarterly* 8 (1971): 223–41.

14. For discussions of weak foundationism, see Bertrand Russell, *Human Knowledge,* part 2, chap. 2, and part 5, chaps. 6 and 7; Nelson Goodman, "Sense and Certainty," *Philosophical Review* 61 (1952): 160–67; Israel Scheffler, *Science and Subjectivity,* chapter 5; and Roderick Firth, "Coherence, Certainty, and Epistemic Priority," *Journal of Philosophy* 61 (1964): 545–57.

15. How good a reason must one have? Presumably some justification accrues from any reason which makes the belief even minimally more likely to be true than not, but considerably more than this would be required to make the justification adequate for knowledge. (See note 3, above.) (The James-

Clifford controversy concerning the "will to believe" is also relevant here. I am agreeing with Clifford to the extent of saying that epistemic justification requires some positive reason in favor of the belief and not just the absence of any reason against.)

16. For a similar use of the notion of epistemic irresponsibility, see Ernest Sosa, "How Do You Know?," *American Philosophical Quarterly* 11 (1974), p. 117.

17. In fact, the premises would probably have to be true as well, in order to avoid Gettier-type counterexamples. But I shall ignore this refinement here.

18. On a Carnap-style *a priori* theory of probability it could, of course, be the case that very general empirical propositions were more likely to be true than not, i.e., that the possible state-descriptions in which they are true outnumber those in which they are false. But clearly this would not make them likely to be true in a sense which would allow the detached assertion of the proposition in question (on pain of contradiction), and this fact seems to preclude such justification from being adequate for knowledge.

19. Armstrong, *Belief, Truth and Knowledge,* chaps. 11–13. Bracketed page references in this section are to this book.

20. Fred I. Dretske, *Seeing and Knowing,* chap. 3, especially pp. 126–39. It is difficult to be quite sure of Dretske's view, however, since he is not concerned in this book to offer a general account of knowledge. Views which are in some ways similar to those of Armstrong and Dretske have been offered by Goldman and by Unger. See Alvin Goldman, "A Causal Theory of Knowing," *The Journal of Philosophy* 64 (1967): 357–72; and Peter Unger, "An Analysis of Factual Knowledge," *The Journal of Philosophy* 65 (1968): 157–70. But both Goldman and Unger are explicitly concerned with the Gettier problem and not at all with the regress problem, so it is hard to be sure how their views relate to the sort of externalist view which is at issue here.

21. On the one hand, Armstrong seems to argue that it is *not* a requirement for knowledge that the believer have "sufficient evidence" for his belief, which sounds like a rejection of the adequate-justification condition. On the other hand, he seems to want to say that the presence of the external relation makes it rational for a person to accept a belief, and he seems (though this is not clear) to have *epistemic* rationality in mind; and there appears to be no substantial difference between saying that a belief is epistemically rational and saying that it is epistemically justified.

22. One way to put this point is to say that whether a belief is likely to be true or whether in contrast it is an accident that it is true depends significantly on how the belief is described. Thus it might be true of one and the same belief that it is "a belief connected in a law-like way with the state of affairs which it describes" and also that it is "a belief adopted on the basis of no apparent evidence"; and it might be likely to be true on the first description and unlikely to be true on the second. The claim here is that it is the believer's own conception which should be considered in deciding whether the belief is

justified. (Something analogous seems to be true in ethics: the moral worth of a person's action is correctly to be judged only in terms of that person's subjective conception of what he is doing and not in light of what happens, willy-nilly, to result from it.)

23. Notice, however, that if beliefs standing in the proper external relation should happen to possess some subjectively distinctive feature (such as being spontaneous and highly compelling to the believer), and if the believer were to notice empirically, that beliefs having this feature were true a high proportion of the time, he would then be in a position to construct a justification for a new belief of that sort along the lines sketched at the end of section II. But of course a belief justified in that way would no longer be basic.

24. I suspect that something like the argument to be given here is lurking somewhere in Sellars's "Empiricism and the Philosophy of Mind" (reprinted in Sellars, *Science, Perception, and Reality*, pp. 127–96), but it is difficult to be sure. A more recent argument by Sellars which is considerably closer on the surface to the argument offered here is contained in "The Structure of Knowledge," his Machette Foundation Lectures given at the University of Texas in 1971, in Hector-Neri Castañeda, ed., *Action, Knowledge, and Reality: Critical Studies in Honor of Wilfrid Sellars* (Indianapolis, 1975), Lecture III, sections 3–4. A similar line of argument was also offered by Neurath and Hempel. See Otto Neurath, "Protocol Sentences," tr. in A. J. Ayer, ed., *Logical Positivism* (New York, 1959), pp. 199–208; and Carl G. Hempel, "On the Logical Positivists' Theory of Truth," *Analysis* 2 (1934–5): 49–59. The Hempel paper is in part a reply to a foundationist critique of Neurath by Schlick in "The Foundation of Knowledge," also translated in Ayer, *Logical Positivism*, pp. 209–27. Schlick replied to Hempel in "Facts and Propositions," and Hempel responded in "Some Remarks on 'Facts' and Propositions," both in *Analysis*, 2 (1934–35): 65–70 and 93–96, respectively. Though the Neurath-Hempel argument conflates issues having to do with truth and issues having to do with justification in a confused and confusing way, it does bring out the basic objection to givenism.

25. Chisholm, "Theory of Knowledge," in Chisholm et al., *Philosophy* (Englewood Cliffs, N.J., 1964), pp. 270ff. [Cf. Chapter 2 above.]

26. Ibid. Bracketed page references in this section will be to this book.

27. Quinton does offer one small bit of clarification here, by appealing to the notion of ostensive definition and claiming in effect that the sort of awareness involved in the intuitive justification of a basic belief is the same as that involved in a situation of ostensive definition. But such a comparison is of little help, for at least two reasons. First, as Wittgenstein, Sellars, and others have argued, the notion of ostensive definition is itself seriously problematic. Indeed, an objection quite analogous to the present one against the notion of a basic belief could be raised against the notion of an ostensive definition; and this objection, if answerable at all, could not only be answered by construing the awareness involved in ostension in such a way as to be of no help to the

foundationist in the present discussion. Second, more straightforwardly, even if the notion of ostensive definition were entirely unobjectionable, there is no need for the sort of awareness involved to be *justified*. If all that is at issue is learning the meaning of a word (or acquiring a concept), then justification is irrelevant. Thus the existence of ostensive definitions would not show how there could be basic beliefs.

28. Notice, however, that to reject an epistemological given does not necessarily rule out other varieties of givenness which may have importance for other philosophical issues. In particular, there may still be viable versions of givenness which pose an obstacle to materialist views in the philosophy of mind. For useful distinctions among various versions of givenness and a discussion of their relevance to the philosophy of mind, see James W. Cornman, "Materialism and Some Myths about Some Givens," *The Monist* 56 (1972): 215–33.

29. Compare the Husserlian notion of a "pre-predicative awareness."

30. It is interesting to note that Quinton seems to offer an analogous critique of givenness in an earlier paper, "The Problem of Perception," reprinted in Robert J. Swartz, ed., *Perceiving, Sensing, and Knowing* (Garden City, New York, 1965), pp. 4497–526; cf. Especially p. 503.

31. For a discussion of such a coherence theory, see my "The Coherence Theory of Empirical Knowledge."

32. I am grateful to my friends Jean Blumenfeld, David Blumenfeld, Hardy Jones, Jeff Pelletier, and Martin Perlmutter for extremely helpful comments on an earlier version of this essay.

4

The Coherence Theory of Knowledge

Keith Lehrer

My research program has been the articulation of a coherence theory of knowledge.[1] The analysis of knowledge embedded in the theory is traditional in form but not in content. It is in the tradition of undefended justified true belief analyses. Such analyses were introduced in my earlier research with Paxson.[2] The salient character of the theory of knowledge is contained in the theory of justification. Justification is the intersection of the subjective, the mental operations of the knower, and the objective, the truth about reality. Justification thus effects what Cohen and I have referred to as the truth connection.[3] The coherence theory of justification on which the coherence theory of knowledge rests effects the truth connection and explains the intersection between the mind and the world. The explanation depends on our conception of ourselves as trustworthy with respect to some matters in some circumstances and untrustworthy about other matters in other circumstances. If we are as trustworthy as we think we are, then we are justified in what we accept.

Reid remarked that it is the first principle of the human mind that our faculties are trustworthy and not fallacious.[4] Our knowledge of the world depends on our capacity to discern when we are trustworthy and when we are not. It is a fundamental feature of human knowledge that we have the capacity to discern when the information we receive by means of our senses is to be trusted, when the probability of veracity is high and the probability of error is small. Such evaluation is essential to human knowledge from the mere possession or exhibition of information. It is not merely the acquisition of information but the certification of that information as trustworthy in terms of background information by the central system that is the hallmark of human

Reprinted with permission of the publisher and the author from *Philosophical Topics* 14 (1986), pp. 5–25.

knowledge. Such knowledge involves, therefore, higher order processing of lower order information. The essential feature of human knowledge is that it is metaknowledge. The essential feature of the human mind is that it is a metamind.

General Features of the Coherence Theory. There are two principal features of the coherence theory of justification that are to be noticed at the outset. First of all, such justification is embedded in a perfectly traditional analysis of knowledge. The analysis which I still advocate, with some variation of detail, is the one I proposed in *Knowledge.*[5] Where 'S' is a variable replaced by the name or description of a knowing subject and 'p' is replaced by a declarative sentence

S knows that p = df (i) p
(ii) S accepts that p
(iii) S is completely justified in accepting that p and
(iv) S is so justified in accepting that p in a way that does not depend essentially on any false statement.

Condition (iv) could be interpreted in such a way as to incorporate the third condition. I shall not be concerned with condition (iv) here, and, consequently, it will simplify exposition to separate condition (iii). I shall also not be concerned here with condition (i), though I shall assume that it is understood. It implies, for instance, that Brand knows that Lehrer wrote the present article only if Lehrer wrote the present article. No matter how convincing the evidence might be that Lehrer wrote the present article, nobody, Brand included, knows that Lehrer wrote it, if Lehrer did not write it. With these admittedly cursory remarks concerning conditions (i) and (iv), I turn to those conditions that are the focus of my present effort.

Acceptance. Condition (ii) is only unusual in that the notion of acceptance rather than belief is employed. It is clear that mere conception or representation that p is not adequate in that a person may conceive that p or represent that p without affirming that p. A special sort of positive attitude toward what is conceived or represented is required for knowledge. Acceptance is the notion that I have adopted for several reasons. In the first, place, the notion is essentially relative to some purpose. What one accepts for one purpose, to be congenial, for example, one may not accept for another, to solve an intellectual or scientific problem. It is acceptance of something for the purposes of obtaining truth and eschewing error with respect to just the thing one accepts that is a condition of knowledge. That notion of acceptance is what is intended by condition (ii). I add the qualification "with respect to just the thing one accepts" because it is possible that accepting something that the evidence indicates is false might serve the general purpose of accepting as much as one can of what is true and accepting as little as one can of what is false. Accepting that one falsehood might be bountifully fecund with respect to accepting other truths. It is clear, however, that accepting something that the evidence indicates is false for such a generally worthwhile purpose is not the sort of acceptance that is required to know that the thing in question is true.

The second reason that acceptance seems to be the appropriate notion is that acceptance does not imply any long term disposition to recall or to act as though the thing accepted is true. In my opinion, belief does not imply this either, but there is a long tradition in philosophy that identifies belief with a long term disposition to act as though the thing believed were true, or, at least, to think that the thing believed is true. It seems clear to me, however, that one may know things for an instant and immediately forget. The experimental evidence concerning short term memory clearly supports this, and anyone who has been told the name of a person at a party knows that what one knows for an instant, the name of a person, may be immediately forgotten.

The third reason for preferring this notion is that acceptance, like judgment, is a notion more closely connected with decision and optionality. A person may decide to accept something, and, though such a decision may lead one to believe something, such a decision does not constitute belief. Note here that if I say to a person, "I accept what you said," I have accepted what the person said whatever I feel about it, but if I say instead "I believe what you said," whether I believed what the person said depends, not on what I said, but on what I feel about it. The locution "I accept . . ." is akin to performatives, but "I believe . . ." is not. One may come to know that what one believes is false when one accepts that it is false and is appropriately justified in doing so. In such instances, belief and acceptance conflict, belief and knowledge conflict. It should be noted, however, that the ordinary use of terms such as "belief" and "accept" is not perfectly consistent. For example, a person may say that she cannot accept something which she knows to be true, but, though this is said, it is clear that the person has accepted what she says she cannot accept. I think this is to be interpreted as hyperbole. We say we cannot do things when they are difficult or painful—even as we do them. "I really can't stay any longer," he says, staying longer.

Justification. Justification divides into a subjective and an objective component. A person may be justified in accepting something in terms of other things he accepts when he is not justified in accepting it in some more objective sense. Both senses of justification may be explicated in terms of coherence with some background system. Whether the person is subjectively or objectively justified depends on the character of the system. I call what a person accepts in the sense specified above an acceptance system, and I say that a person is *subjectively* justified in believing that p if and only if p coheres with the acceptance system of the person. An acceptance system, though motivated by the purpose of obtaining truth and eschewing error in each thing that is accepted, may, given human fallibility, be rife with error. To be objectively justified in accepting that p, the coherence of p with the acceptance system must be sustained when all error is removed. I call the system resulting from the diminution of the acceptance system when all acceptance of false things is expunged the *verific* system of the person. It is a subsystem of the acceptance system.

We thus obtain the following definitions of justification culminating in a definition of complete justification. I use the term *personally* justified to represent subjective justification and the term *verifically* justified to represent objective notion. I acknowledge the technical character of even the defined notions.

(iip) S is *personally* justified in accepting that p if and only if p coheres with the acceptance system of S.

(iiv) S is *verifically* justified in accepting that p if and only if p coheres with the verific system of S.

(iic) S is *completely* justified in accepting that p if and only if S is personally justified in accepting that p and S is verifically justified in accepting that p.

The significance of (iic) is that it expresses the idea that a person is justified in accepting something in the manner required for human knowledge only if there is a coincidence between the background system of things that a person accepts in the attempt to obtain truth and avoid error and the subsystem of that system that would remain when it is cleansed of error. More precisely articulate, the acceptance system of a person is a set of propositions of the form 'S accepts that p', 'S accepts that q', 'S accepts that r' and so forth, articulating the things S accepts in the attempt to accept something true and eschew accepting what is false with respect to just the thing one accepts. The verific system is the subset of the acceptance system that results when all those propositions of the subsystem are deleted if the thing S is said to accept, that is, p, q, or r and so forth, is false.

Coherence and Reasonableness. A number of philosophers have noted that justification is a normative notion. My proposal is that coherence is also implicitly normative. Whether something coheres with some system of a person depends on how reasonable it is for the person to accept it on the basis of the system in question. One achieves coherence by reducing conflict, and so I propose that something coheres with a system of a person if it is more reasonable to accept it than to accept anything with which it conflicts on the basis of the system. For example, it is more reasonable for me to accept that I now see a monitor than that I am hallucinating, and here the former conflicts with the latter. When it is more reasonable to accept one thing than a second with which it conflicts, I shall say that the first beats the second. It is sufficient for something to cohere with a system that it beats anything with which it conflicts with respect to the system. It is not, however, necessary. The reason is that conflict may be a subtle affair. One thing may conflict with a second even though they are logically consistent with each. The notion of conflict may also be explicated in terms of one thing being more reasonable to accept than a second. Thus, it is more reasonable for me to accept that I now see a monitor on the condition that I am not now hallucinating than on the condition that I am now hallucinating, though, in fact, it is not contradictory to suppose that that I really now see a monitor and that I am now hallucinating. People do sometimes see things that are real even when they are hallucinating and also see things that are not real. It is not that the assumption that I am now hallucinating contradicts that I now see a monitor, but the former assumption does render it less reasonable for me to accept that I now see a monitor than the denial of that assumption.

In order to explicate the notion of coherence, I shall employ the undefined locution "it is more reasonable for S to accept that p on the assumption that c than to accept that q on the assumption that d on the basis of system A." Chisholm developed his system taking the simpler locution "it is more reasonable for S to accept p than to accept q" as primitive, and I am clearly indebted to him.[6] I choose the more complicated locution because it makes explicit the fact that our judgments of reasonableness are implicitly relative to some assumptions and system that we unreflectively take for granted. The nonrelativized notion may, of course, be defined in terms of the more complex one by selecting assumptions and a system that are irrelevant to the comparative reasonableness of accepting p and q. Thus, we may say that it is more reasonable for S to accept p than to accept q on the basis of A if and only if it is more reasonable for S to accept p on the assumption that t than to accept q on the assumption that t on the basis of A, where t is some trivial tautology. The reference to A becomes similarly otiose when A is null. It is less clear that the relativized notion may be defined in terms of the nonrelativized notion.

I would then define the notion of conflict or competition as follows.

(vc) p competes with q for S on the basis of system A if and only if it is more reasonable for S to accept p on the assumption that not q than on the assumption that q on the basis of system A.

With conflict defined, it is then possible to define the notion of something beating a competitor, when beating a competitor is sufficient for justification. The definition is as follows.

(vib) p beats q for S on the basis of system A if and only if p competes with q for S on the basis of system A if and only if p competes with q for S on the basis of system A and it is more reasonable for S to accept p than to accept q on the basis of system A.

It is important to notice, however, that beating all competitors, though sufficient for justification, is not necessary. The reason is that some competitors, though not beaten, may be dispensed with in another manner. The suggestion of a skeptic that I am now hallucinating when I accept that I now see a monitor may be beaten. It is more reasonable for me to accept what I do than what he suggests. But the skeptic may be more subtle and only insinuate that I may be hallucinating by noting that people sometimes hallucinate. Now the claim that people sometimes hallucinate does compete with the claim that I am now seeing a monitor. It is more reasonable for me to accept that I now see a monitor on the assumption that people never hallucinate than on the assumption that they sometimes do hallucinate. Moreover, it is not clear that it is more reasonable for me to accept that I see a monitor than that people sometimes hallucinate. I am, of course, quite certain that the former is true, but I am also quite certain that the latter is true.

It would be best to reply to the skeptic in another manner in this case. The right reply would be to neutralize the innuendo rather than refute his claim. For, even though people sometimes hallucinate, I am not hallucinating, and, therefore, his innuendo is beside the point. The competition is not beaten but neutralized. We may define the required notion of neutralization as follows.

(ivn) n neutralizes q as a competitor of p for S on the basis of system A if and only if q competes with p for S on the basis of system A, and n is such that the conjunction of q and n does not compete with p for S on the basis of system A when it is as reasonable for S to accept the conjunction of q and n as to accept q alone on the basis of system A.

This definition is somewhat baroque, but the idea is simple and may be illustrated by our example. If the skeptic attempts to undermine my claim that I see a monitor by pointing out that people sometimes hallucinate, thus suggesting that I may be hallucinating, the way to neutralize the suggestion is for me to point out that I am not hallucinating. The conjunction of the statements that people sometimes hallucinate and that I am not now hallucinating is as reasonable for me to accept as the single statement that people sometimes hallucinate. The conjunction does not compete with the statement that I now see a monitor. Consequently, the statement that I am not now hallucinating neutralizes the competitor that people sometimes hallucinate. The competitor of my claim that I see a monitor is thus neutralized.

It might be objected to this that there is a greater chance of the conjunction being false than a single conjunct. My reply is that the increase in risk of error in accepting the conjunction is negligible and the gain in perspicuity in accepting the conjunction is considerable. The objectives of obtaining truth and avoiding error pull in opposite directions. To avoid error it is always better to withhold acceptance, but that strategy works against the objective of accepting truths. When the acceptance of a more informative statement involves only a negligibly greater risk of error than accepting a less informative one, then it is as reasonable to accept the more informative as the less informative one. I admit that the problem of specifying when a greater risk of error is negligible is difficult.

We may now define coherence and, therefore, indirectly personal and verific justification as follows.

(viic) p coheres with the system A of S if and only if, for every q that competes with p for S on the basis of the system A, q is either beaten or neutralized for S on the basis of the system A.

Personal justification results when the acceptance system is specified and verific justification results when the verific system is specified. This completes the formal analysis of justification. What is now necessary is to say something about what makes it more reasonable to accept one thing rather than another on the basis of a system and to provide the material motivation for adopting this formal account.

Reasonableness and Probability. The foregoing analyses adopt a comparative notion of reasonableness as a basic or primitive notion. What is required for a development of the theory is some indication of characteristics or conditions that make one thing more or less reasonable to accept than another. Without proposing any analysis of reasonableness, I shall propose some criteriological conditions, sufficient conditions, for the application of the notion. The most obvious criterion is one of probability. The acceptance in question has the avoidance of error as an objective and, therefore, the risk of error is a determinant of the reasonableness of such acceptance. The probability of a statement is a measure of the risk of error. The more probable the truth of the statement the smaller the risk of error in accepting it. The risk of error is, however, not the only determinant of reasonableness, and, therefore, the avoidance of error is not the only objective of acceptance. The acceptance of truth is another.

The relation between reasonableness and probability may be articulated by saying that, other things being equal, the more probable a statement is, the more reasonable it is to accept it, and, conversely, the less probable a statement is the less reasonable it is to accept it. I call this the *correspondence* principle. The reason that the qualification, other things being equal, is added is that, as noted, the risk of error is only one determinant of epistemic reasonableness. It is important to notice, however, that the principle is sufficient to solve an epistemological problem of some interest. The lottery paradox of Kyburg is resolved by the principle.[7]

The lottery paradox arises when one assumes that there is some degree of probability less than one that is sufficient for reasonable or justified acceptance. Leaving aside the niceties, the paradox is that once a level of probability less than unity is specified as sufficient for acceptance, the set of accepted statements is a logically inconsistent set. If, for example, one picked .99, then one need only consider a fair lottery with a hundred tickets numbered from one to a hundred, the winning ticket having been selected but not examined, and reflect on the probability that a ticket numbered N has been drawn. Whatever number N is, the probability that it has not been drawn, is .99, and, consequently, by the rule in question, one would be justified in accepting each hypothesis to the effect that ticket number N has not been drawn, where N takes the values from 1 to 100. This set of hypotheses taken together with the background knowledge that one of the tickets numbered 1 to 100 has been drawn is logically inconsistent. Kyburg and others have argued that the paradox depends on a conjunctive principle to the effect that if one is justified in accepting p and one is justified in accepting q, then one is justified in accepting the conjunction of p and q, which they then repudiate.[8] It is important to notice, however, the logical inconsistency of the *set* of accepted statements is not avoided by denying the conjunctivity principle. For this reason, I have argued that to avoid paradox we must deny that any degree of probability less than one is sufficient for justified acceptance. This solution is a consequence of the assumption concerning the relation between probability and reasonable acceptance.

The correspondence principle is sufficient to resolve the paradox. The hypothesis that a given ticket has not been drawn, say the number 6 ticket, has a lower probability on the condition that another ticket, say the number 5 ticket, has not been drawn than on the denial of the latter. The various negative hypotheses are negatively relevant to each. In this instance, other things are obviously equal. Hence, the various negative hypotheses compete with each other, and, since they are all equally probable, no such competitor of a negative hypothesis can be beaten, nor, for that matter, can it be neutralized. This result is, moreover, intuitive rather than *ad hoc*. In a lottery, people may sometimes say that they know that they will not win, especially if the lottery is very large, but surely that is merely hyperbole. Though it is very improbable that they will win, they do not by any means know that they will lose. The reason is that whatever ticket is, in fact, the winner is must as improbable. Indeed, there is exact cognitive symmetry between the winning ticket and any other, that is, there is no relevant difference that is discernible in advance. The analysis yields the result that we do not know that any specific ticket will lose before the winner is revealed.

Probability and the Background System. The second condition concerning probability is that probability is conditional on some background system, in the case of personal justification, the acceptance system. This naturally raises the question of whether conditionalizing on the acceptance system commits us to assigning a probability of one to the statements in the acceptance system. Were this required, the requirement would be that the subject S assign a probability of one to statement of the form 'S accepts that p' which is, perhaps, not an unrealistic requirement. In favor of such a policy, one might appeal to the optionality of acceptance and argue that if a person has decided to accept something, then she might well assign a probability of one to what is thus accepted.

It is, however, not necessary to assign a probability of unity to a statement in order to conditionalize on such probabilities, as Jeffrey has shown.[9] The crux of Jeffrey's proposal is that one can regard certain statements as the focus for assigning probabilities to other statements. To conditionalize on statements, one assigns probabilities to the local statements and then uses those statements as the basis for distributing probabilities to other statements. Formally, supposing A to a focal statement, for example, a statement to the effect that S accepts that p, then the formula for assigning a new probability, p_N, to any other statement H, assuming an original probability assignment, p_O, is as follows.

$$p_N(H) = p_N(A)p_O(H/A) + p_N(\sim A)p_O(H/\sim A).$$

This contrasts with the simpler formula of traditional Bayesian conditionalization which is as follows.

$$p_N(h) = p_O(H/A)$$

The latter formula has the consequence that

$$p_N(A) = p_O(A/A) = 1$$

which amounts to assigning a probability of one to all statements that one uses as the basis of conditionalization.

If one is a fallibilist, and the present author is favorably disposed toward fallibilism, then one might hold that there is always some chance of error, even if negligible, attached to accepting any statement, even the statement that one accepts something or other. Hence, the more complicated Jeffrey formula is the superior formula for articulating probabilities conditional on an acceptance system. Neither a statement to effect that one has accepted something, nor the statement that one has accepted, need be assigned a probability of unity conditional on the acceptance system.

It is especially important to distinguish between accepting a statement and assigning that statement a probability of one. Various probability assignments are reasonable depending on the background beliefs that one has concerning the trustworthiness of the statement one has accepted. Thus, for example, if someone tells me that he will have some work done by tomorrow, the construction of a class schedule, I might accept what he says, that he will have the work done, but what probability I assign to the statement that I accept will depend on how reliable I consider my informant to be. If I see an automobile which I think is a Plymouth, and I accordingly accept that there is a Plymouth before me, the probability that I assign to that statement will depend on how trustworthy I consider myself to be in identifying such an automobile in the present circumstances. Finally, if I merely accept that I bear some external thing buzzing, the probability I assign to the statement I accept will depend on how trustworthy I take my senses to be in their present condition to discern a buzzing in my ears from one in the external world. In general, then, how probable I take something to be, what probability is assigned on the basis of the acceptance system, depends on and is determined by what I accept within that system concerning how trustworthy I am in judging such matters on the basis of the information that I thus possess.

When I take myself to be in circumstances that are highly deceptive, or in which I consider myself to be highly unreliable, then I assign probabilities accordingly. This implies that the acceptance system uses probabilities to assign probabilities. If I accept that it is quite probable that a person in circumstances *C* would err concerning the truth of p, then if I accept that p, I would not assign a very high probabilities to the proposition that p. This assignment of probabilities on the basis of an acceptance system that contains probabilities could, of course, lead to paradox. Kyburg has shown how such paradox may be avoided, and Skyrms has also made such proposals.[10] Ordinary thought about probabilities like ordinary thought about sets proceeds from assumptions that could lead to paradox. Kyburg has shown how such paradox may be avoided, and Skyrms has also made such proposals.[10] Ordinary thought about probabilities like ordinary thought about sets proceeds from assumptions that could lead to paradox. It need not, however.

Coherence and Higher Order Evaluation. The consideration of reliability leads to higher order evaluation and especially to higher order probabilities. Incoming information is evaluated and assigned probabilities depending on what one accepts about the reliability of the information. Those probabilities are also evaluated and assigned probabilities depending on what one accepts about the reliability of the first level probabilities. This is most clear from Keynes' example.[11] Suppose I see a coin that looks quite ordinary lying on a table. I assign a paradox of .5 to the coin falling heads on a fair toss coming to rest on one side or the other. I have not checked on the coin at all carefully, however, and, consequently, I am less certain about the paradox than I might be. By contrast, suppose that I have tossed the coin a few hundred times and checked on it with instruments to insure that it is well balanced with positive results. In that case, I am much more certain that the probability of .5 of falling heads is correct. Thus, I would give much greater weight to the probability on the extended evidence than on the initial evidence. This situation arises, moreover, in much more common cases. In ordinary perception, for example, one accepts perceptual hypotheses about what one sees quickly and without careful observation. Careful observation might not alter the probabilities that one assigns, but it would alter the weight one attaches to the probabilities originally assigned. Thus, it is not only probabilities but the weights that one assigns to those probabilities that influences how reasonable it is to accept a statement.

Probability and Reliability. There is a question that immediately arises about the connection between probability and reliability. Suppose a person is generally reliable in some matter, say judging color with normal vision, who is in unusual circumstances that are highly deceptive, say ones involving the play of colored lights on the objects. As a result, she is not reliable in these circumstances. Here there might seem to be a conflict between reliability and probability. Though she is not in this instance reliable, it is highly probable that her judgment is correct because the circumstances in which she finds herself are, though deceptive, highly improbable. A more fanciful philosophical case, discussed by myself and Cohen, is one in which an evil demon is bent on deceiving us in all of our perceptual judgments. In this case, we would all be very unreliable, though our judgments would have a high probability of being true, because the demon hypothesis, even if true, is a highly improbable one.

This apparent division between reliability and probability is easily explained. The probability is either a prior probability or a conditional probability. The prior probability of error is low in the cases considered because the unusual circumstances have a small prior probability. On the other hand, the conditional probability, the probability of error given what the circumstances are like, is high. Similarly, in both cases we may say that a person has a high general reliability, but, in the special circumstances, her reliability is low. An objection in the evil demon case would seem to be that, given the activity of the evil demon, there is no reliability in the sense of a high frequency of truth. The reply to this is, of course, counterfactual. The people are highly reliable in the absence of the demon.

Higher Level Evaluation and Convergence. The higher level evaluation of probabilities may also be essential to convergence and coherence when it eliminates lower level conflict. There is reason to believe that more than one probability value for a given statement may be suggested by experience. The value that one would assign on the basis of background information may conflict with testimony from another which in turn may conflict with visual perception. This conflict between background information, testimony of others and visual perception may result in three different assignments of probability from differing perspectives. We might designate them as p_B, p_T and p_V. The obvious resolution of the conflict would be a weighted average of the probabilities, $wp_B + wp_T + wp_V = p_1$, where the latter is the probability that integrates the lower level conflict. This weighted average assumes that the weights fall between 0 and 1. Indeed, the weight of 0 may also be neglected as merely a limiting case. A probability to which one gives no weight is a probability that is ignored.

The weighted average involves higher order evaluation of first level probabilities and, once again, is essentially a metamental activity. There is, however, the problem of the conflict recurring at a still higher level. Just as probabilities may be assigned from the point of view of background knowledge, testimony and visual perception, so weights may be assigned from these perspectives. We may find ourselves, therefore, with a matrix of weights as follows, where the weight w_{XY} is the weight that is assigned to the probability of perspective Y from the perspective X.

$$W = \begin{matrix} w_{BB} & w_{BT} & w_{BV} \\ w_{TB} & w_{TT} & w_{TV} \\ w_{VB} & w_{VT} & w_{VV} \end{matrix}$$

which may be multiplied by the vector of original probabilities

$$p^0 = \begin{matrix} p_B{}^0 \\ p_T{}^0 \\ p_V{}^0 \end{matrix}$$

to obtain a new stage one vector of three probabilities, $p_B{}^1$, $p_T{}^1$, $p_V{}^1$, each of which is a weighted average of the original probabilities from one perspective. This is the result of multiplying the matrix times the column vector, Wp^0, of original probabilities. Each stage on probability is an average of the original probabilities from each perspective by the formula

$$Px^1 = w_{XB}p_B{}^0 + w_{XTpT}{}^0 = w_{XVpV}{}^0$$

This process of averaging or aggregating the original, or stage 0 probabilities, to arrive at the new, or stage one probabilities, does not insure convergence. The result of averaging may be three new but divergent probabilities in the column vector p^1. If, however, the new probabilities are again averaged using the same weights, and this process of averaging is repeated, then convergence will be obtained. This means that the probabilities calculated from the three perspectives will converge toward a single probability. Mathematically expressed, the values of the three probabilities in the column vector resulting from multiplying W^N times p^0 converge toward a single probability value as N goes to infinity. That convergence represents the integration of the information articulated in the assignment of weights and original probabilities. Weighted averaging is, therefore, a method for obtaining a coherent probability assignment from conflicting information.[12]

An interest in obtaining integration and coherence from divergence and a conflict suffices as an argument for averaging. There is, however, another argument for averaging, one of consistency. Notice that the refusal to aggregate from one perspective is mathematically equivalent to assigning weights of zero to the other probabilities and averaging. If, therefore, some positive weight is assigned to other probabilities from a perspective, then the refusal to average is inconsistent with the information motivating the assignment of those positive weights. This argument for averaging suffices from stage to stage so long as positive weight is assigned to probabilities from other perspectives. The argument for averaging as opposed to some other use of the weights is given by Wagner in Lehrer and Wagner.[13] The argument is, essentially, that very intuitive axioms lead to the consequence that an aggregation method that yields coherence must be averaging. I have argued that the same axioms used to justify the method described above may also be used to justify the application of Jeffrey's method and Bayes Theorem.[14] It is important to notice that the assumptions that the weights are all positive and that they remain constant through the various stages are sufficient but not necessary for convergence. Convergence will also result when some zero weights are assigned and weights shift from stage to stage. These conditions for convergence are described in detail in Lehrer and Wagner.[15]

It is also important to notice that higher order evaluation and aggregation is here proposed as a normative model. The model permits us to determine how *reasonable* it is to accept some statement or hypothesis. It is, however, testable as an empirical model of the integration of conflicting information. The input of probabilities and weights has a mathematical rate of convergence, some initial inputs converging more rapidly than others. Some will not converge at all. The empirical applicability of the model might, therefore, be tested in terms of the time required to resolve conflicting sources of information. If there was a correspondence between the rate of mathematical convergence and the time required by a subject to resolve conflict, that would confirm the empirical applicability of the model. It is also worth noting that weighted averaging of input is one standard model of information processing at the neural level. This is outlined in some detail by Palm.[16] In general, there is reason to think that higher level processing mimics or resembles lower level processing. Thus, weighted averaging, and, at the higher level, weighted averaging of weighted averages converging toward integrated values, provides us with a unified model of information processing in the human mind.

It should also be noticed that weighted averaging of probabilities may be mathematically interpreted as the probability of probabilities. If a person assigns weights to conflicting probabilities by dividing a unit weight of one among the conflicting probabilities, the set of weights, summing to unity, constitutes a probability vector. The weight thus assigned to a probability could not reasonably be construed as a simple second order probability of the first order probability. The weight may, however, be construed as a proportional or comparative second order probability of the first order probability in comparison to the other first order probabilities. If, moreover, one first obtained second order probabilities of the first order probabilities, one could obtain the comparative second order probabilities assigned to all first level probabilities and assign the ratio of the second order probability over that sum as the weight or comparative probability of a first order probability.

Thus, given the cogency of a theory of second order probability, as proposed by Skyrms for example, weights could be obtained from second order probabilities.[17] There is, moreover, a point of some philosophical interest to be gleaned from this discussion. Hume argued, in his early work, that if we assign probabilities to probabilities, then the probability of first order probability being less than unity would diminish the first order probability. The third order probability of the second order probability being less than unity would diminish the first order probability yet further, and, as we proceed infinitely, the first order probability would be diminished to zero.[18] Hume did not repeat this argument, and it is clearly fallacious on the grounds that diminishing values need not converge toward zero.

In terms of the present model, however, the result of considering the higher order probabilities, even a very low second order probability, may have the consequence of increasing rather than diminishing the first order probability. If, for example, the first order probability of a hypothesis is .6, and the second order probability of that first order probability is low, with the result that a low weight is assigned, .3, for example, averaging may result in a higher first order probability than .6. For example, if another higher first order probability, .8, had a higher second order probability, and correspondingly a greater weight, .7, the averaged first order probability would be .74. In short, if second order probabilities are normalized and used as weights for averaging, we obtain the result that the convergent probability value as aggregation goes to infinity will fall between the lowest and the highest first order probability. The idea that higher order probabilities will result in an infinite regress that will reduce all first order probabilities to zero is as fallacious as Zeno's argument that an infinite regress of motion will reduce all motion to zero and for the same reason. Infinite aggregation may yield positive values.

Justification under risk. The foregoing is a brief account of the theory I have advanced as a coherence theory of human knowledge. Unlike other such models, Dretske's for example, it assumes that justification and knowledge are the result of higher order evaluation of first level information in terms of some acceptance system constituting background information.[19] One part of that acceptance system is an integrated system of probabilities. Input beliefs are evaluated in terms of such probabilities. The probabilities in the acceptance system are personal, those in the verific alternative objective. Personal and objective justification are based, in part, on probabilities, and neither requires a probability of unity. Justification and knowledge are, on this account, compatible with some risk of error, provided that the risk is justified in the quest for truth. This theory of justification might, therefore, be called a theory of justification under risk, and it contrasts with theories like those of Dretske and Armstrong which are theories of justification with certainty.[20] In order to exhibit the advantages of such a theory it will be useful to compare it with other theories of knowledge.

Comparison with the Foundation and Causal Theories. The *Foundation* theory of knowledge and justification is one that postulates certain principles of justification affirming that certain mental states, sensory or belief states, for example, yield justified beliefs. This conception is articulated with precision by Chisholm who also postulates that some very cautious perceptual beliefs are evident.[21] The problem with the foundation theory is that it does not provide us with any explanation of why we should accept the postulates, when, in fact, it is perfectly clear why we accept those postulates. The reason is that we think that the beliefs in question are very probably true or, what is the same, very unlikely to be false. That is why we find the postulates of the foundation theory plausible, why we think, for example, that it is more reasonable to postulate that very cautious perceptual beliefs are evident than to postulate that all perceptual beliefs are evident. The cautious ones are less likely to be in error.

Van Cleve has argued that the foundation theory can provide some justification for the foundationary principles, but such justification does not explain why we accept them initially.[22] The needed explanation depends on a background system of beliefs concerning the conditions under which beliefs are trustworthy and those in which the beliefs are unreliable. This explanation is easy to provide within the context of the coherence theory on the basis of an acceptance system, and, so far as I can ascertain, impossible to provide within the constraints of the foundation theory.

It is, however, important not to confuse the explanation of justification in terms of coherence with inferential justification. If I believe that I see something red, this belief may be justified because it coheres with my acceptance system, but that does not mean that it is justified because I infer it from my acceptance system. On the contrary, the belief may be immediate and noninferential, though the justification of the belief depends on coherence with the acceptance system. Coherence does not depend on inference. Thomas Reid said that there are some beliefs that are justified in themselves without argument.[23] The coherence theory is compatible with this claim. Though argument does not generate the justification, an argument based on beliefs in the

acceptance system may, nevertheless, explain the immediate justification. The explanation explicates a justified it does not generate.

Consider an example. Suppose I argue from the premises that I am
almost always right about the color of objects seen with normal vision
in daylight, that I am in those conditions, and that I believe that I see
something red, to the conclusion that all competitors to the latter belief
are either beaten or neutralized. That would explain why the belief is
justified. Chisholm has noted that such argumentation is ultimately
circular.[24] It is. But that does not prevent it from being explicate. The
circle of our beliefs, what we accept in the interest of obtaining truth
and avoiding error, explains what is justified and what is not. Some of
what is justified does not depend on inference but is immediate. This
immediacy arises within the circle of our beliefs. Within that circle,
particular perceptual beliefs are justified, in part, because they cohere
with general principles and theories, and those theories and principles
are justified, in part, because they cohere with particular perceptual
beliefs. Justification cycles to yield knowledge. Knowledge that is
immediate may supply the premises for the inference of further knowledge, but even immediate knowledge depends on coherence.

Causal Theories: An Objection. Causal theories of knowledge are
vulnerable to another objection, to wit, that they neglect the effect of
background beliefs and higher order evaluation on justification and
knowledge. It is characteristic of such theories, in their primitive
forms, to maintain that beliefs that arise in a particular causal manner,
from perception or from communication, for example, constitute
knowledge simply because of the etiology of the beliefs. This sort of
theory seems to me to be defective. In the first place, a true belief can
arise in circumstances that are highly deceptive, and, in the second
place, a true belief can arise in circumstances that are not deceptive
but believed to be so. In neither instance do those beliefs constitute
knowledge.

Consider a simple belief. A person sees a red patch, and her seeing
a red patch causes her to believe that she sees a red patch. That is a
paradigmatic instance of the appropriate etiology to yield knowledge.
But does she know that she sees a red patch? Her conviction may fall
short of knowledge. Suppose that she is looking at a painted red patch
on a white wall in a room where there appear to be several such red
patches which are quite indistinguishable from each other in quality.
Suppose, moreover, some of the patches are painted red patches while
others are the result of projecting a red light on the wall. If the person
seeing the painted red patch cannot tell a red patch from a spot that is

merely illuminated with red light, then she does not know that she sees a painted red patch when she sees one, for she cannot tell a painted red patch from an illusion in this context. We may imagine, furthermore, that the person has no idea that she is confronted with any illuminated spots. The problem is that the person assumes, incorrectly, that circumstances are those in which she can tell the real thing when she sees one.

Another instance is one in which the circumstances are in no way deceptive but the person believes them to be, she believes, perhaps quite groundlessly, that the circumstances are deceptive, that the wall is illuminated by colored lights. She finds, however, that in spite of her belief about the deceptiveness of the circumstances that she cannot help but believe that she sees a painted red patch, which, in fact, she does. Nature and habit triumph over doubt, but she retains the view that it is no more likely that she is correct in her belief about what she sees than in error. Even though she does see a painted red patch and her seeing causes her to believe that she sees one, she does not *know* that she sees one. She thinks that her beliefs are untrustworthy, and, whether she is right or wrong, this is sufficient to deprive her of knowledge.

Reliabilism: An Objection. These objections may be combined to provide an objection to a somewhat different theory that makes justification a matter of etiology, namely, the reliabilism of Goldman.[25] According to Goldman, justified belief is belief that is an output of a reliable belief forming process. The coherence theory incorporates the idea that one must believe that one is trustworthy with respect to what one accepts and be right in that belief in order to be completely justified in accepting what one does. Reliabilism requires that one be reliable, but it does not entail that one believes that one is. The result is that, according to reliabilism, when one believes one sees something red, as in the second example above, as the result of a reliable belief forming process, though one believes that it is not reliable, then one is justified in believing what one does. It seems, on the contrary, that given the incoherence involved in believing something when one, at the same time, believes that one is not trustworthy in believing it, one is not justified in what one believes.

Now a reliabilist may, as Goldman notes, argue that a belief forming process is not a reliable belief process in the relevant sense if one does not believe that it is. But that seems to me to be in error. There may be belief forming processes that are perfectly reliable even though one does not believe that they are, for example, when one mistakenly believes, as in the second example, that the belief forming process is not reliable. The crux is that one may have a false belief to the effect that some belief is the outcome of a belief forming process that is not reliable. In such a case, the belief forming process is reliable, though one believes that it is not, and, therefore, one is not completely justified in what one believes. The problem is not unreliability. It is coherence. A reliabilist might, of course, add a coherence condition as a condition of justification, but that would be to abandon pure reliabilism for a hybrid theory.

The Coherence Theory: A Reply. The coherence theory treats the examples in the following manner. In the case of the second example, the fact that the person accepts that she is in circumstances that render her belief untrustworthy means that what she believes cannot beat a competitor—the negation of what she believes, for example. In the first case, the person is personally justified in believing that she sees a red patch, but she is not completely justified because of the nature of the verific system. Since it is false that the circumstances are ones in which she is trustworthy in what she accepts, relative to the verific system the person does not accept that she is trustworthy in accepting that she sees a painted red patch. Hence, relative to the verific system, competitors cannot be beaten or neutralized. There is no neutralizing proposition for the competitor that says that she is seeing a white patch illuminated with red light, nor can that competitor be beaten. So the person is not completely justified in believing that she sees a red patch.

It might be objected to this line of argumentation that the coherence theory has incorporated reliabilism as a component and is itself a hybrid theory. The objection would be that an acceptance system might contain nothing concerning when a person is trustworthy and when not, and, therefore, the supposition that it contains assumptions about when we are trustworthy is an ad hoc way of smuggling reliablism into coherence. The reply is twofold. First that, as a matter of fact, people do make assumptions about when they are trustworthy. So the assumption is psychologically well grounded. An analysis of completely justified belief should be based on a realistic psychology of belief.

Coherence, Justification, and Higher Order Evaluation. This brings us to the most fundamental observation concerning the coherence theory of knowledge and justification. It is that human knowledge and justification involve higher order evaluation and certification of information. One may receive information from various sources, but the mere possession of information does not constitute knowledge. The thermometer, as Armstrong noted, contains information, but, contrary to his suggestion, the thermometer does not know anything.[26] The ignorance of the thermometer is a result of the fact that it does not know that the reading is a temperature much less whether it is a correct temperature. The thermometer is an ignorant source of information. To know, one must understand and evaluate the information one receives. To do that, one must have a background system to evaluate the trustworthiness of the incoming information. It is such higher order evaluation of information that yields knowledge. There is no knowledge without such evaluation. Knowledge without meta-knowledge is an impossibility.

Notes

1. Lehrer, K., *Knowledge,* Oxford, 1974, paperback edition, 1979; "The Gettier Problem and the Analysis of Knowledge", in *Justification and Knowledge,* G. Pappas, ed. Dordrecht, 1979; "A Self Profile," in *Keith Lehrer,* R. Bogdan, ed., Dordrecht, 1981, pp. 74–98; and several articles, most recently, "Coherence and the Hierarchy of Method," in *Philosophie als Wissenschaft/ Essays in Scientific Philosophy,* E. Morshcer et al., eds., Salzburg, 1981, 25–56; "Knowledge, Truth and Ontology", in *Language and Ontology,* Dordrecht, 1982, pp. 201–211; and "Justification, Truth and Coherence," with S. Cohen, *Synthese,* 1983, 191–208.

2. Lehrer, K. and Paxson, T. Jr., "Knowledge: Undefeated Justified True Belief," in *Journal of Philosophy,* 1969, pp. 285–297, reprinted in *Essays on Knowledge and Justification,* G. S. Pappas and M. Swain, eds., Ithaca, 1978, pp. 146–154.

3. Cohen and Lehrer, op. cit.

4. Reid, T., *Inquiry and Essays,* R. Beanblossom and K. Lehrer, eds., Bobbs-Merrill, Indianapolis, 1975, new edition, Hackett, Indianapolis, 1984.

5. Lehrer, *Knowledge.*

6. Chisholm, R. M., *Theory of Knowledge,* second edition, Prentice-Hall, Englewood Cliffs, 1976, esp. pp. 6–7.

7. Kyburg, H., Jr., *Probability and the Logic of Rational Belief,* Wesleyan Press, Middleton, 1961, p. 197.

8. Hempel, C. G., "Deductive-Nomological vs. Statistical Explanation", in *Minnesota Studies in the Philosophy of Science,* vol. Iii, H. Feigl and G. Maxwell, eds., Minneapolis, 1962, and Kyburg, H., Jr., "Conjunctivitis," in *Induction, Acceptance, and Rational Belief,* M. Swain, ed., Dordrecht, 1970.

9. Jeffrey, R. C., *The Logic of Decision,* New York, 1965.

10. Kyburg, *Probability,* and Brian Skyrms, *Causal Necessity,* New Haven, 1980.

11. Keynes, J. M., *A Treatise on Probability,* London, 1952.

12. Lehrer, K., and Wagner, C., *Rational Consensus in Science and Society: A Philosophical and Mathematical Study,* Dordrecht, 1981.

13. Ibid.

14. Lehrer, K., "Rationality as Weighted Averaging", *Synthese,* 1983, pp. 283–295.

15. Lehrer and Wagner, *Rational Consensus.*

16. Palm, G. *Neural Assemblies,* Berlin, 1982.

17. Skyrms, B., "Higher Order Degrees of Belief", in *Prospects for Pragmatism—Essays in Honor of F.P. Ramsey,* D. H. Mellor, ed., Cambridge, 1980.

18. D. Hume, *The Treatise of Human Nature,* London, 1739, Book I, Part IV, Section I.

19. Dretske, F., *Knowledge and the Flow of Information,* Oxford, 1981.

20. Armstrong, D., *Belief, Truth, and Knowledge,* London, 1973.

21. Chisholm, R. M., *Theory,* p. 84.

22. Van Cleve, J., "Foundationism, Epistemic Principles, and the Cartesian Circle", *Philosophical Review,* 1979, pp. 55–19.

23. Reid, T. *Inquiry,* passim, e.g., p. 89.

24. Chisholm, R. M., *Perceiving: A Philosophical Study,* Ithaca, 1957.

25. Goldman, A. I., "What is Justified Belief?", in *Justification and Knowledge,* pp. 1–24. In the same volume, see also Swain, M., "Justification and the Basis of Belief", pp. 25–49.

26. Armstrong, D. M. *Belief,* p. 166.

5

The Foundationalism–Coherentism Controversy

Robert Audi

Foundationalism and coherentism each contain significant epistemological truths.[1] Both positions are, moreover, intellectually influential even outside epistemology. But most philosophers defending either position have been mainly concerned to argue for their view and to demolish the other, which they have often interpreted through just one leading proponent. It is not surprising, then, that philosophers in each tradition often feel misunderstood by those in the other. The lack of clarity—and unwarranted stereotyping—about both foundationalism and coherentism go beyond what one would expect from terminological and philosophical diversity: there are genuine obscurities and misconceptions. Because both positions, and especially foundationalism, are responses to the epistemic regress problem, I want to start with that. Once it is seen that this perennial conundrum can take two quite different forms, both foundationalism and coherentism can be better understood.

I. Two Conceptions of the Epistemic Regress Problem

It is widely agreed that the epistemic regress argument gives crucial support to foundationalism. Even coherentists, who reject the argument, grant that the regress problem which generates it is important in motivating their views.[2] There are at least two major contexts—often not distinguished—in which the regress problem arises. Central to one

Reprinted by permission from Robert Audi, *The Structure of Justification*, pp. 117, 125–64. Cambridge: Cambridge University Press, 1993.

is pursuit of the question of how one knows or is justified in believing some particular thing, most typically a proposition about the external world, e.g., that one saw a bear in the woods. This context is often colored by conceiving such questions as skeptical challenges, and this is the conception of them most important for our purposes. The challenges are often spearheaded by "How do you know?" Central to the other main context in which the regress problem arises are questions about what *grounds* knowledge or justification, or a belief taken to be justified or to constitute knowledge, where there is no skeptical purpose, or at least no philosophically skeptical one. Other terms may be used in framing these questions. People interested in such grounds may, for instance, want to know the source, basis, reasons, evidence, or rationale for a belief. We must consider the regress problem raised in both ways. I begin with the former.

Suppose I am asked how I know that *p*, say that there are books in my study. The skeptic, for instance, issues the question as a challenge. I might reply by citing a ground of the belief in question, say *q:* I have a clear recollection of books in my study. The skeptic then challenges the apparent presupposition that I know the ground to hold; after all, if I do have a ground, it seems natural to think that I should be able (at least on reflection) not just to produce it, but also to justify it: how else can I be entitled to take it as a ground? Thus, if "How do you know?" is motivated by a skeptical interest in knowledge, the question of how I know is likely to be reiterated, at least if my ground, *q*, is not self-evident; for unless *q* is self-evident, and in that sense a self-certifying basis for *p*, the questioner—particularly if skeptical—will accept my citing *q* as answering "How do you know that *p*?", only on the assumption that I *also* know that *q*. How far can this questioning reasonably go?

For epistemologists, the problem posed by "How do you know?" and "What justifies you?" is to answer such questions without making one or another apparently inevitable move that ultimately undermines the possibility of knowledge or even of justification. Initially, there seem to be three unpleasant options. The first is to rotate regressively in a vicious circle, say from *p* to *q* as a ground for *p*, then to *r* as a ground for *q*, and then back to *p* as a ground for *r*. The second option is to fall into a vicious regress: from *p* to *q* as a ground for *p*, then to *r* as a ground for *q*, then to *s* as a ground for *r*, and so on to infinity. The third option is to stop at a purported ground, say *s*, that does not constitute knowledge or even justified belief; but the trouble with this is that if one neither knows nor justifiedly believes *s*, it is at best difficult to see how citing *s* can answer the question of how one knows that *p*. The fourth option is to stop with something that is known or justifiedly believed, say *r*, but *not* known on the basis of any further knowledge or justified belief. Here the problem as many see it is that *r*, not believed on any further ground, serves as just an arbitrary way of stopping the regress and is only capriciously taken to be known or justifiedly believed. Thus, *citing r* as a final answer to the chain of queries seems dogmatic. I want to call this difficulty—how to answer, dialectically, questions about how one knows, or about what justifies one—the *dialectical form of the regress problem.*[3]

Imagine, by contrast, that we consider either the entire body of a person's apparent knowledge, as Aristotle seems to have done,[4] or a representative item of apparent knowledge, say my belief that there are books in my study, and ask on what this apparent knowledge is grounded (or based) and whether, if it is grounded on some further belief, *all* our knowledge or justified belief could be so grounded. We are now asking a structural question about knowledge, not requesting a verbal response in defense of a claim to it. No dialectic need even be imagined; we are considering a person's overall knowledge, or some presumably representative item of it, and asking how that body of knowledge is structured or how that item of knowledge is grounded. Again we get a regress problem: how to specify one's grounds without vicious circularity or regress or, on the other hand, stopping with a belief that does not constitute knowledge (or is not justified) or seems only capriciously regarded as knowledge. Call this search for appro-priate grounds of knowledge the *structural form of the regress problem.*

To see how the two forms of the regress problem differ, we can think of them as arising from different ways of asking, "How do you know?" It can be asked with *skeptical force,* as a challenge to people who either claim to know something or (more commonly) presuppose that some belief they confidently hold represents knowledge. Here the question is roughly equivalent to "Show me that you know." It can also be asked with *informational force,* as where someone simply wants to know by what route, such as observation or testimony, one came to know something. Here the question is roughly equivalent to "How is it that you know?" The skeptical form of the question does *not* presuppose that the person in question really has any knowledge, and, asked in this noncommittal way, the question tends to generate the dialectical form of the regress. The informational form of the question typically *does* presuppose that the person knows the proposition in which way we formulate the problem. But it does matter, for at least four reasons.

Knowing versus showing that one knows. First, the dialectical form of the regress problem invites us to think that an adequate answer to "How do you know?" *shows* that we know. This is so particularly in the context of a concern to reply to skepticism. For the skeptic is not interested in the information most commonly sought when people ask how someone knows, say information about the origin of the belief, e.g., in first-hand observation as opposed to testimony. It is, however, far from clear that an adequate answer to the how-question must be an adequate answer to the show-question. If I tell you how I know there were injuries in the accident by citing the testimony of a credible witness who saw it, you may be satisfied; but I have not shown that I know (as I might by taking you to the scene), and the skeptic who, with the force of a challenge, asks how I know will not be satisfied. I have answered the informational form of the question, but not the skeptical form.

First-order versus second-order knowledge. Second, when the regress problem is dialectically formulated, any full non-skeptical answer to "How do you know that p?" will tend to imply an epistemic self-ascription, say "I know that q"; thus, my answer is admissible only if I both have the concept of knowledge—since I would otherwise not understand what I am attributing to myself—*and* am at least dialectically warranted in asserting that I do know that q. If you ask, informationally, how I know that there were injuries, I simply say (for instance) that I heard it from Janet, who saw them. But if you ask, skeptically, how I know it, I will realize that you will not accept evidence I merely *have,* but only evidence I *know;* and I will thus tend to say something to the effect that I *know* that Janet saw the injuries. Since this in effect claims knowledge of knowledge, it succeeds only if I meet the second-order standard for having knowledge that I know she saw this. If, however, the regress problem is structurally formulated, it is sufficient for its solution that there *be* propositions which, whether or not I believe them *prior* to being questioned, are both warranted for me (reasonable for me *to* believe) and together justify the proposition originally in question. For this to be true of me, I need only meet a first-order standard, e.g., by remembering the accident, and thereby be justified in believing that there were injuries.

Having, giving, and showing a solution. Third, and largely implicit in the first two points, the two formulations of the regress problem differ as to what must hold in order for there to *be,* and for S to *give,* an adequate answer to "How do you know?" or "What justifies you?" On the structural formulation, if there *are* warranted propositions of the kind just described, as where I am warranted in believing that there were injuries, the problem (as applied to *p,* the proposition in question) *has* a solution; and if I *cite* them in answering "How do you know that *p?*" I *give* a solution to the problem. The problem has a solution because of the mere existence of propositions warranted for me; and the solution is given, and the problem thus actually solved, by my simply affirming those propositions in answering "How do you know?" By contrast, when the problem is dialectically formulated, it is taken to have a solution only if there not only *are* such propositions, but I can show by *argument* that there are; and to give a solution I must not merely cite these propositions but also show that they are justified and that they in turn justify *p.* Thus, I cannot adequately say how I know there are books in my study by citing my recollection of them unless I can show by argument that it is both warranted and justifies concluding that there are indeed books there. Raising the structural form of the problem presupposes only that if I know that *p,* I have grounds of this knowledge that are expressible in propositions warranted for me; it does not presuppose that I can formulate the grounds or show that they imply knowledge. The structural form thus encourages us to conceive solutions as *propositional,* in the sense that they depend on the evidential propositions warranted for me; the dialectical form encourages conceiving solutions, as *argumental,* because they depend on what *arguments* about the evidence are accessible to me. I must be able to enter the dialectic with good arguments for *p,* not simply to be warranted in believing evidence propositions that justify *p.*

The process of justification versus the property of justification. Fourth, a dialectical formulation, at least as applied to justification (and so, often, to knowledge as at least commonly embodying justification), tends to focus our attention on the *process* of justification, i.e., of justifying a proposition, though the initial question concerns whether the relevant belief has the *property* of justification, i.e., of being justified. The skeptical forms of the questions "How do you know?" and "What justifies you?" tend to start a process of argument; "Show me that you know" demands a response, and what is expected is a process of justifying the belief that *p.* The informational form of these questions tends to direct one to cite a ground, such as clear recollection, and the knowledge or (property of) justification in question may be simply taken to be based on this ground. "By what route (or on what basis) do you know?'' need not start a process (though it may). It implies that providing a good ground—one in virtue of which the belief that *p* has the property of being justified—will fully answer the question. Granted, the epistemologist pursuing the regress problem in either form must use second-order formulations (though in different ways); still, the criteria for knowledge and justified belief tend to differ depending on which approach is dominant in determining those criteria.

If I am correct in thinking that the dialectical and structural formulations of the regress problem are significantly different, which of them is preferable in appraising the foundationalism–coherentism controversy? One consideration is neutrality; we should try to avoid bias toward any particular epistemological theory. The dialectical formulation, however, favors coherentism, or at least non-foundationalism. Let me explain.

Foundationalists typically posit beliefs that are grounded in experience of reason and are direct—and so not grounded through other, mediating beliefs—in two senses. First, they are *psychologically direct:* non-inferential (in the most common sense of that term), and thus not held on the basis of (hence through) some further belief. Second, they are *epistemically direct:* they do not depend (inferentially) for their status as knowledge, or for any justification they have, on other beliefs, justification, or knowledge. The first kind of direct belief has no psychological intermediary of the relevant kind, such as belief. The second kind has no evidential intermediary, such as knowledge of a premise for the belief in question. Roughly, epistemically direct beliefs are not inferentially *based on* other beliefs or knowledge, and this point holds whether or not there is any actual *process* of inference. Now imagine that, in dealing with the dialectical form of the regress problem, say in answering the question of how I know I have reading material for tonight, I cite, as an appropriate ground, my knowing that there are books in my study. In choosing this as an example of knowledge, I express a belief that I do in fact know that there are books in the study. But am I warranted in this *second-order* belief, as I appear to be warranted simply in believing that there *are* books in the study (the former belief is construed as second-order on the assumption that knowing entails believing, and the belief that one knows is thus in some sense a belief about another belief)? Clearly it is far less plausible to claim that my second-order belief that I *know* there are books in my study is epistemically direct than to claim this status for my *perceptual* belief that there *are* books in it; for the latter seems non-inferentially based on my seeing them, whereas the former seems inferential, e.g., based on beliefs about epistemic status. Thus, foundationalists are less likely to seem able to answer the dialectical formulation of the problem, since doing that requires positing direct second-order knowledge (or at least direct, second-order justified belief).

In short, the dialectical form of the problem seems to require foundationalists to posit foundations of a higher order, and a greater degree of complexity, than they are generally prepared to posit. The same point emerges if we note that "How do you know?" can be repeated, and in some fashion answered, indefinitely. Indeed, because this question (or a similar one) is central to the dialectical formulation, that formulation tends to be inimical to foundationalism, which posits at least one kind of natural place to stop the regress: a place at which, even if a skeptical challenge *can* be adequately answered, having an answer to it is not necessary for having knowledge or justified belief.

It might seem, on the other hand, that the structural formulation, which stresses our actual cognitive makeup, is inimical toward coherentism, or at least non-foundationalism. For given our knowledge of cognitive psychology, it is difficult to see how a normal person might *have* anything approaching an infinite chain of beliefs constituting knowings; hence, an infinite chain of answers to "How do you know?" seems out of the question. But this only cuts against an infinite regress approach in epistemology, not against any finistic coherentism, which seems the only kind ever plausibly defended. Indeed, even assuming—as coherentists may grant—that much of our knowledge in fact arises, non-inferentially, from experiential states like seeing, the structural formulation of the problem allows *both* that, as foundationalists typically claim, there is non-inferential knowledge, and that, as coherentists typically claim, non-inferential beliefs are dialectically defensible indefinitely and (when true) capable of constituting knowledge only by virtue of coherence. The structural formulation may not demand that such defenses be available indefinitely; but it also does not preclude this nor even limit the mode of defense to circular reasoning.

I believe, then, that the structural formulation is not significantly biased against coherentism. Nor is it biased in favor of internalism over externalism about justification, where internalism is roughly the view that what justifies a belief, such as a visual impression, is internal in the sense that one can become (in some way) aware of it through reflection or introspection (internal processes), and externalism denies that what justifies a belief is always accessible to one in this sense. The dialectical formulation, by contrast, tends to favor internalism, since it invites us to see the regress problem as solved in terms of what propositions warranted for one are *also* accessible to one in answering "How do you know that *p*?" If the structural formulation is biased against internalism or coherentism, I am not aware of good reasons to think so, and I will work with it here.

II. The Epistemic Regress Argument

If we formulate the regress problem structurally, then a natural way to state the famous epistemic regress argument is along these lines. First, suppose I have knowledge, even if only of something so simple as there being a patter outside my window. Could all my knowledge be inferential? Imagine that this is possible by virtue of an infinite epistemic regress—roughly, an infinite series of knowings, each based (inferentially) on the next. Just assume that a belief constituting inferential knowledge is based on knowledge of some other proposition, or at least on a further belief of another proposition; the further knowledge or belief might be based on knowledge of, or belief about, something still further, and so on. Call this sequence an epistemic chain; it is simply a chain of beliefs, with at least the first constituting knowledge, and each belief linked to the previous one by being based on it. A standard view is that there are just four kinds: an epistemic chain might be infinite or circular, hence in either case unending and in that sense regressive, third, it might terminate with a belief that is not knowledge; and fourth, it might terminate with a belief constituting direct knowledge. The epistemic regress problem is above all to assess these chains as possible sources (or at least carriers) of knowledge or justification.

The foundationalist response to the regress problem is to offer a regress argument favoring the fourth possibility as the only genuine one. The argument can be best formulated along these lines:

1. If one has any knowledge, it occurs in an epistemic chain (possibly including the special case of a single link, such as a perceptual or a priori belief, which constitutes knowledge by virtue of being anchored directly in experience or reason);
2. the only possible kinds of epistemic chains are the four mutually exclusive kinds just sketched;
3. knowledge can occur only in the last kind of chain; hence,
4. if one has any knowledge, one has some direct knowledge.[5]

Some preliminary clarification is in order before we appraise this argument.

First, the conclusion, being conditional, does not presuppose that there *is* any knowledge. This preserves the argument's neutrality with respect to skepticism, as is appropriate since the issue concerns *conceptual* requirements for the possession of knowledge. The argument would have existential import, and so would not be purely conceptual, if it presupposed that there *is* knowledge and hence that at least one knower exists. Second, I take (1) to imply that inferential knowledge depends on at least one epistemic chain for its status *as* knowledge. I thus take the argument to imply the further conclusion that any inferential knowledge one has exhibits (inferential) *epistemic dependence* on some appropriate inferential connection, via some epistemic chain, to some non-inferential knowledge one has. Thus, the argument would show not only that if there is inferential knowledge, there *is* non-inferential knowledge, but also that if there is inferential knowledge, that very knowledge is *traceable* to some non-inferential knowledge as its foundation.

The second point suggests a third. If two epistemic chains should *intersect,* as where a belief that *p* is both foundationally grounded in experience and part of a circular chain, then if the belief is knowledge, that knowledge *occurs in* only the former chain, though the knowledge qua *belief* belongs to both chains. Knowledge, then, does not occur in a chain merely because the belief constituting it does. Fourth, the argument concerns the structure, not the content, of a body of knowledge and of its constituent epistemic chains. The argument may thus be used regardless of what purported items of knowledge one applies it to in any particular person. The argument does not presuppose that in order to have knowledge, there are specific things one must believe, or that a body of knowledge must have some particular content.

A similar argument applies to justification. We simply speak of *justificatory chains* and proceed in a parallel way, substituting justification for knowledge. The conclusion would be that if there are any justified beliefs, there are some non-inferentially justified beliefs, and that if one has any inferentially justified belief, it exhibits (inferential) *justificatory dependence* on an epistemic chain appropriately linking it to some non-inferentially justified belief one has, that is, to a foundational belief. In discussing foundationalism, I shall often focus on justification.

Full-scale assessment of the regress argument is impossible here. I shall simply comment on some important aspects of it to provide a better understanding of foundationalism and of some major objections to it.

Appeal to infinite epistemic chains has seldom seemed to philosophers to be promising. Let me suggest one reason to doubt that human beings are even capable of having infinite sets of beliefs. Consider the claim that we can have an infinite set of arithmetical beliefs, say that 2 is twice 1, that 4 is twice 2, etc. Surely for a finite mind there will be some point or other at which the relevant proposition cannot be grasped. The required formulation (or entertaining of the proposition) would, on the way "toward" infinity, become too lengthy to permit understanding it. Thus, even if we could read or entertain it part by part, when we got to the end we would be unable to remember enough of the first part to grasp and thereby believe what the formulation expresses. Granted, we could believe that the formulation just read expresses *a* truth; but this is not sufficient for believing *the truth* that it expresses. That truth is a specific mathematical statement; believing, of a formulation we cannot even get before our minds, or remember, in toto, that it expresses *some* mathematical truth is not sufficient for believing, or even grasping, the true statement in question. Since we cannot understand the formulation as a whole, we cannot grasp that truth; and what we cannot grasp, we cannot believe. I doubt that any other lines of argument show that we can have infinite sets of beliefs; nor, if we can, is it clear how infinite epistemic chains could account for any of our knowledge. I thus propose to consider only the other kinds of chain.

The possibility of a circular epistemic chain as a basis of knowledge has been taken much more seriously. The standard objection has been that such circularity is vicious, because one would ultimately have to know something on the basis of itself—say *p* on the basis of *q, q* on the basis of *r,* and *r* on the basis of *p.* A standard reply has been that if the circle is wide enough and its content sufficiently rich and coherent, the circularity is innocuous. I bypass this difficult matter, since I believe that coherentism as most plausibly formulated does not depend on circular chains.

The third alternative, namely that an epistemic chain terminates in a belief which is not knowledge, has been at best rarely affirmed; and there is little plausibility in the hypothesis that knowledge can originate through a belief of a proposition *S* does not know. If there are exceptions, it is where, although I do not know that *p*, I am justified, to *some* extent, in believing that *p*, as in making a reasonable estimate that there are at least thirty books on a certain shelf. Here is a different case. Suppose it vaguely seems to me that I hear strains of music. If, on the basis of the resulting, somewhat justified belief that there is music playing, I believe that my daughter has come home, and she has, do I know this? The answer is not clear. But this apparent indeterminacy would not help anyone who claims that knowledge can arise from belief which does not constitute knowledge. For it is equally unclear, and for the same sort of reason, whether my belief that there is music playing is *sufficiently* reasonable—say, in terms of how good my perceptual grounds are—to give me knowledge that music is playing. The stronger our tendency to say that I know she is home, the stronger our inclination to say that I do after all know that there are strains of music in the air. Notice something else. In the only cases where the third kind of chain seems likely to ground knowledge (or justification), there is a degree—apparently a substantial degree—of justification. If there can be an epistemic chain which ends with belief that is not knowledge only because it ends, in this way, with justification, then we are apparently in the general vicinity of knowledge. We seem to be at most a few degrees of justification away. Knowledge is not emerging from nothing, as it were—the picture originally evoked by the third kind of epistemic chain—but from something characteristically much like it: justified true belief. There would thus be a foundation after all: not bedrock, but perhaps ground that is nonetheless firm enough to yield a foundation we can build upon.

The fourth possibility is that epistemic chains which originate with knowledge end in non-inferential knowledge: knowledge not inferentially based on further knowledge (or further justified belief). That knowledge, in turn, is apparently grounded in experience, say in my auditory impression of music or in my intuitive sense that if A is one mile from B, then B is one mile from A. This non-inferential grounding of my knowledge can explain how that knowledge is (epistemically) direct. It arises non-inferentially—and so without any intermediary premise that must be known along the way—from (I shall assume) one of the four classical kinds of foundational material, namely, perception, memory, introspection, and reason.

Such direct grounding in experience also seems to explain why a belief so grounded may be expected to be *true;* for experience seems to connect the beliefs it grounds to the reality they are apparently about, in such a way that what is believed concerning that reality tends to be the case. For empirical beliefs at least, this point seems to explain best why we have those beliefs. Let me illustrate all this. Normally, when I know that there is music playing, it is just because I hear it, and not on the basis of some further belief of mine; hence, the chain grounding my knowledge that my daughter has come home is anchored in my auditory perception, which in turn reflects the musical reality represented by my knowledge that there is music playing. This reality explains both my perception and, by explaining that, indirectly explains my believing the proposition I know on the basis of this perception—that my daughter is home.

The non-inferentially grounded epistemic chains in question may differ in many ways. They differ *compositionally,* in the sorts of beliefs constituting them, and *causally,* in the kind of causal relation holding between one belief and its successor. This relation, for instance, may or may not involve the predecessor belief's being necessary or sufficient for its successor: perhaps, on grounds other than the music, I would have believed my daughter was home; and perhaps not, depending on how many indications of her presence are accessible to me. Such chains also differ *structurally,* in the kind of *epistemic transmission* they exhibit; it may be deductive, as where I infer a theorem from an axiom by rigorous rules of deductive inference, or inductive, as where I infer from the good performance of a knife that others of that kind will also cut well; or the transmission of knowledge or justification may combine deductive and inductive elements. Epistemic chains also differ *foundationally,* in their ultimate grounds, the anchors of the chains; the grounds may, as illustrated, be perceptual or rational, and they may vary in justificational strength.

Different proponents of the fourth possibility have held various views about the character of the *foundational knowledge,* i.e., the beliefs constituting the knowledge that makes up the final link and anchors the chain in experience or reason. Some, including Descartes, have thought that the appropriate beliefs must be infallible, or at least indefeasibly justified.[6] But in fact all that the fourth possibility requires is *non-inferential knowledge,* knowledge not (inferentially) based on other knowledge (or other justified belief). Non-inferential knowledge need not be of self-evident propositions, nor constituted by indefeasibly justified belief, the kind whose justification cannot be defeated. The case of introspective beliefs, which are paradigms of those that are non-inferentially justified, supports this view, and we shall see other reasons to hold it.

III. Fallibilist Foundationalism

The foundationalism with which the regress argument concludes is quite generic and leaves much to be determined, such as how *well* justified foundational beliefs must be if they are to justify a superstructure belief based on them. In assessing the foundationist—coherentist controversy, then, we need a more detailed formulation. The task of this section is to develop one. I start with a concrete example.

As I sit reading on a quiet summer evening, I sometimes hear a distinctive patter outside my open window. I immediately believe that it is raining. It may then occur to me that if I do not bring in the lawn chairs, the cushions will be soaked. But this I do not believe immediately, even if the thought strikes me in an instant; I believe it on the basis of my prior belief that it is raining. The first belief is perceptual, being grounded directly in what I hear. The second is inferential, being grounded not in what I perceive but in what I believe. My belief that it is raining expresses a premise for my belief that the cushions will be soaked. There are many beliefs of both kinds. Perception is a major source of beliefs; and, from beliefs we have through perception, many others arise inferentially. The latter, inferential beliefs are then based on the former, perceptual beliefs. When I see a headlight beam cross my window and immediately believe, perceptually, that a car's light is moving out there, I may, on the basis of that belief, come to believe, inferentially, that someone has entered my driveway. From this proposition in turn I might infer that my doorbell is about to ring; and from that I might infer still further propositions. Assuming that knowledge implies belief, the same point holds for knowledge: much of it is perceptually grounded, and much of it is inferential.[7] There is no definite limit on how many inferences one may draw in such a chain, and people differ in how many they tend to draw. Could it be, however, that despite the apparent obviousness of these points, there really *is* no non-inferential knowledge or belief, even in perceptual cases? If inference can take us forward indefinitely beyond perceptual beliefs, why may it not take us backward indefinitely from them? To see how this might be thought to occur, we must consider more systematically how beliefs arise, what justifies them, and when they are sufficiently well grounded to constitute knowledge.

Imagine that when the rain began I had not trusted my ears. I might then have believed just the weaker proposition that there was a pattering sound and only on that basis, and after considering the situation, come to believe that it was raining. We need not stop here, however. For suppose I do not trust my sense of hearing. I might then believe merely that it *seems* to me that there is a patter, and only on that basis believe that there is such a sound. But surely this cannot go much further, and in fact there is no need to go even this far. Still, what theoretical reason is there to stop? It is not as if we had to articulate all our beliefs. Little of what we believe is at any one time before our minds being inwardly voiced. Indeed, perhaps we can have infinitely many beliefs, as some think we do.[8] But, as I have already suggested, it is simply not clear that a person's cognitive system can sustain an infinite set of beliefs, and much the same can be said regarding a circular cognitive chain.

Even if there could be infinite or circular belief chains, foundationalists hold that they cannot be sources of knowledge or justification. The underlying idea is in part this. If knowledge or justified belief arises through inference, it requires belief of at least one premise; and that belief could produce knowledge or justified belief of a proposition inferred from the premise only if the premise belief is itself an instance of knowledge or is at least justified. But if the premise is justified, it must be so by virtue of *something*—otherwise it would be self-justified, and hence one kind of foundational belief after all. If, however, experience cannot do the justificatory work, then the belief must derive its justification from yet another set of premises, and the problem arises all over again: what justifies that set? In the light of such points, the foundationalist concludes that if any of our beliefs are justified or constitute knowledge, then some of our beliefs are justified, or constitute knowledge, simply because they arise (in a certain way) from experience or reflection (including intuition as a special case of reflection). Indeed, if we construe experience broadly enough to include logical reflection and rational intuition, then experience may be described as the one overall source. In either case, there appear to be at least four basic sources of knowledge and justified belief: perception; consciousness, which grounds, e.g., my knowledge that I am thinking about the structure of justification; reflection, which is, for instance, the basis of my justified belief that if A is older than B and B is older than C, then A is older than C; and memory: I can be justified in believing that, say, I left a light on simply by virtue of the sense of recalling my having done so.[9]

Particularly in the perceptual cases, some foundationalists tend to see experience as a mirror of nature.[10] This seems to some foundationalists a good, if limited, metaphor because it suggests at least two important points: first, that some experiences are *produced* by external states of the world, somewhat as light produces mirror images; and second, that (normally) the experiences in some way *match* their causes, for instance in the color and shape one senses in one's visual field.[11] If one wants to focus on individual perceptual beliefs, one might think of a thermometer model; it suggests both the causal connections just sketched, but also, perhaps even more than the mirrored meta-phor, *reliable* responses to the external world.[12] From this causal–responsiveness perspective, it is at best unnatural to regard perceptual beliefs as inferential. They are not formed by inference from anything else believed but directly reflect the objects and events that cause them.

The most plausible kind of foundationalism will be fallibilist (moderate) in at least the following respects—and I shall concentrate on foundationalism about justification, though much that is said will also hold for foundationalism about knowledge. First, as a purely philosophical thesis about the *structure* of justification, foundationalism should be neutral with respect to skepticism and should not entail that there *are* justified beliefs. Second, if it is fallibilistic, it must allow that a justified belief, even a foundational one, be false. To require here justification of a kind that entails truth is to require that justified foundational beliefs be infallible. Third, superstructure beliefs may be only inductively, hence fallibly, justified by foundational ones and thus (unless they are necessary truths) can be false even when the latter are true. Just as one's warranted beliefs may be fallible, one's inferences may be, also, leading from truth to falsity. If the proposition is sufficiently supported by evidence one justifiedly believes, one may justifiedly hold it on the basis of that evidence, even if one could turn out to be in error. Fourth, a fallibilist foundationalism must allow for *discovering* error or lack of justification, in foundational as well as in superstructure beliefs. Foundational beliefs may be discovered to conflict either with other such beliefs or with sufficiently well-supported superstructure beliefs.

These four points are quite appropriate to the inspiration of the theory as expressed in the regress argument: it requires epistemic unmoved movers, but not unmovable movers. Solid ground is enough, even if bedrock is better. There are also different kinds of bedrock, and not all of them have the invulnerability apparently belonging to beliefs of luminously self-evident truths of logic. Even foundationalism as applied to knowledge can be fallibilistic; for granting that false propositions cannot be known, foundationalism about knowledge does not entail that one's *grounds* for knowledge (at any level) are indefeasible. Perceptual grounds, e.g., may be overridden; and one can fail (or cease) to know a proposition not because it is (or is discovered to be) false, but because one ceases to be justified in believing it.

I take *fallibilist foundationalism,* as applied to justification, to be the inductivist thesis that

I. For any S and any t, (1) the structure of S's body of justified beliefs is, at t, foundational in the sense that any inferential (hence non-foundational) justified beliefs S has depend for their justification on one or more non-inferential (thus in a sense foundational) justified beliefs of S's; (2) the justification of S's foundational beliefs is at least typically defeasible; (3) the inferential transmission of justification need not be deductive; and (4) non-foundationally justified beliefs need not derive *all* of their justification from foundational ones, but only enough so that they would remain justified if (other things remaining equal) any other justification they have (say, from coherence) were eliminated.[13]

This is fallibilist in at least three ways. Foundational beliefs may turn out to be unjustified or false or both; superstructure beliefs may be only inductively, hence fallibly, justified by foundational ones and hence can be false even when the latter are true; and possibility of *discovering* error or lack of justification, even in foundational beliefs, is left open: they may be found to conflict either with other such beliefs or with sufficiently well-supported superstructure beliefs. Even foundationalism as applied to knowledge can forswear infallibility. For although false beliefs cannot be knowledge, what is known can be both contingent—and so might have been false—*and* based on defeasible grounds—and so might cease to be known. We can lose knowledge when our grounds are overridable; hence, even self-knowledge is defeasible.

Since I am particularly concerned to clarify foundationalism in contrast to coherentism, I want to focus on the roles fallibilist foundationalism allows for coherence (conceived in any plausible way) in relation to justification. There are at least two important roles coherence may apparently play.

The first role fallibilist foundationalism allows for coherence—or at least for incoherence—is negative. Incoherence may defeat justification or knowledge, even the justification of a directly justified, hence foundational, belief (or one constituting knowledge), as where my justification for believing I am hallucinating books prevents me from knowing, or remaining justified in believing, certain propositions incoherent with it, say that the books in my study are before me. If this is not ultimately a role for coherence itself—which is the opposite and not merely the absence of incoherence—it *is* a role crucial for explaining points stressed by coherentism. Coherentists have not taken account of the point that incoherence is not merely the absence of coherence and cannot be explicated simply through analyzing coherence, nor accounted for as an epistemic standard only by a coherentist theory (a point to which I shall return); but they have rightly noted, for instance, such things as the defeasibility of the justification of a memorial belief owing to its incoherence with perceptual beliefs, as where one takes oneself to remember an oak tree in a certain spot, yet, standing near the very spot, can find no trace of one. Because fallibilist foundationalism does not require indefeasible justification on the part of the relevant memory belief, there is no anomaly in its defeat by perceptual evidence.

Second, fallibilist foundationalism can employ an *independence principle,* one of a family of principles commonly emphasized by coherentists, though foundationalists need not attribute its truth to coherence. This principle says that the larger the number of independent mutually coherent factors one believes to support the truth of a proposition, the better one's justification for believing it (other things being equal). The principle can explain, e.g., why my justification for believing, from what I hear, that my daughter has come home increases as I acquire new beliefs supporting that conclusion, say that there is a smell of popcorn. For I now have a confirmatory belief which comes through a different sense (smell) and does not depend for its justification on my other evidence beliefs.

Similar principles consistent with foundationalism can accommodate other cases in which coherence apparently enhances justification, for instance where a proposition's explaining, and thereby cohering with, something one justifiably believes, tends to confer some justification on that proposition. Suppose I check three suitcases at the ticket counter. Imagine that as I await them at the baggage terminal I glimpse two on the conveyor at a distance and tentatively believe that they are mine. The propositions that (a) the first is mine, (b) the second is, and (c) these two are side by side—which I am fully justified in believing because I can clearly see how close they are to each other—would be explained by the hypothesis that my three suitcases are now coming off together; and that hypothesis, in turn, derives some justification from its explaining what I already believe. When I believe the further proposition, independent of (a)–(c), that my third suitcase is coming just behind the second, the level of my justification for the hypothesis rises.

Fallibilist foundationalism thus allows for coherence to play a significant though restricted role in explicating justification, and it provides a major place for incoherence in this task. But there remains a strong contrast between the two accounts of justification, as we shall soon see.

IV. Holistic Coherentism

The notion of coherence is frequently appealed to in epistemological and other contexts, but it is infrequently explicated. Despite the efforts that have been made to clarify coherence, explaining what it is remains difficult.[14] It is not mere consistency, though *in*consistency is the clearest case of incoherence. Whatever coherence is, it is a cognitively *internal* relation, in the sense that it is a matter of how one's beliefs (or other cognitive items) are related *to one another,* not to anything outside one's system of beliefs, such as one's perceptual experience. Coherence is sometimes connected with explanation; it is widely believed that propositions which stand in an explanatory relation cohere with one another and that this coherence counts toward that of a person's beliefs of the propositions in question. If the wilting of the leaves is explained by billowing smoke from a chemical fire, then presumably the proposition expressing the first event coheres with the proposition expressing the second (even if the coherence is not obvious and is relative to the context). Probability is also relevant: if the probability of one proposition you believe is raised by that of a second you believe, this at least counts toward the coherence of the first of the beliefs with the second. The relevant notions of explanation and probability are themselves philosophically problematic, but our intuitive grasp of them can still help us understand coherence.

In the light of these points, let us try to formulate a plausible version of coherentism as applied to justification. The central coherentist idea concerning justification is that a belief is justified by its coherence with other beliefs one holds. The unit of coherence may be as large as one's entire set of beliefs, though some may be more significant in producing the coherence than others, say because of differing degrees of their closeness in subject matter to the belief in question. This conception of coherentism would be accepted by a proponent of the circular view, but the thesis I want to explore differs from that view in not being *linear:* it does not take justification for believing that *p,* or knowledge that *p,* to emerge from an inferential line running from premises for *p* to that proposition as a conclusion from them, and from other premises to the first set of premises, and so on until we return to the original proposition as a premise. On the circular view, no matter how wide the circle or how rich its constituent beliefs, there is a line from any one belief in a circular epistemic chain to any other. In practice I may never trace the entire line, as by inferring one thing I know from a second, the second from a third, and so on until I reinfer the first. Still, on this view there is such a line for every belief constituting knowledge.

Coherentism need not, however, be linear, and I believe that the most plausible versions are instead holistic.[15] A moderate version of *holistic coherentism* might be expressed as follows:

> II. For any S and any t, if S has any justified beliefs at t, then, at t, (1) they are each justified by virtue of their coherence with one or more others of S's beliefs; and (2) they would remain justified even if (other things remaining equal) any justification they derive from sources other than coherence were eliminated.

The holism required is minimal, since the unit of coherence may be as small as one pair of beliefs—though it may also be as large as the entire system of S's beliefs (including the belief whose justification is in question, since we may take such partial ''self-coherence'' as a limiting case). But the formulation also applies to the more typical cases of holistic coherentism; in these cases a justified belief coheres with a substantial number of other beliefs, but not necessarily with all of one's beliefs. Some beliefs, like those expressing basic principles of one's thinking, can be justified only by coherence with a large and diverse group of related beliefs. Coherentist theories differ concerning the sense (if any) in which the set of beliefs whose coherence determines the justification of some belief belonging to it must be a ''system.''

To illustrate holistic coherentism, consider a question that evokes a justification. Ken wonders how, from my closed study, I know (or why I believe) that my daughter is home. I say that there is music playing in the house. He next wants to know how I can recognize my daughter's music from behind my closed doors. I reply that what I hear is the wrong sort of thing to come from any nearby house. He then asks how I know that it is not from a passing car. I say that the volume is too steady. He now wonders whether I can distinguish, with my door closed, my daughter's vocal music from the singing of a neighbor in her yard. I reply that I hear an accompaniment. In giving each justification I apparently go only one step along the inferential line: initially, for instance, just to my belief that there is music playing in the house. For my belief that my daughter is home *is* based on this belief about the music. After that, I do not even mention anything that this belief, in turn, is based on; rather, I defend my beliefs as appropriate, in terms of an entire pattern of interrelated beliefs I hold. And I may appeal to many different parts of the pattern. For coherentism, then, beliefs representing knowledge do not lie at one end of a grounded chain; they fit a coherent pattern and are justified through their fitting it in an appropriate way.

Consider a variant of the case. Suppose I had seemed to hear music of neither the kind my daughter plays nor the kind the neighbors play nor the sort I expect from passing cars. The proposition that this is what I hear does not cohere well with my belief that the music is played by my daughter. Suddenly I recall that she was bringing a friend, and I remember that her friend likes such music. I might now be justified in believing that my daughter is home. When I finally hear her voice, I know that she is. The crucial thing here is how, initially, a kind of *incoherence* prevents justification of my belief that she is home, and how, as relevant pieces of the pattern develop, I become justified in believing, and (presumably) come to know, that she is. Arriving at a justified belief, on this view, is more like answering a question by looking up diverse information that suggests the answer than like deducing a theorem from axioms.

Examples like this show how a holistic coherentism can respond to the regress argument *without* embracing the possibility of an epistemic circle (though its proponents need not reject that either). It may deny that there are only the four kinds of possible epistemic chains I have specified. There is apparently another possibility, not generally noted: that the chain terminates with belief which is *psychologically direct* but *epistemically indirect* or, if we are talking of coherentism about justification, *justificationally indirect*. Hence, the last link is, as belief, direct, since it is non-inferential; yet, as knowledge, it is *indirect*, not in the usual sense that it is inferential but rather in the broad sense that the belief constitutes knowledge only by virtue of receiving support from other knowledge or belief. Thus, my belief that there is music playing is psychologically direct because it is simply grounded, causally, in my hearing and is not (inferentially) based on any other belief; yet my *knowledge* that there is music is not epistemically direct. It is epistemically, but not inferentially, based on the coherence of my belief that there is music with my other beliefs, presumably including many that constitute knowledge themselves. It is thus knowledge *through*, though not by inference from, other knowledge—or at least through justified beliefs; hence it is epistemically indirect and thus non-foundational.

There is another way to see how this attack on the regress argument is constructed. The coherentist grants that the belief element *in* my knowledge is non-inferentially grounded in perception and is in that sense direct; but the claim is that the belief constitutes knowledge only by virtue of coherence with my other beliefs. The strategy, then—call it the *wedge strategy*—is to sever the connection foundationalism usually posits between the psychological and the epistemic. In the common cases, foundationalists tend to hold, the basis of one's *knowledge* that *p,* say a perceptual experience, is also the basis of one's belief that *p;* similarly, for justified belief, the basis of its justification is usually also that of the belief itself. For the coherentist using the wedge strategy, the epistemic ground of a belief need not be a psychological ground. Knowledge and justification are a matter of how well the system of beliefs hangs together, not of how well grounded the beliefs are—and they may indeed hang: one could have a body of justified beliefs, at least some of them constituting knowledge, even if *none* of them is justified by a belief or experience in which it is psychologically grounded.

In a sense, of course, coherentism does posit a *kind* of foundation for justification and knowledge: namely, coherence. But so long as coherentists deny that justification and knowledge can be *non-inferentially* grounded in experience or reason, this point alone simply shows that they take justification and knowledge to be based on something (to be supervenient properties, as some would put it). Justification and knowledge are still grounded in the coherence of elements which themselves admit of justification and derive their justification (or status as knowledge) from coherence with other such items rather than from grounding in elements like sensory impressions (say of music), which, though not themselves justified or unjustified, confer justification on beliefs they ground.

Apparently, then, the circularity objection to coherentism can be met by construing the thesis holistically and countenancing psychologically direct beliefs. One could insist that if a non-inferential, thus psychologically direct, belief constitutes knowledge, it *must* be direct knowledge. But the coherentist would reply that in that case there will be two kinds of direct knowledge: the kind the foundationalist posits, which derives from grounding in a basic experiential or rational source, and the kind the coherentist posits, which derives from coherence with other beliefs and not from being based on those sources. This is surely a plausible response.

Is the holistic coherentist trying to have it both ways? Not necessarily. Holistic coherentism can grant that a variant of the regress argument holds for belief, since the only kind of inferential belief chain that it is psychologically realistic to attribute to us is the kind terminating in direct (non-inferential) belief. But even on the assumption that knowledge is constituted by (certain kinds of) beliefs, it does not follow that direct belief which constitutes knowledge is also direct *knowledge*. Epistemic dependence, on this view, does not imply inferential or psychological dependence; hence, a non-inferential belief can depend for its status as knowledge on other beliefs. Thus, the coherentist may grant a kind of *psychological foundationalism*—which says (in part) that if we have any beliefs at all, we have some direct (non-inferential) ones—yet deny epistemological foundationalism, which requires that there be knowledge which is epistemically (and normally also psychologically) direct, if there is any knowledge at all. Holistic coherentism may grant experience and reason the status of psychological foundations of our belief systems, but it denies that they are the basic sources of justification or knowledge.

V. Foundationalism, Coherentism, and Defeasibility

Drawing on our results above, this section considers how fallibilist foundationalism and holistic coherentism differ and, related to that, how the controversy is sometimes obscured by failure to take account of the differences.

There is one kind of case that seems both to favor foundationalism and to show something about justification that coherentism in any form misses. It might seem that coherence theories of justification are decisively refuted by the possibility of *S*'s having, if just momentarily, only a single belief which is nonetheless justified, say that there is music playing. For this belief would be justified without cohering with any others *S* has. But could one have just a single belief? Could I, for instance, believe that there is music playing yet not believe, say, that there are (or could be) musical instruments, melodies, and chords? It is not clear that I could; and foundationalism does not assume this possibility, though the theory may easily be wrongly criticized for implying it. Foundationalism is in fact consistent with *one* kind of coherentism—*conceptual coherentism*. This is a coherence theory of the acquisition of concepts which says that a person acquires concepts, say of musical pieces, only in relation to one another and must acquire an entire family of related concepts in order to acquire any concept.

It remains questionable, however, whether my justification for believing that there is music playing ultimately *derives* from the coherence of the belief with others, i.e., whether coherence is even partly the basis of my justification in holding this belief.[16] Let us first note an important point. Suppose the belief turns out to be *in*coherent with a second, such as my belief that I am standing before the phonograph playing the music yet see no movement of its turntable; now the belief may *cease* to be justified, since if I really hear the phonograph, I should see its turntable moving. But this shows only that the belief's justification is *defeasible*—liable to being either overridden (roughly, outweighed) or undermined—should sufficiently serious incoherence arise. It does not show that the justification derives from coherence. In this case the justification of my belief grounded in hearing may be overridden. My better-justified beliefs, including the belief that a phonograph with a motionless turntable cannot play, may make it more reasonable for me to believe that there is *not* music playing in the house.

The example raises another question regarding the possibility that coherence is the source of my justification, as opposed to incoherence's constraining it. Could incoherence override the justification of my belief if I were not *independently* justified in believing that a proposition incoherent with certain other ones is, or probably is, false, e.g., in believing that if I do not see the turntable moving, then I do not hear music from the phonograph? For if I lacked independent justification, should I not suspend judgment on, or even reject, the other propositions and retain my original belief? And aren't the relevant other beliefs or propositions—those that can override or defeat my justification—precisely the kind for which, directly or inferentially, we have some degree of justification through the experiential and rational sources, such as visual perception of a stockstill turntable? Note that the example shows that these beliefs or propositions need not be a priori; thus it is not open to coherentists to claim that only the a priori is an exception to the thesis that justification is determined by coherence.

A similar question arises regarding the crucial principles themselves. Could incoherence play the defeating role it does if we did not have a kind of foundational justification for principles to the effect that certain kinds of evidences or beliefs override certain other kinds? More generally, can we *use,* or even benefit from, considerations of coherence in acquiring justification, or in correcting mistaken presuppositions of justification, if we do not bring to the various coherent or incoherent patterns principles not derived from those very patterns? If, without such principles to serve as justified standards that guide belief formation and belief revision, we can become justified by coherence, then coherence would seem to be playing the kind of generative role that foundational sources are held to play in producing justification. One could become justified in believing that *p* by virtue of coherence even if one had no justified principles by which one could, for instance, inferentially connect the justified belief that *p* with others that cohere with it.

There is a second case, in which one's justification is simply undermined: one ceases to be justified in believing the proposition in question, though one does not become justified in believing it false. Suppose I seem to see a black cat, yet there no longer appears to be one there if I move five feet to my left. This experience could justify my believing, and lead me to believe, that I might be hallucinating. This belief in turn is to a degree incoherent with, and undermines the justification of, my visual belief that the cat is there, though it does not by itself justify my believing that there is *not* a cat there. Again, however, I am apparently justified, independently of coherence, in believing a proposition relevant to my overall justification for an apparently foundational perceptual belief: namely, the proposition that my seeing the cat there is incoherent with my merely hallucinating it there. The same seems to hold for the proposition that my seeing the cat there coheres with my feeling fur if I extend my hand to the feline focal point of my visual field. Considerations like these suggest that coherence has the role it does in justification only because *some* beliefs are justified independently of it.

Both examples illustrate an important distinction that is often missed.[17] It is between defeasibility and epistemic dependence or, alternatively, between *negative epistemic dependence,* which is a form of defeasibility, and *positive epistemic dependence,* the kind beliefs bear to the source(s) from which they *derive* any justification they have or, if they represent knowledge, their status as knowledge. The defeasibility of a belief's justification by incoherence does not imply that, as coherentists hold, its justification positively depends on coherence. If my garden is my source of food, I (positively) depend on it. The fact that people could poison the soil does not make their nonmalevolence part of my food *source* or imply a (positive) dependence on them, such as I have on the sunshine. Moreover, it is the sunshine that (with rainfall and other conditions) explains both my having the food and the amount I have. The non-malevolence is necessary for, but does not explain, this; it alone, under the relevant conditions of potential for growth, does not even tend to produce food.

So it is with perceptual experience as a source of justification. Foundationalists need not deny that a belief's justification negatively depends on something else, for as we have seen they need not claim that justification must be indefeasible. It may arise, unaided by coherence, from a source like perception; yet it remains defeasible from various quarters—including conflicting perceptions. Negative dependence, however, does not imply positive dependence. The former is determined by the absence of something—defeaters; the latter is determined by the presence of something—justifiers. Justification can be defeasible by incoherence and thus overridden or undermined should incoherence arise, without owing its existence to coherence. Fallibilist foundationalism is not, then, a blend of coherentism, and it remains open just what positive role, if any, it must assign to coherence in explaining justification.

There is a further point that fallibilist foundationalism should stress, and in appraising the point we learn more about both coherentism and justification. If I set out to *show* that my belief is justified—as the dialectical formulation of the regress problem invites one to think stopping the regress of justification requires—I do have to cite propositions that cohere with the one to be shown to be justified for me, say that there is music in my house. In some cases, these are not even propositions one already believes. Often, in defending the original belief, one forms new beliefs, such as the belief one acquires, in moving one's head, that one can vividly see the changes in perspective that go with seeing a black cat. More important, these beliefs are highly appropriate to the *process* of self-consciously justifying one's belief; and the result of that process is twofold: forming the second-order belief that the original belief is justified and showing that the latter is justified. Thus, coherence is important in showing that a belief is justified. In *that* limited sense coherence is a pervasive element in justification: it is pervasive in the process of *justifying,* especially when that is construed as showing that one has justification.

Why, however, should the second-order beliefs appropriate to *showing* that a belief is justified be necessary for its *being* justified? They need not be. Indeed, why should one's simply having a justified belief imply even that one could be justified in holding the second-order beliefs appropriate to showing that it is justified? It would seem that just as a little child can be of good character even if unable to defend its character against attack, one can have a justified belief even if, in response to someone who doubts this, one could not show that one does. Supposing I have the sophistication to form a second-order belief that my belief that there is a cat before me is justified, the latter belief can be justified so long as the former is *true;* and it can be *true* that my belief about the cat is justified even if I am not justified in holding it or am unable to show that it is true. Justifying a second-order belief is a sophisticated process. The process is particularly sophisticated if the second-order belief concerns a special property like the justification of the original belief. Simply being justified in a belief about, say, the sounds around one is a much simpler matter. But confusion is easy here, particularly if the governing context is an imagined dialectic with a skeptic. Take, for instance, the question of how a simple perceptual belief "is justified." The very phrase is ambiguous. The question could be "By what process, say of reasoning, has the belief been (or might it be) justified?" or, on the other hand, "In virtue of what is the belief justified?" These are very different questions. The first invites us to conceive justification as a process of which the belief is a beneficiary, the second to conceive it as a property that a belief has, whether in virtue of its content, its genesis, or others of its characteristics or relations. Both aspects of the notion are important, but unfortunately much of our talk about justification makes it easy to run them together. A justified belief could be one that *has* justification or one that *has been* justified; and a request for someone's justification could be a request for a list of justifying factors or for a recounting of the process by which the person justified the belief.

Once we forswear the mistakes just pointed out, what argument is left to show the (positive) dependence of perceptual justification on coherence? I doubt that any plausible one remains, though given how hard it is to discern what coherence is, we cannot be confident that no plausible argument is forthcoming. Granted, one could point to the oddity of saying things like, "I am justified in believing that there is music playing, but I cannot justify this belief." Why is this odd if not because, when I have a justified belief, I can give a justification for it by appeal to beliefs that cohere with it? But consider this. Typically, in asserting something, say that there were lawsuits arising from an accident, I imply that, in some way or other, I *can* justify what I say, especially if the belief I express is, like this one, not plausibly thought to be grounded in a basic source such as perception. In the quoted sentence I deny that I can justify what I claim. The foundationalist must explain why that is odd, given that I can be justified in believing propositions even when I cannot show that I am (and may not even believe I am). The main point needed to explain this is that it is apparently my *asserting* that my belief is justified, rather than its being so, that gives the appearance that I must be able to give a justification of the belief. Compare "*She* is justified in believing that there is music playing, but (being an intuitive and unphilosophical kind of person) she cannot justify that proposition." This has no disturbing oddity, because the person said to have justification is not the one claiming it. Since she might be shocked to be asked to justify the proposition and might not know how to justify it, this statement might be true of her. We must not stop here, however. There are at least two further points.

First, there is quite a difference between *showing* that one is justified and simply *giving* a justification. I can give my justification for believing that there is music simply by indicating that I hear it. But this does not show that I am justified, at least in the sense of 'show' usual in epistemology. That task requires not just exhibiting what justifies one but also indicating conditions for being justified *and* showing that one meets them. It is one thing to cite a justifier, such as a clear perception; it is quite another to show that it meets a sufficiently high standard to *be* a justifier of the belief it grounds. Certainly skeptics—and probably most coherentists as well—have in mind something more like the latter process when they ask for a justification. Similarly—and this is the second point—where a regress of justification is, for fallibilist foundationalism, stopped by giving a (genuine) justification for the proposition in question, and the regress problem can be considered soluble because such stopping is possible, the skeptic will not countenance any stopping place, and certainly not any solution, that is not dialectically defended by argument showing that one is justified.

To be sure, it may be that at least typically when we do have a justified belief we can give a justification for it. When I justifiedly believe that there is music playing, I surely can give a justification: that I hear it. But I need not *believe* that I hear it *before* the question of justification arises. That question leads me to focus on my circumstances, in which I first had a belief solely about the music. I also had a *disposition,* based on my auditory experience, to form the belief that *I hear* the music, and this is largely why, in the course of justifying that belief, I then *form* the further belief that I do hear it. But a disposition to believe something does not imply an actual belief of it, not even a dispositional one, as opposed to one manifesting itself in consciousness. If I am talking loudly and excitedly in a restaurant, I may be disposed to believe this—so much so that if I merely think of the proposition that I am talking loudly, I will form the belief that I am and lower my voice. But this disposition does not imply that I *already* believe that proposition—if I did, I would not be talking loudly in the first place. In the musical case, I tend to form the belief that I hear the music if, as I hear it, the question of whether I hear it arises; yet I need not have subliminally believed this already. The justification I offer, then, is not by appeal to coherence with other beliefs I already had—such as that I saw the turntable moving—but by reference to what has traditionally been considered a basic source of both justification and knowledge: perception. It is thus precisely the kind of justification that foundationalists are likely to consider appropriate for a non-inferential belief. Indeed, one consideration favoring foundationalism about both justification and knowledge, at least as an account of our everyday epistemic practices, including much scientific practice, is that typically we cease to offer justification or to defend a knowledge claim precisely when we reach a basic source.

VI. Coherence, Foundations, and Justification

There is far more to say in clarifying both foundationalism and coherentism. But if what I have said so far is correct, then we can at least understand their basic thrusts. We can also see how coherentism may respond to the regress argument—in part by distinguishing psychological from epistemic directness. And we can see how foundationalism may reply to the charge that, once made moderate enough to be plausible, it depends on coherence criteria rather than on grounding in experience and reason. The response is in part to distinguish negative from positive epistemic dependence and to argue that foundationalism does not make justification depend positively on coherence, but only negatively on (avoiding) incoherence.

One may still wonder, however, whether fallibilist foundationalism concedes enough to coherentism. Granted that it need not restrict the role of coherence any more than is required by the regress argument, it still denies that coherence is (independently) necessary for justification. As most plausibly develop, fallibilist foundationalism also denies that coherence is a *basic* (non-derivative) source of justification—or at least that if it is, it can produce *enough* justification to render a belief unqualifiedly justified or (given truth and certain other conditions) to make it knowledge. A single drop of even the purest water will not quench a thirst. The moderate holistic coherentism formulated above is parallel in this: while it may grant foundationalism its typical psychological picture of how belief systems are structured, it denies that foundational justification (is independently) necessary for justification and that it is a basic source of justification, except possibly of degrees of justification too slight for knowledge or unqualifiedly justified belief.

The issue here is the difference in the two conceptions of justification. Broadly, foundationalists tend to hold that justification belongs to a belief, whether inferentially or directly, by virtue of its grounding in experience or reason; coherentists tend to hold that justification belongs to a belief by virtue of its coherence with one or more other beliefs. This is apparently a difference concerning basic sources. To be sure, my formulation may make coherentism sound foundationalistic, because justification is grounded not in an inferential relation to premises but in coherence itself, which sounds parallel to experience or reason. But note three contrasts with foundationalism: (1) the source of coherence is *cognitive,* because the coherence is an internal property of the belief system, whereas foundationalism makes no such restriction; (2) coherence is an inferential or at least epistemic generator, in the sense that it arises, with or without one's propositional objects, e.g., from entailment, inductive support, or explanation of one belief or proposition by another, whereas experiential sources and (for pure coherentists) even rational sources are a non-inferential generator of belief (these sources can produce and thereby explain belief, but they do not, according to coherentism, justify it); and (3) *S* has *inferential access* to the coherence-making relations: *S* can wield them in inferentially justifying the belief that *p*, whereas foundationalism does not require such access to its basic sources. Still, I want to pursue just how deep the difference between foundationalism and coherentism is; for once foundationalism is moderately expressed and grants the truth of conceptual coherentism, and once coherentism is (plausibly) construed as consistent with psychological foundationalism, it may appear that the views differ far less than the prevailing stereotypes would have us think.

It should help if we first contrast fallibilist foundationalism with *strong foundationalism* and compare their relation to coherentism. If we use Descartes' version as a model, strong foundationalism is deductivist, takes foundational beliefs as indefeasibly justified, and allows coherence at most a limited generative role. To meet these conditions, it may reduce the basic sources of justification to reason and some form of introspection. Moreover, being committed to the indefeasibility of foundational justification, it would not grant that incoherence can defeat such justification. It would also concede to coherentists, and hence to any independence principle they countenance, at most a minimal positive role, say by insisting that if a belief is supported by two or more independent cohering sources, its justification is increased at most "additively," that is, only by combining the justification transmitted separately from each relevant basic source.

By contrast, what fallibilist foundationalism denies regarding coherence is only that it is a basic (hence sufficient) source of justification. Thus, coherence by itself does not ground justification, and hence the independence principle does not apply to sources that have *no* justification; at most, the principle allows a coherence to raise the level of justification originally drawn from other sources to a level higher than it would reach if those sources did not mutually cohere. Similarly, if inference is a basic source of coherence (as some coherentists seem to believe), it is not a basic source of justification. It may enhance justification, as where one strengthens one's justification for believing someone's testimony by inferring the same point from someone else's. But inference *alone* does not generate justification. Suppose I believe several propositions without a shred of evidence and merely through wishful thinking. I might infer any number of others; yet even if by good luck I arrive at a highly coherent set of beliefs, I do not automatically gain justification for believing any of them. If I am floating in mid-ocean, strengthening my boat with added nails and planks may make it hang together more tightly and thereby make me feel secure; but if nothing indicates my location, there is no reason to expect this work to get me any closer to shore. Coherence may, to be sure, enable me to draw a beautiful map; but if there are no experiences I may rely on to connect it with reality, I may follow it forever to no avail. Even to be justified in *believing* that it will correspond with reality, I must have some experiential source to work from.

A natural coherentist reply is that when we consider examples of justified belief, not only do we always find some coherence, we also apparently find the right sort to account for the justification. This reply is especially plausible if—as I suggest is reasonable—coherentism as usually formulated is modified to include, in the coherence base, *dispositions to believe*. Consider my belief that music is playing. It coheres both with my beliefs about what records are in the house, what music my daughter prefers, my auditory capacities, etc., *and* with many of my dispositions to believe, say to form the belief that no one else in the house would play that music. Since such dispositions can themselves be well grounded, say in justification and, when they produce beliefs, can lead to reasonable inferences. These dispositions are thus appropriate for the coherence base, and including them among generators of coherence is particularly useful in freeing coherentism from implausibly positing all the beliefs needed for the justificational capacities it tends to take to underlie justified belief. We need not "store" beliefs of all the propositions needed for our own system of justified belief; the disposition to believe them is enough. Given this broad conception of coherence, it is surely plausible to take coherence as at least necessary for justified belief. And it might be argued that its justification is based on coherence, not on grounding in experience.

Let us grant both that the musical case does exhibit a high degree of coherence among my beliefs and dispositions to believe and even that the coherence is necessary for the justification of my belief. It does not follow that the justification is based on the coherence. Coherence could still be at best a *consequential necessary condition* for justification, one that holds as a result of the justification itself or what that is based on, as opposed to a *constitutive necessary condition,* one that either expresses part of what *is* for a belief to be justified or constitutes a basic source of it. The relation of coherence to the properties producing it might be analogous to that of heat to friction: a necessary product of it, but not part of what constitutes it.

If coherence is a constitutive necessary condition for justification, and especially if it is a basic source of it, we might expect to find cases in which the experiential and rational sources are absent, yet there is sufficient coherence for justified belief. But this is precisely what we do not easily find, if we ever find it. If I discover a set of my beliefs that intuitively cohere very well yet receive no support from what I believe (or at least am disposed to believe) on the basis of experience or reason, I am not inclined to attribute justification to any of them. To be sure, if the unit of coherence is large enough to include my actual beliefs, then because I have so many that *are* grounded in experience or reason (indeed, few that are not), I will almost certainly not in fact have any beliefs that, intuitively, seem justified yet are not coherent with some of my beliefs so grounded. This complicates assessment of the role of coherence in justification. But we can certainly imagine beings (or ourselves) artificially endowed with coherent sets of beliefs *not* grounded in experience or reason; and when we do, it appears that coherence does not automatically confer justification.

One might conclude, then, that it is more nearly true that coherence is based on justification (or whatever confers justification) than that the latter is based on the former. Further, the data we have so far considered can be explained on the hypothesis that both coherence among beliefs and their justification rest on the beliefs' being grounded (in an appropriate way) in the basic sources. For particularly if a coherence theory of the acquisition of concepts is true, one perhaps cannot have a belief justified by a basic source without having beliefs—or at least dispositions to believe—related in an intimate (and intuitively coherence-generating) way to that belief. One certainly cannot have a justified belief unless no incoherence defeats its justification. Given these two points, it is to be expected that on a fallibilist foundationalism, justification will normally imply coherence, both in the positive sense involving mutual support and in the weak sense of the absence of potential incoherence. There is some reason to think, then, that coherence is not a basic source of justification and is at most a consequentially necessary condition for it.

There is at least one more possibility to be considered, however: that *given* justification from foundational sources, coherence can generate more justification than S would have from those sources alone. If so, we might call coherence a *conditionally basic* source, in that, where there is already some justification from other sources, it can produce new justification. This bears on interpreting the independence principle. It is widely agreed that our justification increases markedly when we take into account independent sources of evidence, as where I confirm that there is music playing by moving closer to enhance my auditory impression and by visually confirming that a phonograph is playing. Perhaps what explains the dramatic increase in my overall justification here is not just "additivity" of foundational justification but also coherence as a further source of justification.

There is plausibility in this reasoning, but it is not cogent. For one thing, there really are no such additive quantities of justification. Perhaps we simply combine degrees of justification, so far as we can, on analogy with combinations of independent probabilities. Thus, the probability of at least one heads on two fair coin tosses is not $\frac{1}{2} + \frac{1}{2}$ (the two independent probabilities), which would give the event a probability of 1 and make it a certainty; the probability is $\frac{3}{4}$, i.e., 1 minus the probability of two tails, which is $\frac{1}{4}$. Insofar as degrees of justification are quantifiable, they combine similarly. Moreover, the relevant probability rules do not seem to depend on coherence; they seem to be justifiable by *a priori* reasoning in the way beliefs grounded in reason are commonly thought to be justifiable, and they appear to be among the principles one must *presuppose* if one is to give an account of how coherence contributes to justification. The (limited) analogy) between probability and justification, then, does not favor coherentism and may well favor foundationalism.

There remains a contrast between, say, having six independent credible witnesses tell me that *p* on separate occasions which I do not connect with one another, and having them do so on a single occasion when I can note the coherence of their stories. In the first case, while my isolated beliefs cohere, I have no belief that they do, nor even a sense of their collective weight. This is not, to be sure, a case of six increments of isolated foundational justification versus a case of six cohering items of evidence. Both cases exhibit coherence; but in the second there is an additional belief (or justified disposition to believe): *that* six independent witnesses agree. Foundationalists as well as coherentists can plausibly explain how this additional belief increases the justification one has in the first case. It would be premature, then, to take cases like this to show that coherence is even a conditionally basic source of justification. It may only reflect other sources of justification, rather than contribute any.

VII. Epistemological Dogmatism and the Sources of Justification

Of the problems that remain for understanding the foundationalism–coherentism controversy, the one most readily clarified by the results of this chapter, is the dogmatism objection. This might be expressed as follows. If one can have knowledge or justified belief without being able to show that one does, and even without a premise from which to derive it, then the way is open to claim just about anything one likes, defending it by cavalierly noting that one can be justified without being able to show that one is. Given the conception of the foundationalism–coherentism controversy developed here, we can perhaps throw some new light on how the charge of dogmatism is relevant to each position.

The notion of dogmatism is not easy to characterize, and there have apparently been few detailed discussions of it in recent epistemological literature.[18] My focus will be dogmatism as an epistemological attitude or stance, not as a trait of personality. I am mainly interested in what it is to hold a belief dogmatically. This is probably the basic notion in any case: a general dogmatic attitude, like the personality trait of dogmatism, is surely in some way a matter of having or tending to have dogmatically held beliefs.[19]

It will be useful to start with some contrasts. Dogmatism in relation to a belief is not equivalent to stubbornness in holding it; for even if a dogmatically held belief cannot be easily given up, one could be stubborn in holding a belief simply from attachment to it, and without the required disposition to defend it or regard it as better grounded than alternatives. For similar reasons, psychological certainty in holding a belief does not entail dogmatism. Indeed, even if one is both psychologically certain of a simple logical truth *and* disposed to reject denials of it with confidence and to suspect even well-developed arguments against it as sophistical, one does not qualify as dogmatic. The content of one's view is important: even moderate insistence on a reasonably disputed matter may bespeak dogmatism; stubborn adherence to the self-evident need not. An attitude that would be dogmatic in holding one belief may not be so in holding another.

Dogmatic people are often closed-minded, and dogmatically held beliefs are often closed-mindedly maintained; but a belief held closed-mindedly need not be held dogmatically: it may be maintained with a guilty realization that emotionally one simply cannot stand to listen to challenges of it, and with an awareness that it might be mistaken. Moreover, although people who hold beliefs dogmatically are often intellectually pugnacious in defending them, or even in trying to win converts, such pugnacity is not sufficient for dogmatism. Intellectual pugnacity is consistent with a keen awareness that one might be mistaken, and it may be accompanied by open-minded argumentation for one's view. Nor need a dogmatically held belief generate such pugnacity; I might be indisposed to argue, whether from confidence that I know or from temperament, and my dogmatism might surface only when I am challenged.

One thing all of these possible conceptions of dogmatism have in common is lack of a second-order component. But that component may well be necessary for a dogmatic attitude, at least of the full-blooded kind. Typically, a dogmatically held belief is maintained with a conviction (often unjustified) to the effect that one is right, e.g., that one knows, is amply justified, is properly certain, or can just see the truth of the proposition in question. Such a second-order belief is not, however, sufficient for a dogmatic attitude. This is shown by certain cases of believing simple logical truths. These can be held both with such a second-order belief and in the stubborn way typical of a dogmatic attitude yet not bespeak a dogmatic attitude. It might be held that in this case they would at least be held *dogmatically;* but if the imagined tenacity is toward, say, the principle that if a = b, and b = c, then a = c, one could not properly call the attitude dogmatic, and we might better speak of maintaining the belief steadfastly rather than dogmatically.

It might be argued, however, that even if the only examples of dogmatism so far illustrated are the second-order ones, there are still two kinds of dogmatism: first- and second-order. It may be enough, for instance, that one be *disposed* to have a certain belief, usually an unwarrantedly positive one, about the status of one's belief that *p.* Imagine that Tom thinks that Mozart is a far greater composer than Haydn, asserts it without giving any argument, and sloughs off arguments to the contrary. If he does not believe, but is disposed to believe on considering the matter, that his belief is, say, obviously correct, then he may qualify as dogmatically holding it. Here, then, there is no actual second-order attitude, but only a disposition to form one upon considering the status of one's belief. I want to grant that this kind of first-order pattern may qualify as dogmatism; but the account of it remains a second-order one, and it still seems that the other first-order cases we have considered, such as mere stubbornness in believing, are not cases of dogmatism. They may exhibit believing dogmatically, but that does not entail dogmatism as an epistemic attitude or trait of character, any more than doing something lovingly entails a loving attitude, or being a loving person. It appears, then, that at least the clear cases of dogmatically holding beliefs imply either second-order attitudes or certain dispositions to form them.

There may be no simple, illuminating way to characterize dogmatism with respect to a belief that *p;* but if there is, the following elements should be reflected at least as typical conditions and should provide the materials needed in appraising the foundationalism–coherentism controversy: (1) confidence that *p,* and significantly greater confidence than one's evidence or grounds warrant; (2) unjustified resistance to taking plausible objections seriously when they are intelligibly posed to one; (3) a willingness, or at least a tendency, to assert the proposition flat-out even in the presence of presumptive reasons to question it, including simply the conflicting views of one or more persons whom *S* sees or should see to be competent concerning the subject matter; and (4) a (second-order) belief, or disposition to believe, that one's belief is clearly true (or certainly true). Note, however, that (i) excessive confidence can come from mere foolhardiness and can be quite

unstable; (ii) resistance to plausible objections may be due to intellectual laziness; (iii) a tendency to assert something flat-out can derive
from mere bluntness; and (iv) a belief that one is right might arise not
from dogmatism but merely from conceit, intellectual mistake (such as
a facile anti-skepticism), or sheer error. Notice also that the notion of
dogmatism is not just psychological, but also epistemic.

Of the four elements highly characteristic of dogmatism the last may
have the best claim to be an unqualifiedly necessary condition, and
perhaps one or more of the others is necessary. The four are probably
jointly sufficient; but this is not self-evident, and I certainly doubt that
we can find any simple condition that is nontrivially sufficient, such as
believing that one knows, or is justified in believing, that p (which one
does believe), while also believing one has no reasons for believing
that p.[20] This condition is not sufficient because it could stem from a
certain view of knowledge and reasons, say a view on which one never
has reasons (as opposed to a basis) for believing simple, self-evident
propositions. The condition also seems insufficient because it could be
satisfied by a person who lacks the first three of the typical conditions
just specified.

Let us work with the full-blooded conception of a dogmatically held
belief summarized by conditions (1)–(4). What, then may we say about
the standard charge that foundationalism is dogmatic, in a sense
implying that it invites proponents to hold certain beliefs dogmatically?
This charge has been leveled on a number of occasions,[21] and some
plausible replies have been made.[22] Given the earlier sections of this
chapter, it should be plain that the charge is more likely to seem cogent
if foundationalism is conceived as answering the dialectical regress
problem, as it has apparently been taken to do by, e.g., Chisholm.[23]
For in this case a (doxastic) stopping place in the regress generated by
'How do you know that p?' will coincide with the assertion of a
second-order belief, such as that I know that q, e.g., that there is a
window before me; and since knowledge claims are commonly justifiable by evidence, flatly stopping the regress in this way will seem
dogmatic. Even if such a claim is justified by one's citing a non-
doxastic state of affairs, such as a visual experience of a window, one
is still asserting the existence of this state of affairs and hence apparently expressing knowledge: making what seems a tacit claim to it,
though not actually claiming to *have* it.

We can formulate various second-order foundationalisms, for instance one which says that if S knows anything, then there is something that S directly knows S knows. But a foundationalist need not

hold such a view, nor would one who does be committed to maintaining that many kinds of belief constitute such knowable foundations, i.e., are knowledge one can know one has, or that every epistemic chain terminates in them. In any event, moderate foundationalists will be disinclined to hold a second-order foundationalism, even if they think that we do in fact have some second-order knowledge. For one thing, if foundational beliefs are only defeasibly justified, it is likely to be quite difficult to know that they are justified, because this requires warrant for attributing certain grounds to the belief and may also require justification for believing that certain defeaters are absent. This is not to deny that there are kinds of knowledge which one may, without having evidence for this, warrantedly and non-dogmatically say one has, for instance where the first-order knowledge is of a simple self-evident proposition. My point is that foundationalism as such, at least in moderate versions, need not make such second-order knowledge (or justification) a condition for the existence of knowledge (or justification) in general.[24]

If we raise the regress problem in the structural form, there is much less temptation to consider foundationalism dogmatic. For there is no presumption that, with respect to anything I know, I non-inferentially know that I know it (and similarly for justification). Granted, on the assumption that by and large I am entitled, without offering evidence, to assert what I directly know, it may seem that even moderate foundationalism justifies me in holding—and expressing—beliefs dogmatically. But this is a mistake. There is considerable difference between what I know or justifiably believe and what I may warrantedly assert without evidence. It is, e.g., apparently consistent with knowing that *p,* say that there is music playing, that I have some reason to doubt that *p;* I might certainly have reason to think others doubt it and that they should not be spoken to as if their objections could not matter. Thus, I might know, through my own good hearing, that *p,* yet be unwarranted in saying that I know it, and warranted, with only moderate confidence, even in saying simply that it is true. Here 'It is true' would *express,* but not *claim,* my knowledge; 'I know it' explicitly claims knowledge and normally implies that I have justification for beliefs about my objective grounds, not just about my own cognitive and perceptual state.

Nothing said here implies that one *cannot* be justified in believing that one holds dogmatically. That one's attitude *in* holding that *p* is not justified does not imply that one's holding that *p* is itself not justified. It might be possible, for all I have said, that in certain cases one might even be justified, overall, in taking a dogmatic attitude toward certain propositions. This will depend on, among other things, the plausibility of the proposition in question and the level of justification one has for believing that one is right. But typically, dogmatic attitudes are not justified, and moderate foundationalism, far from implying otherwise, can readily explain this.

Furthermore, once the defeasibility of foundational beliefs is appreciated, then even if one does think that one may assert the propositions in question without offering evidence, one will not take the attitudes or other stances required for holding a belief dogmatically. As the example of my belief about the music illustrates, most of the time one is likely to be open to counterargument and may indeed tend to be no more confident than one's grounds warrant. To be sure, fallibilism alone, even when grounded in a proper appreciation of defeasibility, does not preclude dogmatism regarding many of one's beliefs. But it helps toward this end, and it is natural for moderate foundationalists to hold a fallibilistic outlook on their beliefs, especially their empirical beliefs, and to bear it in mind in framing an overall conception of human experience.

If foundationalism has been uncritically thought to encourage dogmatism, coherentism has often been taken to foster intellectual openness. But this second stereotypic conception may be no better warranted than the first. Much depends, of course, on the kind of coherentism and on the temperament of its proponent. Let us consider these points in turn.

What makes coherentism seem to foster tolerance is precisely what leads us to wonder how it can account for knowledge (at least without a coherence theory of truth). For as coherentists widely grant, there are definitely many coherent systems of beliefs people might in principle have; hence, to suppose that mine embodies knowledge and thus truth, or even justification and thus a presumption of truth, while yours does not, is prima facie unwarranted. But the moment the view is developed to yield a plausible account of knowledge of the world (an external notion), say by requiring a role for observation beliefs and other cognitively spontaneous beliefs, as some coherentists do, or by requiring beliefs accepted on the basis of a desire to believe truth and avoid error, as others do,[25] it becomes easy to think—and one can be warranted in thinking—that one's beliefs are more likely to constitute knowledge, or to be justified, than someone else's, especially if the other person(s) holds views incompatible with one's own. Indeed, while coherentism makes it easy to see how counterargument can be launched from a wide range of opposing viewpoints, it also provides less in the way of foundational appeals by which debates may be settled—and pretentions quashed. Is one likely to be less dogmatic where one thinks one can always encounter reasoned opposition from someone with a different coherent belief system, right or wrong, than where one believes one can be decisively shown to be mistaken by appeal to foundational sources of knowledge and justification? The answer is not clear; in any given case it will depend on a number of variables, including the temperament of the subject and the propositions in question. And could not my confidence that, using one or another coherent resource, I can always continue to argue for my view generate overconfidence just as much as my thinking that I (defeasibly) know something through experience or reason? Indeed, if coherence is as vague a notion as it seems, it seems quite possible both to exaggerate the extent of its support for one's own beliefs and underestimate the degree of coherence supporting an opposing belief. It turns out that coherentism can also produce dogmatism, even if its proponents have tended to be less inclined toward it than some foundationalists.

If there has been such a lesser inclination, it may be due to temperament, including perhaps a greater sympathy with skepticism, as much as to theoretical commitments. In any case, whether one dogmatically holds certain of one's beliefs surely does depend significantly on whether one is dogmatic in temperament or in certain segments of one's outlook. It may be that the tendency to seek justification in large patterns runs stronger in coherentists than in foundationalists, and that the latter tend more than the former to seek it instead in chains of argument or of inference. If so, this could explain a systematic difference in the degree of dogmatism found in the two traditions. But these tendencies are only contingently connected with the respected theories. Foundationalism can account for the justificatory importance of large patterns, and coherentists commonly conceive argument and inference as prime sources of coherence. One can also wax dogmatic in insisting that a pattern is decisive in justification, as one can dogmatically assert that a single perceptual belief is incontrovertibly veridical.

One source of the charge of dogmatism, at least as advanced by philosophers, is of course the sense that skepticism is being flatly denied. Moreover, the skeptic in us tends to think that any confident assertion of a non-self-evident, non-introspective proposition is dogmatic. On this score, foundationalism is again likely to seem dogmatic if it is conceived as an answer to the dialectical regress formulation. For it may then seem to beg the question against skepticism. But again, foundationalism is not committed to the existence of any knowledge or justified belief; and even a foundationalist who maintains that there is some need not hold that we directly know that there is. Granted, foundationalists are more likely to say, at some point or other, that skepticism is just wrong than are coherentists, who (theoretically) can always trace new justificatory paths through the fabric of their beliefs. But if this is true, it has limited force: perhaps in some such cases foundationalists would be warranted in a way that precludes being dogmatic, and perhaps coherentists are in effect repeating themselves in a way consistent with dogmatic reassertion of the point at issue.

It turns out, then, that fallibilist foundationalism is not damaged by the dogmatism objection and coherentism is not immune to it. Far from being dogmatic, fallibilist foundationalism implies that even where one has a justified belief one cannot show to be justified, one may (and at least normally can) *give* a justification for it. As to coherentism, it, too, may be a refuge for dogmatists, at least those clever enough to find a coherent pattern by which to rationalize the beliefs they dogmatically hold.

Conclusion

The foundationalism–coherentism controversy cannot be settled in a single essay. But we can now appreciate some often neglected dimensions of the issue. One dimension is the formulation of the regress problem itself; another is the distinction between defeasibility and epistemic dependence; still another is that between consequential and constitutive necessary conditions; and yet another is between an unqualifiedly and a conditionally basic source. Even if coherence is neither a constitutive necessary condition for justification nor even a conditionally basic source of it, there is still reason to consider it important for justification. It may even be a *mark* of justification, a common effect of the same causes as it were, or a virtue with the same foundations. Coherence is certainly significant as suggesting a negative constraint on justification; for incoherence is a paradigm of what defeats justification.

I have argued at length for the importance of the regress problem. It matters considerably whether we conceive the problem dialectically or structurally, at least insofar as we cast foundationalism and coherentism in terms of their capacity to solve it. Indeed, while both coherentism and foundationalism can be made plausible on either conception, coherentism is perhaps best understood as a response to the problem *in* some dialectical formulation, and foundationalism is perhaps best understood as a response to it in some structural form. Taking account of both formulations of the regress problem, I have suggested plausible versions of both foundationalism and coherentism. Neither has been established, though fallibilist foundationalism has emerged as the more plausible of the two. In clarifying them, I have stressed a number of distinctions: between the process and the property of justification, between dispositional beliefs and dispositions to believe, between epistemologically and psychologically foundational beliefs, between defeasibility and epistemic dependence, between constitutive and consequential necessary conditions for justification, and between unqualified and conditionally basic sources of it. Against this background, we can see how fallibilist foundationalism avoids some of the objections commonly thought to refute foundationalism, including its alleged failure to account for the defeasibility of most and perhaps all of our justification, and for the role of coherence in justification. Indeed, fallibilist foundationalism can even account for coherence as a mark of justification; the chief tension between the two theories concerns not whether coherence is necessary for justification, but whether it is a basic source of it.

It is appropriate in closing to summarize some of the very general considerations supporting a fallibilist foundationalism, since that is a position which some have apparently neglected—or supposed to be a contradiction in terms—and others have not distinguished from coherentism. First, the theory provides a plausible and reasonably straightforward solution to the regress problem. It selects what seems the best option among the four and does not interpret that option in a way that makes knowledge or justification either impossible, as the skeptic would have it, or too easy to achieve, as they would be if they required no grounds at all or only grounds obtainable without the effort of observing, thinking, or otherwise taking account of experience. Second, in working from the experiential and rational sources it takes as epistemically basic, fallibilist foundationalism (and in its most plausible versions) accords with reflective common sense: the sorts of beliefs it takes as non-inferentially justified, or as constituting non-inferential knowledge, are pretty much those that, on reflection, we think people are justified in holding, or in supposing to be knowledge, without any more than the evidence of the senses or of intuition.

Third, fallibilist foundationalism is psychologically plausible, in two major ways: the account it suggests of the experiential and inferential genesis of many of our beliefs apparently fits what is known about their origins and development; and, far from positing infinite or circular belief chains, whose psychology is at least puzzling, it allows a fairly simple account of the structure of cognition. Beliefs arise both from experience and from inference; some serve to unify others, especially those based on them; and their relative strengths, their changes, and their mutual interactions are all explicable within the moderate foundationalist assumptions suggested. Fourth, the theory serves to integrate our epistemology with our psychology and even biology, particularly in the crucial case of perceptual beliefs. What causally explains why we hold them—sensory experience—is also what justifies them.

From an evolutionary point of view, moreover, many of the kinds of beliefs that the theory (in its most plausible versions) takes to be non-inferentially justified—introspective and memorial beliefs as well as perceptual ones—are plainly essential to survival. We may need a map, and not merely a mirror, of the world to navigate it; but if experience does not generally mirror reality, we are in no position to move to the abstract level on which we can draw a good map. If a mirror without a map is insufficiently discriminating, a map without a mirror is insufficiently reliable. Experience that does not produce beliefs cannot guide us; beliefs not grounded in experience cannot be expected to be true.

Finally, contrary to the dogmatism charge, the theory helps to explain cognitive pluralism. Given that different people have different experiences, and that anyone's experiences change over time, people should be expected to differ from one another in their non-inferentially justified beliefs and, in their own case, across time; and given that logic does not dictate what is to be inferred from one's premises, people should be expected to differ considerably in their inferential beliefs as well. Logic does, to be sure, tell us what *may* be inferred; but it neither forces inferences nor, when we draw them, selects which among the permissible ones we will make. Particularly in the case of inductive inference, say where we infer a hypothesis as the best explanation of some puzzling event, our imagination comes into play; and even if we were to build from the same foundations as our neighbors, we would often produce quite different superstructures.

A properly qualified foundationalism, then, has much to recommend it and exhibits many of the virtues that have been commonly thought to be characteristic only of coherentist theories. Fallibilist foundation-alism can account for the main connections between coherence and justification, and it can provide principles of justification to explain how justification that can be plausibly attributed to coherence can also be traced—by sufficiently complex and sometimes inductive paths—to basic sources in experience and reason.[26]

Notes

1. For recent statements of foundationalism see, e.g., R. M. Chisholm, *Theory of Knowledge* (Englewood Cliffs, N.J.: Prentice-Hall, 1977 and 1989), and, especially, "A Version of Foundationalism," *Midwest Studies in Philosophy* V (1980); William P. Alston, "Two Types of Foundationalism," *The Journal of Philosophy* LXXXIII, 7 (1976); Paul K. Moser, *Empirical Justification* (Dordrecht and Boston: D. Reidel, 1985); and Richard Foley, *The Theory of Epistemic Rationality* (Cambridge, Mass.: Harvard University Press, 1987); and Chapter 1, this volume. For detailed statements of coherentism, see, e.g., Wilfrid Sellars, "Givenness and Explanatory Coherence," *The Journal of Philosophy* LXX (1973); Keith Lehrer, *Knowledge* (Oxford: Oxford University Press, 1974); Gilbert Harman, *Thought* (Princeton, N.J.: Princeton University Press, 1975); and Laurence BonJour, *The Structure of Empirical Knowledge* (Cambridge, Mass.: Harvard University Press, 1985). For useful discussions of the controversy between foundationalism and coherentism, see C. F. Delaney, "Foundations of Empirical Knowledge—Again," *The New Scholasticism* L, 1 (1976), which defends a kind of foundationalism; and Brand Blanshard, "Coherence and Correspondence," in *Philosophical Interrogations*, edited by Sydney and Beatrice Rome (New York: Holt, Rinehart & Winston, 1964), which defends his earlier views against objections by critics quoted in the same chapter.

2. BonJour, e.g., says that the regress problem is "perhaps the most crucial in the entire theory of knowledge" (op. cit., p. 18); and he considers it the chief motivation for foundationalism (p. 17) and regards the failure of foundationalism as "the main motivation for a coherence theory" (p. 149).

3. Chisholm seems to raise the problem in this way when he says, "If we try Socratically to formulate our justification for any particular claim to know ('My justification for thinking that I know that *A* is the fact that *B*'), and if we are relentless in our inquiry ('and my justification for thinking that I know that *B* is the fact that *C*'), we will arrive, sooner or later, at a kind of stopping place ('but my justification for thinking that I know that *N* is simply the fact that *N*'). An example of *N* might be the fact that I seem to remember having been here before or that something now looks blue to me" (*Theory of Knowledge*, 1966, p. 2); cf. 2nd ed., 1977, esp. pp. 19–20. In these and other passages Chisholm seems to be thinking of the regress problem, dialectically and taking

a foundational belief to be second order. To be sure, he is talking about justification of any "claim to know"; but this and similar locutions—such as "knowledge claim"—have often been taken to apply to expressions of first-order knowledge, as where one says that it is raining, on the basis of perceptions which one would normally take to yield knowledge that it is.

4. See *Posterior Analytics*, Bk 3. Having opened Bk. 1 with the statement that "All instruction given or received by way of argument proceeds from pre-existent knowledge" (71a1–2), and thereby established a concern with the structure and presuppositions of knowledge, Aristotle formulated the regress argument as a response to the question of what is required for the existence of (what he called scientific) knowledge (72b4–24). (The translation is by W. D. Ross.)

5. The locus classicus of this argument is the *Posterior Analytics*, Bk. II. But while Aristotle's version agrees with the one given here insofar as his main conclusion is that "not all knowledge is demonstrative," he also says, "since the regress must end in immediate truths, those truths must be inde-monstrable" (72b19–24), whereas I hold that direct knowledge does *not* require indemonstrability. There might be appropriate premises; *S*'s foundational belief is simply not based on them (I also question the validity of the inference in the second quotation, but I suspect that Aristotle had independent grounds for its conclusion).

6. In Meditation I, e.g., Descartes says that "reason already persuades me that I ought no less carefully to withhold my assent from matters which are not entirely certain and indubitable than from those which appear to me manifestly to be false" (from the Haldane and Ross translation).

7. That knowing a proposition implies believing it is not uncontroversial, but most epistemologists accept the implication. For defense of the implication see, e.g., Harman, op. cit., and my *Belief, Justification, and Knowledge* (Belmont, Calif.: Wadsworth, 1988).

8. See, e.g., Richard Foley, "Justified Inconsistent Beliefs," *American Philosophical Quarterly* 16 (1979). I have criticized the infinite-belief view in "Believing and Affirming," *Mind* XCI (1982).

9. It should be noted that memory is different from the other three in this: it is apparently not a *basic* source of knowledge, as it is of justification; i.e., one cannot know something from memory unless one has *come* to know it in some other mode, e.g., through perception. This is discussed in ch. 2 of my *Belief, Justification, and Knowledge*. Cf. Carl Ginet, *Knowledge, Perception, and Memory* (Dordrecht and Boston: D. Reidel, 1973).

10. The view that such experience is a mirror of nature is criticized at length by Richard Rorty in *Philosophy and the Mirror of Nature* (Princeton, N.J.: Princeton University Press, 1979). He has in mind, however, a Cartesan version of foundationalism, which is not the only kind and implies features of the "mirror" that are not entailed by the metaphor used here.

11. This does not entail that there are *objects* in the visual field which have

their own phenomenal colors and shapes; the point is only that there is some sense in which experiences *characterized* by color and shape (however that is to be analyzed) represent the colors and shapes apparently instantiated in the external world.

12. This metaphor comes from D. M. Armstrong. See esp. *Belief, Truth and Knowledge* (Cambridge, Mass.: Cambridge University Press, 1973). His theory of justification and knowledge is reliabilist, in taking both to be analyzable in terms of their being produced or sustained by reliable processes (such as tactile belief-production), those that (normally) yield true beliefs more often than false. Foundationalism may, but need not be reliabilist; and this chapter is intended to be neutral with respect to the choice between reliabilist and internalist views. For further discussion of internalism see Paul K. Moser, *Knowledge and Evidence* (Cambridge and New York.: Cambridge University Press, 1989), and R. M. Chisholm, *The Theory of Knowledge,* 3rd ed. (Englewood Cliffs, N.J.: Prentice-Hall, 1989).

13. Clause (4) requires 'other-things-equal' because removal of justification from one source can affect justification from another even without being a basis of the latter justification; and the *level* of justification in question I take to be (as in the counterpart formulation of coherentism) approximately that appropriate to knowledge. The formulation should hold, however, for any given level.

14. For references to the main contemporary accounts, especially those by Lehrer and BonJour, see John B. Bender, *The Current Status of the Coherence Theory* (Dordrecht and Boston: Kluwer, 1989).

15. This applies to Sellars, Lehrer, and BonJour and is evident in the works cited in note 1. Their coherentist positions are not linear.

16. With this question in mind, it is interesting to read Donald Davidson, "A Coherence Theory of Truth and Knowledge," in Dieter Henrich, ed., *Kant oder Hegel* (Stuttgart, 1976). Cf. Jaegwon Kim, "What is 'Naturalized Epistemology'?" *Philosophical Perspectives* 2 (1988).

17. This distinction seems to have been often missed, e.g., in Hilary Kornblith, "Beyond Foundationalism and the Coherence Theory," *Journal of Philosophy* LXXVII (1980).

18. One exception is David Shatz's "Foundationalism, Coherentism, and the Levels Gambit," *Synthese* 55, 1 (1983).

19. This suggestion may be controversial: an epistemic virtue theorist might argue that the trait is most basic and colors the attitude, and that these together are the basis for classifying beliefs as held dogmatically or otherwise. Most of my points will be neutral with respect to this priority issue.

20. Shatz, op. cit., p. 107, attributes a similar suggestion to me (from correspondence), and it is appropriate to suggest here why I do not mean to endorse it.

21. The dogmatism charge has been brought by, e.g., Bruce Aune in *Knowledge, Mind and Nature* (New York: Random House, 1967), pp. 41–3,

and, by implication, by James Cornman and Keith Lehrer in *Philosophical Problems and Arguments,* 2nd ed. (New York: Macmillan, 1974), pp. 60–1. Alston goes so far as to say that "It is the aversion to dogmatism, to the apparent arbitrariness of putative foundations, that leads many philosophers to embrace some form of coherence or contextualist theory . . ." (op. cit., pp. 182–83).

22. See Alston, op. cit., for a reply (which supports mine) to the dogmatism charge.

23. A formulation of the regress problem by Chisholm is cited in note 3. For a contrasting formulation see Anthony Quinton, *The Nature of Things* (London: Routledge & Kegan Paul 1973), p. 119. Quinton, it is interesting to note, is sympathetic to the kind of moderate foundationalism that would serve as an answer to the problem in his formulation.

24. It is natural to read Descartes as holding a second-order foundationalism; but if he did, he was at least not committed to it by even his strong foundationalism. That requires indefeasible foundations, but it is his commitment to vindicating knowledge in the face of skepticism that apparently commits him to our having second-order knowledge. Similar points hold for Aristotle, who indeed my have taken our second-order knowledge to be at least limited; he said, e.g., "It is hard to be sure whether one knows or not; for it is hard to be sure whether one's knowledge is based on the basic truths appropriate to each attribute—the differentia of true knowledge" (*Posterior Analytics* 76a26–28).

25. I have in mind, for the observation requirement, BonJour, op. cit., and, for the motivational requirement, Keith Lehrer, e.g., in *Knowledge.*

26. This chapter draws substantially on my "Foundationalism, Coherentism, and Epistemological Dogmatism," *Philosophical Perspectives* 2 (1988), 407–59 (edited by James E. Toberlin). I thank Louis P. Pojman for many helpful comments on an earlier draft of much of the material and for permission to use selected passages from my two chapters in his book *The Theory of Knowledge: Contemporary Readings* (Belmont, Calif.: Wadsworth, 1992).

6

Reliabilism and Intellectual Virtue

Ernest Sosa

Externalism and reliabilism go back at least to the writings of Frank Ramsey early in this century.[1] The generic view has been developed in diverse ways by David Armstrong, Fred Dretske, Alvin Goldman, Robert Nozick, and Marshall Swain.[2]

A. Generic Reliabilism

Generic reliabilism might be put simply as follows:

> S's belief that p at t is justified if it is the outcome of a process of belief acquisition or retention which is reliable, or leads to a sufficiently high preponderance of true beliefs over false beliefs.

That simple statement of the view is subject to three main problems: the generality problem, the new evil-demon problem, and the meta-incoherence problem (to give it a label). Let us consider these in turn.

The generality problem for such reliabilism is that of how to avoid processes with only one output ever, or one artificially selected so that if a belief were the output of such a process it would indeed be true; for every true belief is presumably the outcome of some such too-specific processes, so that if such processes are allowed, then every true belief would result from a reliable process and would be justified. But we must also avoid processes that are too generic, such as perception (period), which surely can produce not only justified beliefs but also unjustified ones, even if perception is on the whole a reliable process of belief acquisition for normally circumstanced humans.[3]

Reprinted by permission from Ernest Sosa, *Knowledge in Perspective*, pp. 131–145. Cambridge: Cambridge University Press, 1991.

The evil-demon problem for reliabilism is not Descartes's problem, of course, but it is a relative. What if twins of ours in another possible world were given mental lives just like ours down to the most minute detail of experience or thought, etc., though they were also totally in error about the nature of their surroundings, and their perceptual and inferential processes of belief acquisition accomplished very little except to sink them more and more deeply and systematically into error? Shall we say that we are justified in our beliefs while our twins are not? They are quite wrong in their beliefs, of course, but it seems somehow very implausible to suppose that they are unjustified.[4]

The meta-incoherence problem is in a sense a mirror image of the new evil-demon problem, for it postulates not a situation where one is internally justified though externally unreliable, but a situation where one is internally unjustified though externally reliable. More specifically, it supposes that a belief (that the President is in New York) which derives from one's (reliable) clairvoyance is yet *not* justified if either (a) one has a lot of ordinary evidence against it, and none in its favor; or (b) one has a lot of evidence against one's possessing such a power of clairvoyance; or (c) one has good reason to believe that such a power could not be possessed (e.g., it might require the transmission of some influence at a speed greater than that of light); or (d) one has no evidence for or against the general possibility of the power, or of one's having it oneself, nor does one even have any evidence either for or against the proposition that one believes as a result of one's power (that the President is in New York).[5]

B. Goldman's Reliabilisms

How might reliabilism propose to meet the problems specified? We turn first to important work by Goldman, who calls his theory "Historical Reliabilism," and has the following to say about it:

> The theory of justified belief proposed here, then, is an *Historical* or *Genetic* theory. It contrasts with the dominant approach to justified belief, an approach that generates what we may call (borrowing a phrase from Robert Nozick) *Current Time-Slice* theories. A Current Time-Slice theory makes the justificational status of a belief wholly a function of what is true of the cognizer *at the time* of the belief. An Historical theory makes the justificational status of a belief depend on its prior history. Since my Historical theory emphasizes the reliability of the belief-generating processes, it may be called *Historical Reliabilism*.[6]

The insights of externalism are important, and Goldman has been perceptive and persistent in his attempts to formulate an appropriate and detailed theory that does them justice. His proposals have stimulated criticism, however, among them the three problems already indicated.

Having appreciated those problems, Goldman in his book[7] moves beyond Historical Reliabilism to a view we might call rule reliabilism, and, in the light of further problems,[8] has made further revisions in the more recent "Strong and Weak Justification." The earlier theory, however, had certain features designed to solve the new evil-demon problem, features absent in the revised theory. Therefore, some other solution is now required, and we do now find a new proposal.

Under the revised approach, we now distinguish between two sorts of justification:

A belief is *strongly justified* if and only if it is well formed, in the sense of being formed by means of a process that is truth-conducive in the possible world in which it is produced, or the like.

A belief is *weakly justified* if and only if it is blameless though ill-formed, in the sense of being produced by an unreliable cognitive process which the believer does not believe to be unreliable, and whose unreliability the believer has no available way of determining.[9]

Notice, however, that it is at best in a *very* weak sense that a subject with a "weakly justified" belief is thereby "blameless." For it is not even precluded that the subject take that belief to be very ill-formed, so long as he is in error about the cognitive process that produces it. That is to say, S might hold B, and believe B to be an output of P, and hold P to be an epistemically unreliable process, while in fact it is not P but the equally unreliable P′ that produces B. In this case S's belief B would be weakly justified, so long as S did not believe P′ to be unreliable, and had no available means of determining its unreliability. But it seems at best extremely strained to hold S epistemically "blameless" with regard to holding B in such circumstances, where S takes B to derive from a process P so unreliable, let us suppose, as to be epistemically vicious.

The following definition may perhaps give us a closer approach to epistemic blamelessness.

A belief is *weakly justified (in the modified sense)* if and only if it is blameless though ill-formed, in the sense of being produced by an

unreliable cognitive process while the believer neither takes it to be thus
ill-formed nor has any available way of determining it to be ill-formed.

With these concepts, the Historical Reliabilist now has at least the
beginnings of an answer both for the evil-demon problem and for the
meta-incoherence problem. About the evil demon's victims, those
hapless twins of ours, we can now say that though their beliefs are
very ill-formed—and are not knowledge even if by luck they, some of
them, happen to be true—still there is a sense in which they are
justified, as justified as our corresponding beliefs, which are indistinguishable from theirs so far as concerns only the "insides" of our
respective subjectivities. For we may now see their beliefs to be
weakly justified, in the modified sense defined above.[10]

About the meta-incoherence cases, moreover, we can similarly
argue that, in some of them at least, the unjustified protagonist with
the wrong (or lacking) perspective on his own well-formed (clairvoyant) belief can be seen to be indeed unjustified, for he can be seen as
subjectively unjustified through lack of an appropriate perspective on
his belief: either because he positively takes the belief to be ill-formed,
or because he "ought" to take it to be ill-formed given his total picture
of things, and given the cognitive processes available to him.

Consider now the following definition:

> A belief is *meta-justified* if and only if the believer does place it in
> appropriate perspective, at least in the minimal sense that the believer
> neither takes it to be ill-formed nor has any available way of determining
> it to be ill-formed.

Then any belief that is weakly justified (again, sticking to the unmodified sense) will be meta-justified, but there can be meta-justified beliefs
which are not weakly justified. Moreover, no strongly justified belief
will be weakly justified, but a strongly justified belief can be meta-
justified. Indeed one would wish one's beliefs to be not only strongly
justified but also meta-justified. And what one shares with the victim
of the evil demon is of course not weak justification. For if, as we
suppose, our own beliefs are strongly justified, then our own beliefs
are not weakly justified. What one shares with the evil demon's victim
is rather meta-justification. The victim's beliefs and our beliefs are
equally meta-justified.

Does such meta-justification—embedded thus in weak justification—enable answers both for the new evil-demon problem and for the

problem of meta-incoherence? Does the victim of the evil demon share with us meta-justification, unlike the meta-incoherent? The notion of weak justification does seem useful as far as it goes, as is the allied notion of meta-justification, but we need to go a bit deeper,[11] which may be seen as follows.

C. Going Deeper

Beliefs are states of a subject, which need not be occurrent or conscious, but may be retained even by someone asleep or unconscious, and may also be acquired unconsciously and undeliberately, as are acquired our initial beliefs, presumably, whether innate or not, especially if deliberation takes time. Consider now a normal human with an ordinary set of beliefs normally acquired through sensory experience from ordinary interaction with a surrounding physical world. And suppose a victim in whom evil demons (perhaps infinitely many) implant beliefs in the following way. The demons cast dice, or use some other more complex randomizer, and choose which beliefs to implant at random and in ignorance of what the other demons are doing. Yet, by amazing coincidence, the victim's total set of beliefs is identical to that of our normal human. Now let's suppose that the victim has a beautifully coherent and comprehensive set of beliefs, complete with an epistemic perspective on his object-level beliefs. We may suppose that the victim has meta-justification for his object-level beliefs (e.g., for his belief that there is a fire before him at the moment), at least in the minimal sense defined above: he does not believe such beliefs to derive from unreliable processes, nor has he any available means of determining that they do. Indeed, we may suppose that he has an even stronger form of meta-justification, as follows:

> S has meta-justification, in the stronger sense, for believing that p iff (a) S has weaker meta-justification for so believing, and (b) S has meta-beliefs which positively attribute his object beliefs in every case to some faculty or virtue for arriving at such beliefs in such circumstances, and further meta-beliefs which explain how such a faculty or virtue was acquired, and how such a faculty or virtue, thus acquired, is bound to be reliable in the circumstances as he views them at the time.

And the victim might even be supposed to have a similar meta-meta-perspective, and a similar meta-meta-meta-perspective, and so on, for many more levels of ascent than any human would normally climb. So everything would be brilliantly in order as far as such meta-reasoning is concerned, meta-reasoning supposed flawlessly coherent and comprehensive. Would it follow that the victim was internally and subjectively justified in every reasonable sense or respect? Not necessarily, or so I will now try to show.

Suppose the victim has much sensory experience, but that all of this experience is wildly at odds with his beliefs. Thus he believes he has a splitting headache, but he has no headache at all; he believes he has a cubical piece of black coal before him, while his visual experience is as if he had a white and round snowball before him. And so on. Surely there is then something internally and subjectively wrong with this victim, something "epistemically blameworthy." This despite his beliefs being weakly justified, in the sense defined by Goldman, and despite his beliefs being meta-justified in the weaker and stronger senses indicated above.

Cartesians and internalists (broadly speaking) should find our victim to be quite conceivable. More naturalistic philosophers may well have their doubts, however, about the possibility of a subject whose "experience" and "beliefs" would be so radically divergent. For these there is a different parable. Take our victim to be a human, and suppose that the demon damages the victim's nervous system in such a way that the physical inputs to the system have to pass randomizing gates before the energy transmitted is transformed into any belief. Is there not something internally wrong with this victim as well, even though his beliefs may be supposed weakly and meta-justified, as above?

It may be replied that the "internal" here is not internal in the right sense. What is internal in the right sense must remain restricted to the subjectivity of the subject, to that which pertains to the subject's psychology; it must not go outside of that, even to the physiological conditions holding in the subject's body; or at least it must not do so under the aspect of the physiological, even if in the end it is the physiological (or something physical anyhow) that "realizes" everything mental and psychological.

Even if we accept that objection, however, a very similar difficulty yet remains for the conception of the blameless as the weakly justified or meta-justified (in either the weaker or the stronger sense). For it may be that the connections among the experiences and beliefs of the victim are purely random, as in the example above. True, in that example the randomness derives from the randomizing behavior of the demons involved. But there is no reason why the randomizing may not be brought inside. Thus, given a set of experiences or beliefs, there may be many alternative further beliefs that might be added by the subject, and there may be no rational mechanism that selects only one to be added. It may be rather that one of the many alternatives pops in at random: thus it is a radically random matter which alternative further belief is added in any specific case. Our evil demon's victim, though damaged internally in that way, so that his inner mental processes are largely random, may still by amazing coincidence acquire a coherent and comprehensive system of beliefs that makes him weakly justified and even meta-justified, in both the weaker and stronger senses indicated above. Yet is there not something still defective in such a victim, something that would preclude our holding him to be indiscernible from us in all internal respects of epistemic relevance?

Consider again the project of defining a notion of weak justification, however, a notion applicable to evil-demon victims in accordance with our intuitions; or that of defining a notion of meta-justification as above, one applicable equally to the victims and to ourselves in our normal beliefs. These projects may well be thought safe from the fact that a victim might be internally defective in ways that go beyond any matter of weak or meta-justification. Fair enough. But then of course we might have introduced a notion of superweak justification, and provided sufficient conditions for it as follows:

> S is superweakly justified in a certain belief if (1) the cognitive process that produces the belief is unreliable, but (2) S has not acquired that belief as a result of a deliberate policy of acquiring false beliefs (a policy adopted perhaps at the behest of a cruel master, or out of a deep need for epistemic self-abasement).

Someone may propose that a similarity between the victim of the evil demon on one side and ourselves on the other is that we all are superweakly justified in our object-level beliefs in fires and the like. And this is fair and true enough. But it just does not go very far, not far enough. There is much else that is epistemically significant to the comparison between the victim and ourselves, much else that is left out of account by the mere notion of superweak justification. Perhaps part of what is left out is what the notion of weak justification would enable us to capture, and perhaps the notion of meta-justification, especially its stronger variant, would enable us to do even better. Even these stronger notions fall short of what is needed for fuller illumination, however, as I have tried to show above through the

victims of randomization, whether demon-derived or internally derived. In order to deal with the new evil-demon problem and with the
problem of meta-incoherence we need a stronger notion than either
that of the weakly justified or that of the meta-justified, a stronger
notion of the internally or subjectively justified.

D. A Stronger Notion of the "Internally Justified": Intellectual Virtue

Let us define an intellectual virtue or faculty as a competence in virtue
of which one would mostly attain the truth and avoid error in a certain
field of propositions F, when in certain conditions C. Subject S believes
proposition P at time t out of intellectual virtue only if there is a field
of propositions F, and there are conditions C, such that: (a) P is in F;
(b) S is in C with respect to P; and (c) S would most likely be right if S
believed a proposition X in field F when in conditions C with respect
to X. Unlike Historical Reliabilism, this view does not require that
there be a cognitive process leading to a belief in order for that belief
to enjoy the strong justification required for constituting knowledge.
Which is all to the good, since requiring such a process makes it
hard to explain the justification for that paradigm of knowledge, the
Cartesian cogito. There is a truth-conducive "faculty" through which
everyone grasps their own existence at the moment of grasping.
Indeed, what Descartes noticed about this faculty is its infallible
reliability. But this requires that the existence which is grasped at a
time t be existence at that very moment t. Grasp of earlier existence,
no matter how near to the present, requires not the infallible cogito
faculty, but a fallible faculty of memory. If we are to grant the cogito
its due measure of justification, and to explain its exceptional epistemic
status, we must allow faculties which operate instantaneously in the
sense that the outcome belief is about the very moment of believing,
and the conditions C are conditions about what obtains at that very
moment—where we need place no necessary and general requirements
about what went before.

By contrast with Historical Reliabilism, let us now work with intellectual virtues for faculties, defining their presence in a subject S
by requiring

that, concerning propositions X in field F, once S were in conditions C
with respect to X, S would most likely attain the truth and avoid error.

In fact a faculty or virtue would normally be a fairly stable disposition on the part of a subject *relative to an environment*. Being in conditions C with respect to proposition X would range from just being conscious and entertaining X—as in the case of "I think" or "I am"—to seeing an object O in good light at a favorable angle and distance, and without obstruction, etc.—as in "This before me is white and round." There is no restriction here to processes or to the internal. The conditions C and the field F may have much to do with the environment external to the subject: thus a moment ago we spoke of a C that involved seeing an external object in good light at a certain distance, etc.—all of which involves factors external to the subject.

Normally, we could hope to attain a conception of C and F which at best and at its most explicit will still have to rely heavily on the assumed nature of the subject and the assumed character of the environment. Thus it may appear to you that there is a round and white object before you and you may have reason to think that in conditions C (i.e., for middle-sized objects in daylight, at arm's length) you would likely be right concerning propositions in field F (about their shapes and colors). But of course there are underlying reasons why you would most likely be right about such questions concerning such objects so placed. And these underlying reasons have to do with yourself and your intrinsic properties, largely your eyes and brain and nervous system; and they have to do also with the medium and the environment more generally, and its contents and properties at the time. A fuller, more explicit account of what is involved in having an intellectual virtue or faculty is therefore this:

Because subject S has a certain inner nature (I) and is placed in a certain environment (E), S would most likely be right on any proposition X in field F relative to which S stood in conditions C. S might be a human; I might involve possession of good eyes and a good nervous system including a brain in good order; E might include the surface of the earth with its relevant properties, within the parameters of variation experienced by humans over the centuries, or anyhow by subject S within his or her lifetime or within a certain more recent stretch of it; F might be a field of propositions specifying the colors or shapes of an object before S up to a certain level of determination and complexity (say greenness and squareness, but not chartreuseness or chiliagonicity); and C might be the conditions of S's seeing such an object in good light at arm's length and without obstructions.

If S believes a proposition X in field F, about the shape of a facing surface before him, and X is false, things might have gone wrong at interestingly different points. Thus the medium might have gone wrong unknown to the subject, and perhaps even unknowably to the subject; or something within the subject might have changed significantly: thus the lenses in the eyes of the subject might have become distorted, or the optic nerve might have become defective in ways important to shape recognition. If what goes wrong lies in the environment, that might prevent the subject from knowing what he believes, even if his belief were true, but there is a sense in which the subject would remain subjectively justified or anyhow virtuous in so believing. It is the sense of internal virtue that seems most significant for dealing with the new evil-demon argument and with the meta-incoherence objection. Weak justification and meta-justification are just two factors that bear on internal value, but there are others surely, as the earlier examples were designed to show—examples in which the experience/belief relation goes awry, or in which a randomizer gate intervenes. Can something more positive be said in explication of such internal intellectual virtue?

Intellectual virtue is something that resides in a subject, something relative to an environment—though in the limiting case, the environment may be null, as perhaps when one engages in armchair reflection and thus comes to justified belief.

> A subject S's intellectual virtue V relative to an "environment" E may be defined as S's disposition to believe correctly propositions in a field F relative to which S stands in conditions C, in "environment" E.

It bears emphasis first of all that to be in a certain "environment" is *not* just a matter of having a certain spatio-temporal location, but is more a matter of having a complex set of properties, only some of which will be spatial or temporal. Secondly, we are interested of course in non-vacuous virtues, virtues which are not possessed simply because the subject would never be in conditions C relative to the propositions in F, or the like, though there may be no harm in allowing vacuous virtues to stand as trivial, uninteresting special cases.

Notice now that, so defined, for S to have a virtue V relative to an environment E at a time t, S does not have to be *in* E at t (i.e., S does not need to have the properties required). Further, suppose that, while outside environment E and while not in conditions C with respect to a proposition X in F, S still retains the virtue involved, *relative to E,* because the following ECF conditional remains true of S:

> (ECF) That if in E and in C relative to X, in F, then S would most likely be right in his belief or disbelief of X.

If S does so retain that virtue in that way, it can only be due to some components or aspects of S's intrinsic nature I, for it is S's possessing I together with being in E and in C with respect to X in F that fully explains and gives rise to the relevant disposition on the part of S, namely the disposition to believe correctly and avoid error regarding X in F, when so characterized and circumstanced.

We may now distinguish between (a) possession of the virtue (relative to E) in the sense of possession of the disposition, i.e., in the sense that the appropriate complex and general conditional (ECF) indicated above is true of the subject with the virtue, and (b) possession of a certain ground or basis of the virtue, in the sense of possessing an inner nature I from which the truth of the ECF conditional derives in turn. Of course one and the same virtue might have several different alternative possible grounds or bases. Thus the disposition to roll down an incline if free at its top with a certain orientation, in a certain environment (gravity, etc.), may be grounded in the sphericity and rigidity of an object, or alternatively it may be grounded in its cylindricality and rigidity. Either way, the conditional will obtain and the object will have the relevant disposition to roll. Similarly, Earthians and Martians may both be endowed with sight, in the sense of having the ability to tell colors and shapes, etc., though the principles of the operation of Earthian sight may differ widely from the principles that apply to Martians, which would or might presumably derive from a difference in the inner structure of the two species of being.

What now makes a disposition (and the underlying inner structure or nature that grounds it) an intellectual virtue? If we view such a disposition as defined by a C-F pair, then a being might have the disposition to be right with respect to propositions in field F but not relative to another environment E'. Such virtues, then, i.e., such C-F dispositions, might be virtuous only relative to an environment E and not relative to a different environment E'. And what makes such a disposition a virtue relative to an environment E seems now as obvious as it is that having the truth is an epistemic desideratum, and that being so constituted that one would most likely attain the truth in a certain field in a certain environment, when in certain conditions vis-à-vis propositions in that field, is so far as it goes an epistemic desideratum, an intellectual virtue.

What makes a subject intellectually virtuous? What makes her inner nature meritorious? Surely we can't require that a being have all merit and virtue before it can have any. Consider then a subject who has a minimal virtue of responding, thermometer-like, to environing food, and suppose him to have the minimal complexity and sophistication required for having beliefs at all—so that he is not literally just a thermometer or the like. Yet we suppose him further to have no way of relating what he senses, and his sensing of it, to a wider view of things that will explain it all, that will enable him perhaps to make related predictions and exercise related control. No, this ability is a relatively isolated phenomenon to which the subject yields with infant-like un-selfconscious simplicity. Suppose indeed the subject is just an infant or a higher animal. Can we allow that he knows of the presence of food when he has a correct belief to that effect? Well, the subject may of course have reliable belief that there is something edible there, without having a belief as reliable as that of a normal, well-informed adult, with some knowledge of food composition, basic nutrition, basic perception, etc., and who can at least implicitly interrelate these matters for a relatively much more coherent and complete view of the matter and related matters. Edibility can be a fairly complex matter, and how we have perceptual access to that property can also be rather involved, and the more one knows about the various factors whose interrelation yields the perceptible edibility of something before one, presumably the more reliable one's access to that all-important property.

Here then is one proposal on what makes one's belief that-p a result of enough virtue to make one internally justified in that belief. First of all we need to relativize to an assumed environment, which need not be the environment that the believer actually is in. What is required for a subject S to believe that-p out of sufficient virtue relative to environment E is that the proposition that-p be in a field F and that S be in conditions C with respect to that proposition, such that S would not be in C with respect to a proposition in F while in environment E, without S being most likely to believe correctly with regard to that proposition; and further that by comparison with epistemic group G, S is not grossly defective in ability to detect thus the truth in field F; i.e., it cannot be that S would have, by comparison with G:

(a) only a relatively very low probability of success,
(b) in a relatively very restricted class F,
(c) in a very restricted environment E,
(d) in conditions C that are relatively infrequent.

where all this relativity holds with respect to fellow members of G and to their normal environment and circumstances. (There is of course some variation from context as to what the relevant group might be when one engages in discussion of whether or not some subject knows something or is at least justified in believing it. But normally a certain group will stand out, with humanity being the default value.)

E. Intellectual Virtue Applied

Consider now again the new evil-demon problem and the problem of meta-incoherence. The crucial question in each case seems to be that of the internal justification of the subject, and this in turn seems not a matter of the virtue and total internal justification of that subject relative to an assumed group G and environment E, which absent any sign to the contrary one would take to be the group of humans in a normal human environment for the sort of question under consideration. Given these assumptions, the victim of the evil demon is virtuous and internally justified in every relevant respect, and not just in the respects of enjoying superweak, weak, and meta justification; for the victim is supposed to be just like an arbitrarily selected normal human in all cognitively relevant internal respects. Therefore, the internal structure and goings on in the victim must be at least up to par, in respect of how virtuous all of that internal nature makes the victim, relative to a normal one of us in our usual environment for considering whether we have a fire before us or the like. For those inclined towards mentalism or towards some broadly Cartesian view of the self and her mental life, this means at a minimum that the experience-belief mechanisms must not be random, but must rather be systematically truth-conducive, and that the subject must attain some minimum of coherent perspective on her own situation in the relevant environment, and on her modes of reliable access to information about that environment. Consider next those inclined towards naturalism, who hold the person to be either just a physical organism, or some physical part of an organism, or to be anyhow constituted essentially by some such physical entity; for these it would be required that the relevant physical being identical with or constitutive of the subject, in the situation in question, must not be defective in cognitively relevant internal respects; which would mean, among other things, that the subject would acquire beliefs about the colors or shapes of facing surfaces only under appropriate prompting at the relevant surfaces of the relevant visual organs (and not, e.g., through direct manipulation of the brain by some internal randomizing device).[12]

We have appealed to an intuitive distinction between what is intrinsic or internal to a subject or being, and what is extrinsic or external. Now when a subject receives certain inputs and emits as output a certain belief or a certain choice, that belief or choice can be defective either in virtue of an internal factor or in virtue of an external factor (or, of course, both). That is to say it may be that everything inner, intrinsic, or internal to the subject operates flawlessly and indeed brilliantly, but that something goes awry—with the belief, which turns out to be false, or with the choice, which turns out to be disastrous—because of some factor that, with respect to that subject, is outer, extrinsic, or external.[13]

In terms of that distinction, the victim of the demon may be seen to be internally justified, just as internally justified as we are, whereas the meta-incoherent are internally unjustified, unlike us.

My proposal is that justification is relative to environment. Relative to our actual environment A, our automatic experience-belief mechanisms count as virtues that yield much truth and justification. Of course relative to the demonic environment D such mechanisms are not virtuous and yield neither truth nor justification. It follows that relative to D the demon's victims are not justified, and yet *relative to A their beliefs are justified.* Thus may we fit our surface intuitions about such victims: that they lack knowledge but not justification.

In fact, a fuller account should distinguish between "justification" and "aptness"[14] as follows:

(a) The "justification" of a belief B requires that B have a basis in its inference or coherence relations to other beliefs in the believer's mind—as in the "justification" of a belief derived from deeper principles, and thus "justified," or the "justification" of a belief adopted through cognizance of its according with the subject's principles, including principles as to what beliefs are permissible in the circumstances as viewed by that subject.

(b) The "aptness" of a belief B relative to an environment E requires that B derive from what relative to E is an intellectual virtue, i.e., a way of arriving at belief that yields an appropriate preponderance of truth over error (in the field of propositions in question, in the sort of context involved).

As far as I can see, however, the basic points would remain within the more complex picture as well. And note that "justification" itself would then amount to a sort of inner coherence, something that the demon's victims can obviously have despite their cognitively hostile environment, but also something that will earn them praise relative to that environment only if it is an inner drive for greater and greater explanatory comprehensiveness, a drive which leads nowhere but to a more and more complex tissue of falsehoods. If we believe our world not to be such a world, then we can say that, relative to our actual environment A, "justification" as inner coherence earns its honorific status, and is an intellectual virtue, dear to the scientist, the philosopher, and the detective. Relative to the demon's D, therefore, the victim's belief may be inapt and even unjustified—if "justification" is essentially honorific—or if "justified" simply because coherent then, relative to D, that justification may yet have little or no cognitive worth. Even so, relative to our environment A, the beliefs of the demon's victim may still be both apt and valuably justified through their inner coherence.

The epistemology I defend—virtue perspectivism—is distinguished from generic reliabilism in three main respects:

(a) Virtue perspectivism requires not just any reliable mechanism of belief acquisition for belief that can qualify as knowledge; it requires the belief to derive from an intellectual virtue or faculty.

(b) Virtue perspectivism distinguishes between aptness and justification of belief, where a belief is apt if it derives from a faculty or virtue, but is justified only if it fits coherently within the epistemic perspective of the believer—perhaps by being connected to adequate reasons in the mind of the believer in such a way that the believer follows adequate or even impeccable intellectual procedure. This distinction is used as one way to deal with the new evil-demon problem.

(c) Virtue perspectivism distinguishes between animal and reflective knowledge. For animal knowledge one needs only belief that is apt and derives from an intellectual virtue or faculty. By contrast, reflective knowledge always requires belief that not only is apt but also has a kind of justification, since it must be belief that fits coherently within the epistemic perspective of the believer. This distinction is used earlier in this chapter to deal with the meta-incoherence problem, and it also opens the way to a solution for the generality problem.

Notes

1. Frank Ramsey, *The Foundations of Mathematics and Other Logical Essays* (London: Routledge & Kegan Paul, 1931).

2. David Armstrong, *Belief, Truth and Knowledge* (Cambridge University Press, 1973); Fred Dretske, "Conclusive Reasons," *Australian Journal of Philosophy* 49 (1971);1–22; Alvin Goldman, "What Is Justified Belief?" in George Pappas, ed., *Justification and Knowledge* (Dordrecht: D. Reidel, 1979); Robert Nozick, *Philosophical Explanations* (Cambridge, Mass.: Harvard University Press, 1981), chapter 3; Marshall Swain, *Reasons and Knowledge* (Ithaca, N.Y.: Cornell University Press, 1981).

3. This problem is pointed out by Goldman himself (*op. cit.*, p. 12), and is developed by Richard Feldman in "Reliability and Justification," *The Monist* 68 (1985): 159–74.

4. This problem is presented by Keith Lehrer and Stewart Cohen in "Justification, Truth, and Coherence," *Synthese* 55 (1983): 191–207.

5. This sort of problem is developed by Laurence BonJour in "Externalist Theories of Empirical Knowledge," in *Midwest Studies in Philosophy, Vol. 5: Studies in Epistemology*, ed. P. French et al. (Minneapolis: University of Minnesota Press, 1980).

6. See Goldman, "What Is Justified Belief?" pp. 13–14.

7. Alvin Goldman, *Epistemology and Cognition* (Cambridge, Mass.: Harvard University Press, 1986); *idem.*, "Strong and Weak Justification," in *Philosophical Perspectives, Vol. 2: Epistemology (1988):* 51–71.

8. Some of these are pointed out in my "Beyond Scepticism, to the Best of our Knowledge," *Mind* 97 (1988): 153–88.

9. Goldman, "Strong and Weak Justification," p. 56.

10. I will use the modified sense in what follows because it seems clearly better as an approach to blamelessness; but the substance of the critique to follow would apply also to the unmodified sense of weakly justified belief.

11. Though, actually, it is not really clear how these notions will deal with part (d) of the problem of meta-incoherence: cf. Goldman, *Epistemology and Cognition*, pp. 111–12.

12. As for the generality problem, my own proposed solution appears in Chapter 16 of Sosa, *Knowledge in Perspective* (Cambridge: Cambridge University Press, 1991).

13. This sort of distinction between the internal virtue of a subject and his or her (favorable or unfavorable) circumstances is drawn in "How Do You Know?"—Chapter 2 in Sosa, *Knowledge in Perspective*. There knowledge is relativized to epistemic community, though not in a way that imports any subjectivism or conventionalism, and consequences are drawn for the circumstances within which praise or blame is appropriate (see especially the first part of Section II).

14. For this sort of distinction, see, e.g., "Methodology and Apt Belief," Chapter 14 in Sosa, *Knowledge in Perspective*. The more generic distinction between external and internal justification may be found in "The Analysis of 'Knowledge That *P*'," Chapter 1 in Sosa, *Knowledge in Perspective*.

7

A Contextualist Theory of Epistemic Justification

David B. Annis

I. Foundationalism, Coherentism, and Contextualism

Foundationalism is the theory that every empirical statement which is justified ultimately must derive at least some of its justification from a special class of basic statements which have at least some degree of justification independent of the support such statements may derive from other statements. Such *minimal* foundationalism does not require certainty or incorrigibility; it does not deny the revisability of *all* statements, and it allows an important role for intrasystematic justification or coherence.[1] The main objections to foundationalism have been (a) the denial of the existence of basic statements and (b) the claim that even if such statements were not mythical, such an impoverished basis would never justify all the various statements we normally take to be justified.

Opposed to foundationalism has been the coherence theory of justification. According to coherentism a statement is justified if and only if it coheres with a certain kind of system of statements. Although there has been disagreement among coherentists in explaining what coherence is and specifying the special system of statements, the key elements in these explanations have been consistency, connectedness, and comprehensiveness. The chief objection to the theory has been that coherence within a consistent and comprehensive set of state-

Reprinted from the *American Philosophical Quarterly* 15 (1978): 213–19, by permission of the author and the editor. Copyright 1978, *American Philosophical Quarterly*.

ments is not sufficient for justification.[2] Theorists of epistemic justification have tended to stress foundationalism and coherentism and in general have overlooked or ignored a third kind of theory, namely, *contextualism*. The contextualist denies that there are basic statements in the foundationalist's sense and that coherence is sufficient for justification. According to contextualism both theories overlook contextual parameters essential to justification. In what follows I develop a version of a contextualist theory.[3]

II. The Basic Model—Meeting Objections

The basic model of justification to be developed here is that of a person's being able to meet certain objections. The objections one must meet and whether or not they are met are relative to certain goals. Since the issue is that of epistemic justification, the goals are epistemic in nature. With respect to one epistemic goal, accepting some statement may be reasonable, whereas relative to a different goal it may not be. Two of our epistemic goals are having true beliefs and avoiding having false beliefs. Other epistemic goals such as simplicity, conservation of existing beliefs, and maximization of explanatory power will be assumed to be subsidiary to the goals of truth and the avoidance of error.[4]

Given these goals, if a person S claims that some statement h is true, we may object (A) that S is not in a position to know that h or (B) that h is false. Consider (A). Suppose we ask S how he knows that h and he responds by giving us various reasons $e_1, e_2 \ldots , e_n$ for the truth of h. We may object that one of his reasons e_i-e_n is false, e_i-e_n does not provide adequate support for h, S's specific reasoning from e_i-e_n and i does not provide adequate support for h. These objections may be raised to his reasons for e_i-e_n as well as to his responses to our objections.

There are also cases where a person is not required to give reasons for his claim that h is true. If S claims to see a brown book across the room, we usually do not require reasons. But we may still object that the person is not in a position to know by arguing, for example, that the person is not reliable in such situations. So even in cases where we do not in general require reasons, objections falling into categories (A) or (B) can be raised.

But it would be too strong a condition to require a person to be able to meet all *possible* objections falling into these categories. In some distant time new evidence may be discovered as the result of advances in our scientific knowledge which would call into question the truth of some statement *h*. Even though we do not in fact have that evidence now, it is logically possible that we have it, so it is a possible objection to *h* now. If the person had to meet the objection, he would have to be in a different and better epistemic position than the one he is presently in, that is, he would have to have new evidence in order to respond to the objection. The objectors also would have to be in a better position to raise the objection. But the objections to be raised and answered should not require the participants to be in a new epistemic position. What is being asked is whether the person in his present position is justified in believing *h*. Thus the person only has to answer *current* objections, that is, objections based on the current evidence available.

Merely uttering a question that falls into one of our categories does not make it an objection *S* must answer. To demand a response the objection must be an expression of a *real* doubt. According to Peirce, doubt is an uneasy and dissatisfied state from which we struggle to free ourselves. Such doubt is the result of "some surprising phenomenon, some experience which either disappoints an expectation, or breaks in upon some habit of expectation."[5] As Dewey puts it, it is only when "jars, hitches, breaks, blocks . . . incidents occasioning an interruption of the smooth straight forward course of behavior" occur that doubt arises.[6] Thus for *S* to be held accountable for answering an objection, it must be a manifestation of a real doubt where the doubt is occasioned by a real life situation. Assuming that the subjective probabilities a person assigns reflect the person's actual epistemic attitudes and that these are the product of his confrontation with the world, the above point may be expressed as follows. *S* is not required to respond to an objection if *in general* it would be assigned a low probability by the people questioning *S*.

If an objection must be the expression of a real doubt caused by the jars of a real life situation, then such objections will be primarily *local* as opposed to *global*. Global objections call into question the totality of beliefs held at a certain time or a whole realm of beliefs, whereas local objections call into question a specific belief. This is not to say that a real situation might not occur that would prompt a global objection. If having experienced the nuclear radiation of a third world war, there were a sudden and dramatic increase in the error rate of perceptual beliefs of the visual sort, we would be more hesitant about them as a class.

It must be assumed that the objecting audience has the epistemic

goals of truth and the avoidance of error. If they were not critical truth
seekers, they would not raise appropriate objections. To meet an
objection i, S must respond in such a way as to produce within the
objecting group a general but not necessarily universal rejection of i or
at least the general recognition of the diminished status of i as an
objection. In the latter case S may, for example, point out that although
i might be true, it only decreases the support of e_i (one of his reasons
for believing h) a very small amount, and hence he is still justified in
believing h. There are of course many ways in which S can handle an
objection. He might indicate that it is not of the type (A) or (B) and so
is not relevant. He may respond that it is just an *idle* remark not
prompted by real doubt; that is, there is no reason for thinking that it
is true. He may ask the objector for his reasons, and he can raise any
of the objections of the type (A) or (B) in response. Again the give and
take is based on real objections and responses.

III. The Social Nature of Justification

When asking whether S is justified in believing h, this has to be
considered relative to an *issue-context*. Suppose we are interested in
whether Jones, an ordinary non-medically trained person, has the
general information that polio is caused by a virus. If his response to
our question is that he remembers the paper reporting that Salk said it
was, then this is good enough. He has performed adequately given the
issue-context. But suppose the context is an examination for the M.D.
degree. Here we expect a lot more. If the candidate simply said what
Jones did, we would take him as being very deficient in knowledge.
Thus relative to one issue-context a person may be justified in believing
h but not justified relative to another context.

The issue-context is what specific issue involving h is being raised.
It determines the level of understanding and knowledge that S must
exhibit, and it determines an appropriate objector-group. For example
in the context of the examination for the M.D. degree, the appropriate
group is not the class of ordinary non-medically trained people, but
qualified medical examiners.

The importance (value or utility) attached to the outcome of accepting h when it is false or rejecting h when it is true is a component of
the issue-context. Suppose the issue is whether a certain drug will help
cure a disease in humans without harmful effects. In such a situation
we are much more demanding than if the question were whether it

would help in the case of animals. In both cases the appropriate objector-group would be the same, namely, qualified researchers. But they would require quite a bit more proof in the former case. Researchers do in fact strengthen or weaken the justificatory conditions in relation to the importance of the issue. If accepting h when h is false would have critical consequences, the researcher may increase the required significance level in testing h.

Man is a social animal, and yet when it comes to the justification of beliefs philosophers tend to ignore this fact. But this is one contextual parameter that no adequate theory of justification can overlook. According to the contextualist model of justification sketched above, when asking whether some person S is justified in believing h, we must consider this relative to some specific issue-context which determines the level of understanding and knowledge required. This in turn determines the appropriate objector-group. For S to be justified in believing h relative to the issue-context, S must be able to meet all current objections falling into (A) and (B) which express a real doubt of the qualified objector-group where the objectors are critical truth seekers. Thus social information—the beliefs, information, and theories of others—plays an important part in justification, for it in part determines what objections will be raised, how a person will respond to them, and what responses the objectors will accept.

Perhaps the most neglected component in justification theory is the *actual* social practices and norms of justification of a culture or community of people. Philosophers have looked for universal and a priori principles of justification. But consider this in the context of scientific inquiry. There certainly has been refinement in the methods and techniques of discovery and testing in science. Suppose that at a time t in accordance with the best methods then developed for discovery and testing in a scientific domain by critical truth seekers, S accepts theory T. It is absurd to say that S is not justified in accepting T since at a later time a refinement of those techniques would lead to the acceptance of a different theory. Thus relative to the standards at t, S is justified in accepting T.

The same conclusion follows if we consider a case involving two different groups existing at the same time instead of two different times as in the above example. Suppose S is an Earth physicist and accepts T on the basis of the best methods developed by Earth physicists at t. Unknown to us the more advanced physicists on Twin Earth reject T. S is still justified in accepting T.

To determine whether S is justified in believing h we must consider the actual standards of justification of the community of people to which he belongs. More specifically we determine whether S is justified in believing h by specifying an issue-context raised within a community of people G with certain social practices and norms of justification. This determines the level of understanding and knowledge S is expected to have and the standards he is to satisfy. The appropriate objector-group is a subset of G. To be justified in believing h, S must be able to meet their objections in a way that satisfies their practices and norms.

It follows that justification theory must be *naturalized*. In considering the justification of beliefs we cannot neglect the actual social practices and norms of justification of a group. Psychologists, sociologists, and anthropologists have started this study, but much more work is necessary.[7]

The need to naturalize justification theory has been recognized in recent philosophy of science. Positivists stressed the *logic* of science— the structure of theories, confirmation, explanation—in abstraction from science as actually carried on. But much of the main thrust of recent philosophy of science is that such an approach is inadequate. Science as *practiced* yields justified beliefs about the world. Thus the study of the actual practices, which have changed through time, cannot be neglected. The present tenor in the philosophy of science is thus toward a historical and methodological realism.[8]

From the fact that justification is relative to the social practices and norms of a group, it does not follow that they cannot be criticized nor that justification is somehow subjective. The practices and norms are epistemic and hence have as their goals truth and the avoidance of error. Insofar as they fail to achieve these goals they can be criticized. For example the Kpelle people of Africa rely more on the authority of the elders than we do. But this authority could be questioned if they found it led to too many false perceptual beliefs. An objection to a practice must of course be real; that is, the doubt must be the result of some jar or hitch in our experience of the world. Furthermore such objections will always be local as opposed to global. Some practice or norm and our experiences of the world yield the result that another practice is problematic. A real objection presupposes some other accepted practice. This however does not commit us to some form of subjectivism. Just as there is no theory-neutral observation language in science, so there is no standard-neutral epistemic position that one can adopt. But in neither case does it follow that objectivity and rational criticism are lost.[9]

IV. The Regress Argument

Philosophers who have accepted foundationalism have generally of-fered a version of the infinite regress argument in support of it. Two key premises in the argument are the denial of a coherence theory of justification and the denial that an infinite sequence of reasons is sufficient to justify a belief. But there is another option to the conclu-sion of the argument besides foundationalism. A contextualist theory of the sort offered above stops the regress and yet does not require basic statements in the foundationalist's sense.

Suppose that the Joneses are looking for a red chair to replace a broken one in their house. The issue-context is thus not whether they can discern subtle shades of color. Nor is it an examination in physics where the person is expected to have detailed knowledge of the transmission of light and color perception. Furthermore nothing of great importance hinges on a correct identification. Mr. Jones, who has the necessary perceptual concepts and normal vision, points at a red chair a few feet in front of him and says "here is a red one." The appropriate objector-group consists of normal perceivers who have general knowledge about the standard conditions of perception and perceptual error. In such situations which we are all familiar with, generally, there will be no objections. His claim is accepted as justified. But imagine that someone objects that there is a red light shining on the chair so it may not be red. If Jones cannot respond to this objection when it is real, then he is not in an adequate cognitive position. But suppose he is in a position to reply that he knows about the light and the chair is still red since he saw it yesterday in normal light. Then we will accept his claim.

A belief is *contextually basic* if, given an issue-context, the appro-priate objector-group does not require the person to have reasons for the belief in order to be in a position to have knowledge. If the objector-group requires reasons, then it is not basic in the context. Thus in the first situation above Jones's belief that there is a red chair here is contextually basic, whereas it is not basic in the second situ-ation.

Consider the case either where the objector-group does not require *S* to have reasons for his belief that *h* in order to be in a position to have knowledge and where they accept his claim, or the case where they require reasons and accept his claim. In either case there is no regress of reasons. If an appropriate objector-group, the members of which are critical truth seekers, have no real doubts in the specific issue-context, then the person's belief is justified. The belief has withstood the test of verifically motivated objectors.

V. Objections to the Theory

There are several objections to the contextualist theory offered, and their main thrust is that the conditions for justification imposed are too stringent. The objections are as follows. First according to the theory offered, to be justified in believing *h* one must be able to meet a restricted class of objections falling into categories (A) and (B). But this ignores the distinction between *being* justified and *showing* that one is justified. To be justified is just to satisfy the principles of justification. To show that one is justified is to demonstrate that one satisfies these principles, and this is much more demanding.[10] For example *S* might have evidence that justifies his belief that *h* even though he is not able to articulate the evidence. In this case *S* would not be able to show that he was justified.

Second, if to be justified in believing *h* requires that one be able to meet the objection that *h* is false, then the theory ignores the distinction between truth and justification. A person can be justified in believing a statement even though it is false.

Finally the theory requires *S* to be in a position to answer all sorts of objections from a variety of perspectives. But this again is to require too much. For example assume that two scientists in different countries unaware of each other's work perform a certain experiment. The first scientist, S_1, gets one result and concludes that *h*. The second scientist, S_2, does not get the result (due to incorrect measurements). To require of S_1 that he be aware of S_2's experiment and be able to refute it is to impose an unrealistic burden on him in order for his belief to be justified. It is to build a *defeasibility* requirement into the justification condition. One approach to handling the Gettier problem has been to add the condition that in order to have knowledge, besides having justified true belief, the justification must not be defeated. Although there have been different characterizations of defeasibility, a core component or unrestricted version has been that a statement *i* defeats the justification evidence *e* provides *h* just in case *i* is true and the conjunction of *i* and *e* does not provide adequate support for *h*.[11] But according to the contextualist theory presented in order for *S* to be justified in believing *h*, he must be able to meet the objection that there is defeating evidence.

In reply to the first objection, the theory offered does not ignore the distinction between being justified and showing that one is justified. It is not required of *S* that he be able to state the standards of justification and demonstrate that he satisfies them. What is required is that he be able to meet real objections. This may *sometimes* require him to discuss standards, but not always. Furthermore the example given is not a counterexample since it is not a case of justified belief. Consider a case where relative to an issue-context we would expect *S* to have reasons for his belief that *h*. Suppose when asked how he knows or what his reasons are he is not able to say anything. We certainly would not take him as justified in his belief. We may not be able to articulate all our evidence for *h*, but we are required to do it for some of the evidence. It is not enough that we have evidence for *h*; it must be *taken* by us as evidence, and this places us "in the logical space of reasons, of justifying and being able to justify what one says."[12]

The first point in response to the next objection is that *epistemic* justification makes a claim to knowledge. To be *epistemically* justified in believing *h* is to be in a position to know *h*. Furthermore if the goals of epistemic justification are truth and the avoidance of error, then one *ought not* accept false statements. From an epistemic point of view to do so is objectionable. Hence the falsity of *h* at least counts against the person's being justified.

However, the contextualist account offered does not ignore the distinction between truth and justification. Meeting an objection does not entail showing the objection is false. It only requires general agreement on the response. So the objection may still be true. Thus *S* may be justified in believing *h* since he can meet the objection when *h* is in fact false. Furthermore an objection in order to require a response has to be the expression of a real doubt. Since it is possible for verifically motivated objectors not to be aware of the falsity of *h*, this objection will not be raised, so *S* may be justified in believing *h* even though it is false.

The situation is complex, however, since there are cases where the falsity of *h* implies *S* is not justified in believing *h*. Suppose that Jones is at a party and wonders whether his friend Smith is there. Nothing of great importance hinges on his presence; he simply wonders whether he is there. Perhaps he would not mind a chat with Smith. He looks about and asks a few guests. They have not seen him there. In such a situation Jones is justified in believing Smith is not there.

Imagine now that Jones is a police officer looking for Smith, a suspected assassin, at the party. Merely looking about casually and checking with a few guests is certainly not adequate. If Smith turns out to be hiding in one of the closets, we will not conclude that Jones was justified in his belief only it turned out false. He displayed gross negligence in not checking more thoroughly. There are cases where relative to an issue-context we require the person S to put himself in such an epistemic position that h will not turn out to be false. In this case the falsity of h is *non-excusable*. To be justified in believing h in non-excusable cases, S must be able to meet the objection that h is false. This is not required in excusable cases.

Assume that h is some very complicated scientific theory and S puts himself in the very best evidential position at the time. Even if the truth of h is very important, the falsity of h is excusable. The complexity of the issue and the fact that S put himself in the best position possible excuses S from the falsity of h, so he is still justified. But not all excusable cases involve a complex h nor being in the best position possible. Suppose that Smith has an identical twin brother but the only living person who knows this is the brother. Furthermore there are no records that there was a twin brother. If Jones returns a book to Smith's house and mistakenly gives it to the brother (where the issue-context is simply whether he returned the borrowed book and nothing of great importance hinges on to whom he gave it), he is still justified in his belief that he gave it to his friend Smith. Although Jones could have put himself in a better position (by asking questions about their friendship), there was no reason for him in the context to check further. People did not generally know about the twin brother, and Smith did not notice any peculiar behavior. Given the issue-context, members of the appropriate objector-group would not *expect* Jones to check further. So he evinces no culpability when his belief turns out to be false. Excusability thus depends on the issue-context and what the appropriate objector-group, given their standards of justification and the information available, expect of S.

Part of assimilating our epistemic standards, as is the case with both legal and moral standards, is learning the conditions of excusability. Such conditions are highly context-dependent, and it would be extremely difficult if not impossible to formulate rules to express them. In general we learn the conditions of excusability case by case. One need only consider moral and legal negligence to realize the full complexity of excuses, an area still to be studied despite Austin's well-known plea a number of years ago.

In response to the third objection it should be noted that epistemic justification is not to be taken lightly. Accepting *h* in part determines what other things I will believe and do. Furthermore I can infect the minds of others with my falsehoods and thus affect their further beliefs and actions. So to be epistemically justified requires that our claims pass the test of criticism. This point has motivated some philosophers to build a defeasibility requirement into the conditions of justification.[13]

The contextualist theory presented above, however, does not do this. There may be a defeating statement *i*, but *S* need meet this objection only if the objector-group raises it. For them to raise it, *i* must be the expression of real doubt. But it is perfectly possible for verifically motivated people to be unaware of *i*.

Furthermore the concept of epistemic excusability applies to defeating evidence. Suppose there is defeating evidence *i*. *S* may still be justified in his belief that *h* in the issue-context, even though he is unable to meet the objection. Relative to the issue-context, the appropriate objector-group with their standards of justification and available information may not expect of *S* that he be aware of *i*. Perhaps the issue involving *h* is very complicated. Thus his failure to meet the defeating evidence is excusable.

In the experiment case we can imagine issue-contexts where we would expect the first scientist to know of the experiment of the other scientist. But not all issue-contexts demand this. Nevertheless we may still require that he be in a position to say something about the other experiment if informed about it. For example he might indicate that he knows the area well, has performed the experiment a number of times and gotten similar results, it was performed under carefully controlled conditions, so he has every reason for believing that the experiment is replicable with similar results. Thus there must be something wrong with the other experiment. Requiring the scientist to be able to respond in the *minimal* way seems not to be overly demanding.

VI. Summary

Contextualism is an alternative to the traditional theories of foundationalism and coherentism. It denies the existence of basic statements in the foundationalist's sense (although it allows contextually basic statements), and it denies that coherence as it traditionally has been explained is sufficient for justification. Both theories overlook contextual parameters essential to justification, such as the issue-context and thus the value of *h*, social information, and social practices and norms of justification. In particular, the social nature of justification cannot be ignored.

Notes

1. For a discussion of minimal foundationalism see William P. Alston, "Has Foundationalism Been Refuted?"; James W. Cornman, "Foundationalism versus Nonfoundational Theories of Empirical Justification"; David B. Annis, "Epistemic Foundationalism." (See concluding bibliography for full data.)

2. Recent discussions of coherentism are found in Keith Lehrer, *Knowledge*, chaps. 7–8; Nicholas Rescher, "Foundationalism, Coherentism, and the Idea of Cognitive Systematization." and his *The Coherence Theory of Truth*. Criticism of Lehrer's coherence theory is to be found in Cornman, "Foundational Versus Nonfoundational Theories of Empirical Justification," and in my review of Lehrer in *Philosophia* 6 (1976): 209–13. Criticism of Rescher's version is found in Mark Pastin's "Foundationalism Redux," unpublished, an abstract of which appears in the *The Journal of Philosophy* 61 (1974): 709–10.

3. Historically the key contextualists have been Peirce, Dewey, and Popper. But contextualist hints, suggestions, and theories are also to be found in Robert Ackermann, *Belief and Knowledge;* Bruce Aune, *Knowledge, Mind and Nature;* John Austin *Sense and Sensibilia* (London, 1962); Isaac Levi, *Gambling with Truth* (New York, 1967); Stephen Toulmin, *The Uses of Argument* (London, 1958) and *Human Understanding* (Princeton, New Jersey, 1972); Carl Wellman, *Challenge and Response: Justification in Ethics* (Carbondale, Illinois, 1971); F. L. Will, *Induction and Justification;* Ludwig Wittgenstein, *Philosophical Investigations* (New York, 1953) and *On Certainty*.

4. For a discussion of epistemic goals see Levi, *Gambling with Truth*.

5. C. S. Peirce, *Collected Papers,* vol. 6, ed. Charles Hartshorne and Paul Weiss (Harvard, 1965), p. 469.

6. John Dewey, *Knowing and the Known* (Boston, 1949), p. 315. See also Wittgenstein's *On Certainty*.

7. See, for example, Michael Cole et al., *The Cultural Context of Learning and Thinking* (New York, 1971).

8. For a discussion of the need to naturalize justification theory in the philosophy of science, see Frederick Suppe, "Afterword—1976" in the 2nd edition of his *The Structure of Scientific Theories* (Urbana, Illinois, 1977).

9. See Frederick Suppe's "The Search for Philosophic Understanding of Scientific Theories" and his "Afterword—1976" in *The Structure of Scientific Theories* for a discussion of objectivity in science and the lack of a theory-neutral observation language.

10. Alston discusses this distinction in "Has Foundationalism Been Re-

futed?'' See also his ''Two Types of Foundationalism,'' and ''Self-Warrant: A Neglected Form of Privileged Access.''

11. The best discussion of defeasibility is Marshall Swain's ''Epistemic Defeasibility.''

12. Wilfrid Sellars, *Science, Perception and Reality,* p. 13. Carl Ginet, ''What Must Be Added to Knowing to Obtain Knowing That One Knows?,'' *Synthese* 21 (1970): 163–86.

8

Pragmatism, Relativism, and Irrationalism

Richard Rorty

Part I: Pragmatism

"Pragmatism"is a vague, ambiguous, and overworked word. Nevertheless, it names the chief glory of our country's intellectual tradition. No other American writers have offered so radical a suggestion for making our future different from our past, as have James and Dewey. At present, however, these two writers are neglected. Many philosophers think that everything important in pragmatism has been preserved and adapted to the needs of analytic philosophy. More specifically, they view pragmatism as having suggested various holistic corrections of the atomistic doctrines of the early logical empiricists. This way of looking at pragmatism is not wrong, as far as it goes. But it ignores what is most important in James and Dewey. Logical empiricism was one variety of standard, academic, neo-Kantian, epistemologically-centered philosophy. The great pragmatists should not be taken as suggesting an holistic variation of this variant, but rather as breaking with the Kantian epistemological tradition altogether. As long as we see James or Dewey as having "theories of truth" or "theories of knowledge" or "theories of morality" we shall get them wrong. We shall ignore their criticisms of the assumption that there ought to *be* theories about such matters. We shall not see how radical their thought was—how deep was their criticism of the attempt, common to Kant, Husserl, Russell, and C. I. Lewis, to make philosophy into a foundational discipline.

Reprinted by permission from *Proceedings and Addresses of the American Philosophical Association* 53 (1980), pp. 719–38.

One symptom of this incorrect focus is a tendency to overpraise Peirce. Peirce is praised partly because he developed various logical notions and various technical problems (such as the counterfactual conditional) which were taken up by the logical empiricists. But the main reason for Peirce's undeserved apotheosis is that his talk about a general theory of signs looks like an early discovery of the importance of language. For all his genius, however, Peirce never made up his mind what he wanted a general theory of signs *for,* nor what it might look like, nor what its relation to either logic or epistemology was supposed to be. His contribution to pragmatism was merely to have given it a name, and to have stimulated James. Peirce himself remained the most Kantian of thinkers—the most convinced that philosophy gave us an all-embracing ahistorical context in which every other species of discourse could be assigned its proper place and rank. It was just this Kantian assumption that there was such a context, and that epistemology or semantics could discover it, against which James and Dewey reacted. We need to focus on this reaction if we are to recapture a proper sense of their importance.

This reaction is found in other philosophers who are currently more fashionable than James or Dewey—for example, Nietzsche and Heidegger. Unlike Nietzsche and Heidegger, however, the pragmatists did not make the mistake of turning against the community which takes the natural scientist as its moral hero—the community of secular intellectuals which came to self-consciousness in the Enlightenment. James and Dewey rejected neither the Enlightenment's choice of the scientist as moral example, nor the technological civilization which science had created. They wrote, as Nietzsche and Heidegger did not, in a spirit of social hope. They asked us to liberate our new civilization by giving up the notion of "grounding" our culture, our moral lives, our politics, our religious beliefs, upon "philosophical bases." They asked us to give up the neurotic Cartesian quest for certainty which had been one result of Galileo's frightening new cosmology, the quest for "enduring spiritual values" which had been one reaction to Darwin, and the aspiration of academic philosophy to form a tribunal of pure reason which had been the neo-Kantian response to Hegelian historicism. They asked us to think of the Kantian project of grounding thought or culture in a permanent ahistorical matrix as *reactionary.* They viewed Kant's idealization of Newton, and Spencer's of Darwin, as just as silly as Plato's idealization of Pythagoras, and Aquinas' of Aristotle.

Emphasizing this message of social hope and liberation, however, makes James and Dewey sound like prophets rather than thinkers. This would be misleading. They had things to say about truth, knowledge, and morality, even though they did not have *theories* of them, in the sense of sets of answers to the textbook problems. In what follows, I shall offer three brief sloganistic characterications of what I take to be their central doctrine.

My first characterization of pragmatism is that it is simply antiessentialism applied to notions like "truth," "knowledge," "language," "morality," and similar objects of philosophical theorizing. Let me illustrate this by James's definition of "the true" as "what is good in the way of belief." This has struck his critics as not to the point, as unphilosophical, as like the suggestion that the essence of aspirin is that it is good for headaches. James's point, however, was that there *is* nothing deeper to be said: truth is not the sort of thing which *has* an essence. More specifically, his point was that it is no use being told that truth is "correspondence to reality." Given a language and a view of what the world is like, one can, to be sure, pair off bits of the language with bits of what one takes the world to be in such a way that the sentences one believes true have internal structures isomorphic to relations between things in the world. When we rap out routine undeliberated reports like "This is water," "That's red," "That's ugly," "That's immoral," our short categorical sentences can easily be thought of as pictures, or as symbols which fit together to make a map. Such reports do indeed pair little bits of language with little bits of the world. Once one gets to negative universal hypotheticals, and the like, such pairing will become messy and *ad hoc*, but perhaps it can be done. James's point was that carrying out this exercise will not enlighten us about why truths are good to believe, or offer any clues as to why or whether our present view of the world is, roughly, the one we should hold. Yet nobody would have asked for a "theory" of truth if they had not wanted answers to these latter questions. Those who want truth to have an essence want knowledge, or rationality, or inquiry, or the relation between thought and its object, to have an essence. Further, they want to be able to use their knowledge of such essences to criticize views they take to be false, and to point the direction of progress toward the discovery of more truths. James thinks these hopes are vain. There are no essences anywhere in the area. There is no wholesale, epistemological way to direct, or criticize, or underwrite, the course of inquiry.

Rather, the pragmatists tell us, it is the vocabulary of practise rather than of theory, of action rather than contemplation, in which one can say something useful about truth. Nobody engages in epistemology or semantics because he wants to know how "This is red" pictures the world. Rather, we want to know in what sense Pasteur's views of disease picture the world accurately and Paracelsus' inaccurately, or what exactly it is that Marx pictured more accurately than Machiavelli. But just here the vocabulary of "picturing" fails us. When we turn from individual sentences to vocabularies and theories, critical terminology naturally shifts from metaphors of isomorphism, symbolism, and mapping to talk of utility, convenience, and likelihood of getting what we want. To say that the parts of properly analyzed true sentences are arranged in a way isomorphic to the parts of the world paired with them sounds plausible if one thinks of a sentence like "Jupiter has moons." It sounds slightly less plausible for "The earth goes round the sun," less still for "There is no such thing as natural motion," and not plausible at all for "The universe is infinite." When we want to praise or blame assertions of the latter sort of sentence, we show how the decision to assert them fits into a whole complex of decisions about what terminology to use, what books to read, what projects to engage in, what life to live. In this respect they resemble such sentences as "Love is the only law" and "History is the story of class struggle." The whole vocabulary of isomorphism, picturing, and mapping is out of place here, as indeed is the notion of being true *of objects*. If we ask what objects these sentences claim to be true of, we get only unhelpful repetitions of the subject terms—"the universe," "the law," "history." Or, even less helpfully, we get talk about "the facts," or "the way the world is." The natural approach to such sentences, Dewey tells us, is not "Do they get it right?", but more like "What would it be like to believe that? What would happen if I did? What would I be committing myself to?" The vocabulary of contemplation, looking, *theoria,* deserts us just when we deal with theory rather than observation, with programming rather than input. When the contemplative mind, isolated from the stimuli of the moment, takes large views, its activity is more like deciding what to *do* than deciding that a representation is accurate. James's dictum about truth says that the vocabulary of practice is uneliminable, that no distinction of kind separates the sciences from the crafts, from moral reflection, or from art.

So a second characterization of pragmatism might go like this: there is no epistemological difference between truth about what ought to be and truth about what is, nor any metaphysical difference between facts and values, nor any methodological difference between morality and science. Even nonpragmatists think Plato was wrong to think of moral philosophy as discovering the essence of goodness, and Mill and Kant wrong in trying to reduce moral choice to rule. But every reason for saying that they were wrong is a reason for thinking the epistemological tradition wrong in looking for the essence of science, and in trying to reduce rationality to rule. For the pragmatists, the pattern of all inquiry—scientific as well as moral—is deliberation concerning the relative attractions of various concrete alternatives. The idea that in science or philosophy we can substitute "method" for deliberation between alternative results of speculation is just wishful thinking. It is like the idea that the morally wise man resolves his dilemmas by consulting his memory of the Idea of the Good, or by looking up the relevant article of the moral law. It is the myth that rationality consists in being constrained by rule. According to this Platonic myth, the life of reason is not the life of Socratic conversation but an illuminated state of consciousness in which one never needs to ask if one has exhausted the possible descriptions of, or explanations for, the situation. One simply arrives at true beliefs by obeying mechanical procedures.

Traditional, Platonic, epistemologically-centered philosophy is the search for such procedures. It is the search for a way in which one can avoid the need for conversation and deliberation and simply tick off the way things are. The idea is to acquire beliefs about interesting and important matters in a way as much like visual perception as possible—by confronting an object and responding to it as programmed. This urge to substitute *theoria* for *phronesis* is what lies behind the attempt to say that "There is no such thing as natural motion" pictures objects in the same way as does "The cat is on the mat." It also lies behind the hope that some arrangement of objects may be found which is pictured by the sentence "Love is better than hate," and the frustration which ensues when it is realized that there may be no such objects. The great fallacy of the tradition, the pragmatists tell us, is to think that the metaphors of vision, correspondence, mapping, picturing, and representation which apply to small, routine assertions will apply to large and debatable ones. This basic error begets the notion that where there are no objects to correspond to we have no hope of rationality, but only taste, passion, and will. When the pragmatist attacks the notion of truth as accuracy of representation he is thus attacking the traditional distinctions between reason and desire, reason and appetite, reason and will. For none of these distinctions make sense unless reason is thought of on the model of vision, unless we persist in what Dewey called "the spectator theory of knowledge."

The pragmatist tells us that once we get rid of this model, we see that the Platonic idea of the life of reason is impossible. A life spent representing objects accurately would be spent recording the results of calculations, reasoning through sorites, calling off the observable properties of things, construing cases according to unambiguous criteria, getting things right. Within what Kuhn calls "normal science," or any similar social context, one can, indeed, live such a life. But conformity to *social* norms is not good enough for the Platonist. He wants to be constrained not merely by the disciplines of the day, but by the ahistorical and nonhuman nature of reality itself. This impulse takes two forms—the original Platonic strategy of postulating novel *objects* for treasured propositions to correspond to, and the Kantian strategy of finding *principles* which are definatory of the essence of knowledge, or representation, or morality, or rationality. But this difference is unimportant compared to the common urge to escape the vocabulary and practices of one's own time and find something ahistorical and necessary to cling to. It is the urge to answer questions like "Why believe what I take to be true?" "Why do what I take to be right?" by appealing to something *more* than the ordinary, retail, detailed, concrete reasons which have brought one to one's present view. This urge is common to nineteenth-century idealists and contemporary scientific realists, to Russell and to Husserl; it is definatory of the Western philosophical tradition, and of the culture for which that tradition speaks. James and Dewey stand with Nietzsche and Heidegger in asking us to abandon that tradition, and that culture.

Let me sum up by offering a third and final characterization of pragmatism: it is the doctrine that there are no constraints on inquiry save conversational ones—no wholesale constraints derived from the nature of the objects, or of the mind, or of language, but only those retail constraints provided by the remarks of our fellow-inquirers. The way in which the properly-programmed speaker cannot help believing that the patch before him is red has *no* analogy for the more interesting and controversial beliefs which provoke epistemological reflection. The pragmatist tells us that it is useless to hope that objects will constrain us to believe the truth about them, if only they are approached with an unclouded mental eye, or a rigorous method, or a perspicuous language. He wants us to give up the notion that God, or evolution, or some other underwriter of our present world-picture, has programmed us as machines for accurate verbal picturing, and that philosophy brings self-knowledge by letting us read our own program. The only sense in which we are constrained to truth is that, as Peirce suggested, we can make no sense of the notion that the view which can survive all objections might be false. But objections—conversational constraints—cannot be anticipated. There is no method for knowing *when* one has reached the truth, or when one is closer to it than before.

I prefer this third way of characterizing pragmatism because it seems to me to focus on a fundamental choice which confronts the reflective mind: that between accepting the contingent character of starting-points, and attempting to evade this contingency. To accept the contingency of starting-points is to accept our inheritance from, and our conversation with, our fellow-humans as our only source of guidance. To attempt to evade this contingency is to hope to become a properly-programmed machine. This was the hope which Plato thought might be fulfilled at the top of the divided line, when we passed beyond hypotheses. Christians have hoped it might be attained by becoming attuned to the voice of God in the heart, and Cartesians that it might be fulfilled by emptying the mind and seeking the indubitable. Since Kant, philosophers have hoped that it might be fulfilled by finding the a priori structure of any possible inquiry, or language, or form of social life. If we give up this hope, we shall lose what Nietzsche called "metaphysical comfort," but we may gain a renewed sense of community. Our identification with our community—our society, our political tradition, our intellectual heritage—is heightened when we see this community as *ours* rather than *nature's, shaped* rather than *found*, one among many which men have made. In the end, the pragmatists tell us, what matters is our loyalty to other human beings clinging together against the dark, not our hope of getting things right. James, in arguing against realists and idealists that "the trail of the human serpent is over all," was reminding us that our glory is in our participation in fallible and transitory human projects, not in our obedience to permanent nonhuman constraints.

Part II: Relativism

"Relativism" is the view that every belief on a certain topic, or perhaps about *any* topics, is as good as every other. No one holds this view. Except for the occasional cooperative freshman, one cannot find anybody who says that two incompatible opinions on an important topic are equally good. The philosophers who get *called* "relativists" are those who say that the grounds for choosing between such opinions are less algorithmic than had been thought. Thus one may be attacked as a relativist for holding that familiarity of terminology is a criterion of theory-choice in physical science, or that coherence with the institutions of the surviving parliamentary democracies is a criterion in social philosophy. When such criteria are invoked, critics say that the resulting philosophical position assumes an unjustified primacy for "our conceptual framework," or our purposes, or our institutions. The position in question is criticized for not having done what philosophers are employed to do: explain why our framework, or culture, or interests, or language, or whatever, is at last on the right track—in touch with physical reality, or the moral law, or the real numbers, or some other sort of object patiently waiting about to be copied. So the real issue is not between people who think one view as good as another and people who do not. It is between those who think our culture, or purpose, or intuitions cannot be supported except conversationally, and people who still hope for other sorts of support.

If there *were* any relativists, they would, of course, be easy to refute. One would merely use some variant of the self-referential arguments Socrates used against Protagoras. But such neat little dialectical strategies only work against lightly sketched fictional characters. The relativist who says that we can break ties among serious and incompatible candidates for belief only by "nonrational" or "noncognitive" considerations is just one of the Platonist or Kantian philosopher's imaginary playmates, inhabiting the same realm of fantasy as the solipsist, the skeptic, and the moral nihilist. Disillusioned, or whimsical, Platonists and Kantians occasionally play at being one or another of these characters. But when they do they are never offering relativism or skepticism or nihilism as a serious suggestion about how we might do things differently. These positions are adopted to make *philosophical* points—that is, moves in a game played with fictitious opponents, rather than fellow-participants in a common project.

The association of pragmatism with relativism is a result of a confusion between the pragmatist's attitude toward *philosophical* theories with his attitude towards *real* theories. James and Dewey are, to be sure, metaphilosophical relativists, in a certain limited sense. Namely: they think there is no way to choose, and no point in choosing, between incompatible philosophical theories of the typical Platonic or Kantian type. Such theories are attempts to ground some element of our practices on something external to these practices. Pragmatists think that any such philosophical grounding is, apart from elegance of execution, pretty much as good or as bad as the practice it purports to ground. They regard the project of grounding as a wheel that plays no part in the mechanism. In this, I think, they are quite right. No sooner does one discover the categories of the pure understanding for a Newtonian age than somebody draws up another list that would do nicely for an Aristotelian or an Einsteinian one. No sooner does one draw up a categorical imperative for Christians than somebody draws up one which works for cannibals. No sooner does one develop an evolutionary epistemology which explains why our science is so good than somebody writes a science-fiction story about bug-eyed and monstrous evolutionary epistemologists praising bug-eyed and monstrous scientists for the survival value of their monstrous theories. The reason this game is so easy to play is that none of these philosophical theories have to do much hard work. The real work has been done by the scientists who developed the explanatory theories by patience and genius, or the societies which developed the moralities and institutions in struggle and pain. All the Platonic or Kantian philosopher does is to take the finished first-level product, jack it up a few levels of abstraction, invent a metaphysical or epistemological or semantical vocabulary into which to translate it, and announce that he has *grounded* it.

"Relativism" only seems to refer to a disturbing view, worthy of being refuted, if it concerns *real* theories, not just philosophical theories. Nobody really cares if there are incompatible alternative formulations of a categorical imperative, or incompatible sets of categories of the pure understanding. We *do* care about alternative, concrete, detailed cosmologies, or alternative concrete, detailed proposals for political change. When such an alternative is proposed, we debate it, not in terms of categories or principles but in terms of the various concrete advantages and disadvantages it has. The reason relativism is talked about so much among Platonic and Kantian philosophers is that they think being relativistic about philosophical theories—attempts to "ground" first-level theories—leads to being relativistic about the first-level theories themselves. If anyone really believed that the worth of a theory depends upon the worth of its philosophical grounding, then indeed they would be dubious about physics, or democracy, until relativism in respect to philosophical theories had been overcome. Fortunately, almost nobody believes anything of the sort.

What people do believe is that it would be good to hook up our views about democracy, mathematics, physics, God, and everything else, into a coherent story about how everything hangs together.

Getting such a synoptic view often does require us to change radically our views on particular subjects. But this holistic process of readjustment is just muddling through on a large scale. It has nothing to do with the Platonic-Kantian notion of grounding. That notion involves finding constraints, demonstrating necessities, finding immutable principles to which to subordinate oneself. When it turns out that suggested constraints, necessities, and principles are as plentiful as blackberries, nothing changes except the attitude of the rest of culture towards the philosophers. Since the time of Kant, it has become more and more apparent to nonphilosophers that a really professional philosopher can supply a philosophical foundation for just about anything. This is one reason why philosophers have, in the course of our century, become increasingly isolated from the rest of culture. Our proposals to guarantee this and clarify that have come to strike our fellow-intellectuals as merely comic.

Part III: Irrationalism

My discussion of relativism may seem to have ducked the real issues. Perhaps nobody is a relativist. Perhaps "relativism" is *not* the right name for what so many philosophers find so offensive in pragmatism. But surely there *is* an important issue around somewhere. There is indeed an issue, but it is not easily stated, nor easily made amenable to argument. I shall try to bring it into focus by developing it in two different contexts, one microcosmic and the other macrocosmic. The microcosmic issue concerns philosophy in one of its most parochial senses—namely, the activities of the American Philosophical Association. Our Association has traditionally been agitated by the question of whether we should be free-wheeling and edifying, or argumentative and professional. For my purposes, this boils down to an issue about whether we can be pragmatists and still be professionals. The macrocosmic issue concerns philosophy in the widest sense—the attempt to make everything hang together. This is the issue between Socrates on the one hand and the tyrants on the other—the issue between lovers of conversation and lovers of self-deceptive rhetoric. For my purposes, it is the issue about whether we can be pragmatists without betraying Socrates, without falling into irrationalism.

I discuss the unimportant microcosmic issue about professionalism first because it is sometimes confused with the important issue about irrationalism, and because it helps focus that latter issue. The question of whether philosophy professors should edify agitated our Association in its early decades. James thought they should, and was dubious about the growing professionalization of the discipline. Arthur Lovejoy, the great opponent of pragmatism, saw professionalization as an unmixed blessing. Echoing what was being said simultaneously by Russell in England and by Husserl in Germany, Lovejoy urged the sixteenth annual meeting of the APA to aim at making philosophy into a science. He wanted the APA to organize its program into well-structured controversies on sharply defined problems, so that at the end of each convention it would be agreed who had won.[1] Lovejoy insisted that philosophy could either be edifying and visionary *or* could produce "objective, verifiable, and clearly communicable truths," but not both. James would have agreed. He too thought that one could *not* be both a pragmatist and a professional. James, however, saw professionalization as a failure of nerve rather than as a triumph of rationality. He thought that the activity of making things hang together was *not* likely to produce "objective, verifiable, and clearly communicable truths," and that this did not greatly matter.

Lovejoy, of course, won this battle. If one shares his conviction that philosophers should be as much like scientists as possible, then one will be pleased at the outcome. If one does not, one will contemplate the APA in its seventy-sixth year mindful of Goethe's maxim that one should be careful what one wishes for when one is young, for one will get it when one is old. Which attitude one takes will depend upon whether one sees the problems we discuss today as permanent problems for human thought, continuous with those discussed by Plato, Kant, and Lovejoy—or as modern attempts to breathe life into dead issues. On the Lovejoyan account, the gap between philosophers and the rest of high culture is of the same sort as the gap between physicists and laymen. The gap is not created by the artificiality of the problems being discussed, but by the development of technical and precise ways of dealing with real problems. If one shares the pragmatists' anti-essentialism, however, one will tend to see the problems about which philosophers are now offering "objective, verifiable, and clearly communicable" solutions as historical relics, left over from the Enlightenment's misguided search for the hidden essences of knowledge and morality. This is the point of view adopted by many of our fellow-intellectuals, who see us philosophy professors as caught in a time-warp, trying to live the Enlightenment over again.

I have reminded you of the parochial issue about professionalization not in order to persuade you to one side or the other, but rather to exhibit the source of the anti-pragmatist's passion. This is his conviction that conversation necessarily aims at agreement and at rational consensus, that we converse in order to make further conversation unnecessary. The anti-pragmatist believes that conversation only makes sense if something like the Platonic theory of Recollection is right—if we all have natural starting-points of thought somewhere within us, and will recognize the vocabulary in which they are best formulated once we hear it. For only if something like that is true will conversation have a natural goal. The Enlightenment hoped to find such a vocabulary—nature's own vocabulary, so to speak. Lovejoy— who described himself as an "unredeemed *Aufklärer*"—wanted to continue the project. Only if we had agreement on such a vocabulary, indeed, could conversation be reduced to argumentation—to the search for "objective, verifiable, and clearly communicable" solutions to problems. So the anti-pragmatist sees the pragmatist's scorn for professionalism as scorn for consensus, for the Christian and democratic idea that every human has the seeds of truth within. The pragmatist's attitude seems to him elitist and dilettantish, reminiscent of Alcibiades rather than of Socrates.

Issues about relativism and about professionalization are awkward attempts to formulate this opposition. The real and passionate opposition is over the question of whether loyalty to our fellow-humans presupposes that there is something permanent and unhistorical which explains *why* we should continue to converse in the manner of Socrates, something which guarantees convergence to agreement. Because the anti-pragmatist believes that without such an essence and such a guarantee the Socratic life makes no sense, he sees the pragmatist as a cynic. Thus the microcosmic issue about how philosophy professors should converse leads us quickly to the macrocosmic issue: whether one can be a pragmatist without being an irrationalist, without abandoning one's loyalty to Socrates.

Questions about irrationalism have become acute in our century because the sullen resentment which sins against Socrates, which withdraws from conversation and community, has recently become articulate. Our European intellectual tradition is now abused as "merely conceptual" or "merely ontic" or as "committed to abstractions." Irrationalists propose such rubbishy pseudo-epistemological notions as "intuition" or "an inarticulate sense of tradition" or "thinking with the blood" or "expressing the will of the oppressed classes." Our tyrants and bandits are more hateful than those of earlier times because, invoking such self-deceptive rhetoric, they pose as intellectuals. Our tyrants write philosophy in the morning and torture in the afternoon; our bandits alternately read Hölderlin and bomb people into bloody scraps. So our culture clings, more than ever, to the hope of the Enlightenment, the hope that drove Kant to make philosophy formal and rigorous and professional. We hope that by formulating the *right* conceptions of reason, of science, of thought, of knowledge, of morality, the conceptions which express their *essence,* we shall have a shield against irrationalist resentment and hatred.

Pragmatists tell us that this hope is vain. On their view, the Socratic virtues—willingness to talk, to listen to other people, to weigh the consequences of our actions upon other people—are *simply* moral virtues. They cannot be inculcated nor fortified by theoretical research into essence. Irrationalists who tell us to think with our blood cannot be rebutted by better accounts of the nature of thought, or knowledge, or logic. The pragmatists tell us that the conversation which it is our moral duty to continue is *merely* our project, the European intellectual's form of life. It has no metaphysical nor epistemological guarantee of sucess. Further (and this is the crucial point) *we do not know what "success" would mean except simply "continuance."* We are not conversing because we have a goal, but because Socratic conversation is an activity which is its *own* end. The anti-pragmatist who insists that agreement is its goal is like the basketball player who thinks that the reason for playing the game is to make baskets. He mistakes an essential moment in the course of an activity for the end of the activity. Worse yet, he is like a basketball fan who argues that all men by nature desire to play basketball, or that the nature of things is such that balls can go through hoops.

For the traditional, Platonic or Kantian philosopher, on the other hand, the possibility of *grounding* the European form of life—of showing it to be more than European, more than a contingent human project—seems the central task of philosophy. He wants to show that sinning against Socrates is sinning against our nature, not just against our community. So he sees the pragmatist as an irrationalist. The charge that pragmatism is "relativistic" is simply his first unthinking expression of disgust at a teaching which seems cynical about our deepest hopes. If the traditional philosopher gets beyond such epithets, however, he raises a question which the pragmatist must face up to: the *practical* question of whether the notion of "conversation" *can* substitute for that of "reason." "Reason," as the term is used in the Platonic and Kantian traditions, is interlocked with the notions of truth as correspondence, of knowledge as discovery of essence, of morality as obedience to principle, all the notions which the pragmatist tries to deconstruct. For better or worse, the Platonic and Kantian vocabularies are the ones in which Europe has described and praised the Socratic virtues. It is not clear that we know how to describe these virtues without those vocabularies. So the deep suspicion which the pragmatist inspires is that, like Alcibiades, he is essentially frivolous— that he is commending uncontroversial common goods while refusing to participate in the only activity which can preserve those goods. He seems to be sacrificing our common European project to the delights of purely negative criticism.

The issue about irrationalism can be sharpened by noting that when the pragmatist says "All that can be done to explicate 'truth', 'knowledge', 'morality', 'virtue' is to refer us back to the concrete details of the culture in which these terms grew up and developed," the defender of the Enlightenment takes him to be saying "Truth and virtue are simply what a community agrees that they are." When the pragmatist says "We have to take truth and virtue as whatever emerges from the conversation of Europe," the traditional philosopher wants to know what is so special about Europe. Isn't the pragmatist saying, like the irrationalist, that *we* are in a privileged situation simply by being *us*? Further, isn't there something terribly dangerous about the notion that truth can only be characterized as "the outcome of doing more of what we are doing now"? What if the "we" is the Orwellian state? When tyrants employ Lenin's blood-curdling sense of "objective" to describe their lies as "objectively true," what is to prevent them from citing Peirce in Lenin's defense?[2]

The pragmatist's first line of defense against this criticism has been created by Habermas, who says that such a definition of truth works only for the outcome of *undistorted* conversation, and that the Orwellian state is the paradigm of distortion. But this is *only* a first line, for we need to know more about what counts as "undistorted." Here Habermas goes transcendental and offers principles. The pragmatist, however, must remain ethnocentric and offer examples. He can only say: "undistorted" means employing *our* criteria of relevance, where *we* are the people who have read and pondered Plato, Newton, Kant, Marx, Darwin, Freud, Dewey, etc. Milton's "free and open encounter," in which truth is bound to prevail, must itself be described in terms of examples rather than principles—it is to be more like the Athenian market-place than the council-chamber of the Great King, more like the twentieth century than the twelfth, more like the Prussian Academy in 1925 than in 1935. The pragmatist must avoid saying, with

Peirce, that truth is *fated* to win. He must even avoid saying that truth *will* win. He can only say, with Hegel, that truth and justice lie in the direction marked by the successive stages of European thought. This is not because he knows some "necessary truths" and cites these examples as a result of this knowledge. It is simply that the pragmatist knows no better way to explain his convictions than to remind his interlocutor of the position they both are in, the contingent starting points they both share, the floating, ungrounded conversations of which they are both members. This means that the pragmatist cannot answer the question "What is so special about Europe?" save by saying "Do you have anything non-European to suggest which meets *our* European purposes better?" He cannot answer the question "What is so good about the Socratic virtues, about Miltonic free encounters, about undistorted communication?" save by saying "What else would better fulfill the purposes *we* share with Socrates, Milton, and Habermas?"

To decide whether this obviously circular response is enough is to decide whether Hegel or Plato had the proper picture of the progress of thought. Pragmatists follow Hegel in saying that "philosophy is its time grasped in thought." Anti-pragmatists follow Plato in striving for an escape from conversation to something atemporal which lies in the background of all possible conversations. I do not think one can decide between Hegel and Plato save by meditating on the past efforts of the philosophical tradition to escape from time and history. One can see these efforts as worthwhile, getting better, worth continuing. Or one can see them as doomed and perverse. I do not know what would count as a noncircular metaphysical or epistemological or semantical argument for seeing them in either way. So I think that the decision has to be made simply by reading the history of philosophy and drawing a moral.

Nothing that I have said, therefore, is an argument in favor of pragmatism. At best, I have merely answered various superficial criticisms which have been made of it. Nor have I dealt with the central issue about irrationalism. I have not answered the deep criticism of pragmatism which I mentioned a few minutes ago: the criticism that the Socratic virtues cannot, as a practical matter, be defended save by Platonic means, that without some sort of metaphysical comfort nobody will be able *not* to sin against Socrates. William James himself was not sure whether this criticism could be answered. Exercising his own right to believe, James wrote: "If this life be not a real fight in which something is eternally gained for the universe by success, it is no better than a game of private theatricals from which we may withdraw at will." "It *feels*," he said, "like a fight."

For us, footnotes to Plato that we are, it *does* feel that way. But if James's own pragmatism were taken seriously, if pragmatism became central to our culture and our self-image, then it would no longer feel that way. We do not know how it *would* feel. We do not even know whether, given such a change in tone, the conversation of Europe might not falter and die away. We just do not know. James and Dewey offered us no guarantees. They simply pointed to the situation we stand in, now that both the Age of Faith and the Enlightenment seem beyond recovery. They grasped our time in thought. We did not change the course of the conversation in the way they suggested we might. Perhaps we are still unable to do so; perhaps we never shall be able to. But we can nevertheless honor James and Dewey for having offered what very few philosophers have succeeded in giving us: a hint of how our lives might be changed.

Notes

1. See A. O. Lovejoy, "On Some Conditions of Progress in Philosophical Inquiry," *The Philosophical Review*, XXVI (1917): 123– 163 (especially the concluding pages). I owe the reference to Lovejoy's paper to Daniel J. Wilson's illuminating "Professionalization and Organized Discussion in the American Philosophical Association, 1900–1922," *Journal of the History of Philosophy*, XVII (1979): 53–69.

2. I am indebted to Michael Williams for making me see that pragmatists have to answer this question.

Part II

The Gettier Problem

9

Is Justified True Belief Knowledge?

Edmund Gettier

Various attempts have been made in recent years to state necessary and sufficient conditions for someone's knowing a given proposition. The attempts have often been such that they can be stated in a form similar to the following:[1]

(a) S knows that P   *IFF*   (i) *P* is true,
                              (ii) S believes that P, and
                              (iii) S is justified in believing that P.

For example, Chisholm has held that the following gives the necessary and sufficient conditions for knowledge:[2]

(b) S knows that P   *IFF*   (i) S accepts P.
                              (ii) S has adequate evidence for P, and
                              (iii) P is true.

Ayer has stated the necessary and sufficient conditions for knowledge as follows:[3]

(c) S knows that P   *IFF*   (i) P is true,
                              (ii) S is sure that P is true, and
                              (iii) S has the right to be sure that P
                                    is true.

I shall argue that (a) is false in that the conditions stated therein do not constitute a *sufficient* condition for the truth of the proposition that S

Reprinted from *Analysis* 23 (1963): 121–23, by permission of the author and the publisher, Basil Blackwell. Copyright 1963, Edmund Gettier.

knows that P. The same argument will show that (b) and (c) fail if "has adequate evidence for" or "has the right to be sure that" is substituted for "is justified in believing that" throughout.

I shall begin by noting two points. First, in that sense of "justified" in which S's being justified in believing P is a necessary condition of S's knowing that P, it is possible for a person to be justified in believing a proposition which is in fact false. Second, for any proposition P, if S is justified in believing P and P entails Q and S deduces Q from P and accepts Q as a result of this deduction, then S is justified in believing Q. Keeping these two points in mind, I shall now present two cases in which the conditions stated in (a) are true for some proposition, though it is at the same time false that the person in question knows that proposition.

Case I

Suppose that Smith and Jones have applied for a certain job. And suppose that Smith has strong evidence for the following conjunctive proposition:

(d) Jones is the man who will get the job, and Jones has ten coins in his pocket.

Smith's evidence for (d) might be that the president of the company assured him that Jones would in the end be selected, and that he, Smith, had counted the coins in Jones's pocket ten minutes ago. Proposition (d) entails:

(e) The man who will get the job has ten coins in his pocket.

Let us suppose that Smith sees the entailment from (d) to (e) and accepts (e) on the grounds of (d), for which he has strong evidence. In this case, Smith is clearly justified in believing that (e) is true.

But imagine, further, that unknown to Smith, he himself, not Jones, will get the job. And, also, unknown to Smith, he himself has ten coins in his pocket. Proposition (e) is then true, though proposition (d), from which Smith inferred (e), is false. In our example, then, all of the following are true: *(i)* (e) is true, *(ii)* Smith believes that (e) is true, and *(iii)* Smith is justified in believing that (e) is true. But it is equally clear that Smith does not *know* that (e) is true; for (e) is true in virtue of the number of coins in Smith's pocket, while Smith does not know how many coins are in Smith's pocket, and bases his belief in (e) on a count of the coins in Jones's pocket, whom he falsely believes to be the man who will get the job.

Case II

Let us suppose that Smith has strong evidence for the following proposition:

(f) Jones owns a Ford.

Smith's evidence might be that Jones has at all times in the past within Smith's memory owned a car, and always a Ford, and that Jones has just offered Smith a ride while driving a Ford. Let us imagine, now, that Smith has another friend, Brown, of whose whereabouts he is totally ignorant. Smith selects three place names quite at random and constructs the following three propositions:

(g) Either Jones owns a Ford, or Brown is in Boston.
(h) Either Jones owns a Ford, or Brown is in Barcelona
(i) Either Jones owns a Ford, or Brown is in Brest-Litovsk.

Each of these propositions is entailed by (f). Imagine that Smith realizes the entailment of each of these propositions he has constructed by (f), and proceeds to accept (g), (h), and (i) on the basis of (f). Smith has correctly inferred (g), (h), and (i) from a proposition for which he has strong evidence. Smith is therefore completely justified in believing each of these three propositions. Smith, of course, has no idea where Brown is.

But imagine now that two further conditions hold. First, Jones does *not* own a Ford, but is at present driving a rented car. And second, by the sheerest coincidence, and entirely unknown to Smith, the place mentioned in proposition (h) happens really to be the place where Brown is. If these two conditions hold, then Smith does *not* know that (h) is true, even though *(i)* (h) *is* true, *(ii)* Smith does believe that (h) is true, and *(iii)* Smith is justified in believing that (h) is true.

These two examples show that definition (a) does not state a *sufficient* condition for someone's knowing a given proposition. The same cases, with appropriate changes, will suffice to show that neither definition (b) nor definition (c) does so either.

Notes

1. Plato seems to be considering some such definition at *Theaetetus* 201, and perhaps accepting one at *Meno* 98.
2. Roderick M. Chisholm, *Perceiving: A Philosophical Study,* p. 16.
3. A. J. Ayer, *The Problem of Knowledge.*

10

An Alleged Defect in Gettier Counterexamples

Richard Feldman

A number of philosophers have contended that Gettier counterexamples to the justified true belief analysis of knowledge all rely on a certain false principle. For example, in their recent paper "Knowledge Without Paradox,"[1] Robert G. Meyers and Kenneth Stern argue that "Counterexamples of the Gettier sort all turn on the principle that someone can be justified in accepting a certain proposition h on evidence p even though p is false."[2] They contend that this principle is false, and hence that the counterexamples fail. Their view is that one proposition, p, can justify another, h, only if p is true. With this in mind, they accept the justified true belief analysis.

D. M. Armstrong defends a similar view in *Belief, Truth and Knowledge*.[3] He writes:

> This simple consideration seems to make redundant the ingenious argument of . . . Gettier's . . . article . . . Gettier produces counterexamples to the thesis that justified true belief is knowledge by producing true beliefs based on justifiably believed grounds, . . . but where these grounds are in fact *false*. But because possession of such grounds could not constitute possession of *knowledge*. I should have thought it obvious that they are too weak to serve as suitable grounds.[4]

Thus he concludes that Gettier's examples are defective because they rely on the false principle that false propositions can justify one's belief in other propositions. Armstrong's view seems to be that one

Reprinted from the *Australasian Journal of Philosophy* 52 (1974): 68–69, by permission of the author and the editor. Copyright 1974, *Australasian Journal of Philosophy*.

proposition, p, can justify another, h, only if p is known to be true (unlike Meyers and Stern, who demand only that p in fact be true).[5]

I think, though, that there are examples very much like Gettier's that do not rely on this allegedly false principle. To see this, let us first consider one example in the form in which Meyers and Stern discuss it, and then consider a slight modification of it.

> Suppose Mr. Nogot tells Smith that he owns a Ford and even shows him a certificate to that effect. Suppose, further, that up till now Nogot has always been reliable and honest in his dealings with Smith. Let us call the conjunction of all this evidence m. Smith is thus justified in believing that Mr. Nogot who is in his office owns a Ford (r) and, consequently, is justified in believing that someone in his office owns a Ford (h).[6]

As it turns out, though, m and h are true but r is false. So, the Gettier example runs, Smith has a justified true belief in h, but he clearly does not know h.

What is supposed to justify h in this example is r. But since r is false, the example runs afoul of the disputed principle. Since r is false, it justifies nothing. Hence, if the principle is false, the counterexample fails.

We can alter the example slightly, however, so that what justifies h for Smith is true and he knows that it is. Suppose he deduces from m its existential generalization:

> (n) There is someone in the office who told Smith that he owns a Ford and even showed him a certificate to that effect, and who up till now has always been reliable and honest in his dealings with Smith.

(n), we should note, is true and Smith knows that it is, since he has correctly deduced it from m, which he knows to be true. On the basis of n Smith believes h—someone in the office owns a Ford. Just as the Nogot evidence, m, justified r—Nogot owns a Ford—in the original example, n justifies h in this example. Thus Smith has a justified true belief in h, knows his evidence to be true, but still does not know h.

I conclude that even if a proposition can be justified for a person only if his evidence is true, or only if he knows it to be true, there are still counterexamples to the justified true belief analysis of knowledge of the Gettier sort. In the above example, Smith reasoned from the proposition m, which he knew to be true, to the proposition n, which he also knew, to the truth h; yet he still did not know h. So some examples, similar to Gettier's, do not "turn on the principle that someone can be justified in accepting a certain proposition . . . even though [his evidence] is false.''[7]

Notes

1. *The Journal of Philosophy* 70 (March 22, 1973): 147–60.
2. Ibid., p. 147.
3. (1973).
4. Ibid., p. 152.
5. Armstrong ultimately goes on to defend a rather different analysis.
6. Meyers and Stern, ''Knowledge Without Paradox,'' p. 151.
7. Ibid., p. 147.

11

The Gettier Problem

John Pollock

1. Introduction

It is rare in philosophy to find a consensus on any substantive issue, but for some time there was almost complete consensus on what is called 'the justified true belief analysis of knowing'. According to that analysis:

S knows P if and only if:
(1) P is true;
(2) S believes P; and
(3) S is justified in believing P.

In the period immediately preceding the publication of Gettier's [1963] landmark article "Is justified true belief knowledge?", this analysis was affirmed by virtually every writer in epistemology. Then Gettier published his article and single-handedly changed the course of epistemology. He did this by presenting two clear and undeniable counterexamples to the justified true belief analysis. Recounting the example given in chapter nine, consider Smith who believes falsely but with good reason that Jones owns a Ford. Smith has no idea where Brown is, but he arbitrarily picks Barcelona and infers from the putative fact that Jones owns a Ford that either Jones owns a Ford or Brown is in Barcelona. It happens by chance that Brown is in Barcelona, so this disjunction is true. Furthermore, as Smith has good reason to believe that Jones owns a Ford, he is justified in believing this disjunction. But as his evidence does not pertain to the true disjunct of the disjunction,

Reprinted by permission from John Pollock, *Contemporary Theories of Knowledge*, pp. 180–93. Copyright 1986 by Rowman & Littlefield.

we would not regard Smith as *knowing* that either Jones owns a Ford or Brown is in Barcelona.

Gettier's paper was followed by a spate of articles attempting to meet his counterexamples by adding a fourth condition to the analysis of knowing. The first attempts to solve the Gettier problem turned on the observation that in Gettier's examples, the epistemic agent arrives at his justified true belief by reasoning from a false belief. That suggested the addition of a fourth condition something like the following:

S's grounds for believing P do not include any false beliefs.[1]

It soon emerged, however, that further counterexamples could be constructed in which knowledge is lacking despite the believer's not inferring his belief from any false beliefs. Alvin Goldman (1976) constructed the following example. Suppose you are driving through the countryside and see what you take to be a barn. You see it in good light and from not too great a distance, it looks the way barns look, and so on. Furthermore, it is a barn. You then have justified true belief that it is a barn. But in an attempt to appear more opulent than they are, the people around here have taken to constructing very realistic barn facades that cannot readily be distinguished from the real thing when viewed from the highway. There are many more barn facades around than real barns. Under these circumstances we would not agree that you know that what you see is a barn, even though you have justified true belief. Furthermore, your belief that you see a barn is not in any way inferred from a belief about the absence of barn facades. Most likely the possibility of barn facades is something that will not even have occurred to you, much less have played a role in your reasoning.

We can construct an even simpler perceptual example. Suppose S sees a ball that looks red to him, and on that basis he correctly judges that it is red. But unbeknownst to S, the ball is illuminated by red lights and would look red to him even if it were not red. Then S does not know that the ball is red despite his having a justified true belief to that effect. Furthermore, his reason for believing that the ball is red does not involve his believing that the ball is not illuminated by red lights. Illumination by red lights is related to his reasoning only as a defeater, not as a step in his reasoning. These examples, of other related examples,[2] indicate that justified true belief can fail to be knowledge because of the truth values of propositions that do not play a direct role in the reasoning underlying the belief. This observation led to a number of "defeasibility" analyses of knowing.[3] The simplest defeasibility analysis would consist of adding a fourth condition requiring that there be no true defeaters. This might be accomplished as follows:

> There is no true proposition Q such that if Q were added to S's beliefs then he would no longer be justified in believing P.[4]

But Keith Lehrer and Thomas Paxson (1969) presented the following counterexample to this simple proposal:

> Suppose I see a man walk into the library and remove a book from the library by concealing it beneath his coat. Since I am sure the man is Tom Grabit, whom I have often seen before when he attended my classes, I report that I know that Tom Grabit has removed the book. However, suppose further that Mrs. Grabit, the mother of Tom, has averred that on the day in question Tom was not in the library, indeed, was thousands of miles away, and that Tom's identical twin brother, John Grabit, was in the library. Imagine, moreover, that I am entirely ignorant of the fact that Mrs. Grabit has said these things. The statement that she has said these things would defeat any justification I have for believing that Tom Grabit removed the book, according to our present definition of defeasibility. . . .
>
> The preceding might seem acceptable until we finish the story by adding that Mrs. Grabit is a compulsive and pathological liar, that John Grabit is a fiction of her demented mind, and that Tom Grabit took the book as I believed. Once this is added, it should be apparent that I did know that Tom Grabit removed the book. (p. 228)

A natural proposal for handling the Grabit example is that in addition to there being a true defeater there is a true defeater defeater, and that restores knowledge. For example, in the Grabit case it is true that Mrs. Grabit reported that Tom was not in the library but his twin brother John was there (a defeater), but it is also true that Mrs. Grabit is a compulsive and pathological liar and John Grabit is a fiction of her demented mind (a defeater defeater). It is difficult, however, to construct a precise principle that handles these examples correctly by appealing to true defeaters and true defeater defeaters. It will not do to amend the above proposal as follows:

> If there is a true proposition Q such that if Q were added to S's beliefs then he would no longer be justified in believing P, then there is also a

true proposition R such that if Q and R were both added to S's beliefs then he would be justified in believing P.

The simplest difficulty for this proposal is that adding R may add new reasons for believing P rather than restoring the old reasons. It is not trivial to see how to formulate a fourth condition incorporating defeater defeaters. I think that such a fourth condition will ultimately provide the solution to the Gettier problem, but no proposal of this sort has been worked out in the literature.[5] I will pursue this further in the next section.

2. Objective Epistemic Justification

The Gettier problem has spawned a large number of proposals for the analysis of knowledge. As the literature on the problem has developed, the proposals have become increasingly complex in the attempt to meet more and more complicated counterexamples to simpler analyses. The result is that even if some very complex analysis should turn out to be immune from counterexample, it would seem *ad hoc*. We would be left wondering why we employ any such complicated concept. I will suggest that our concept of knowledge is actually a reasonably simple one. The complexities required by increasingly complicated Gettier-type examples are not complexities in the concept of knowledge, but instead reflect complexities in the structure of our epistemic norms.

In the discussion of externalism I commented on the distinction between subjective and objective senses of 'should believe' and how that pertains to epistemology. The subjective sense of 'should believe' concerns what we should believe given what we actually do believe (possibly incorrectly). The objective sense of 'should believe' concerns what we should believe given what is in fact true. But what we should believe given what is true is just the truths, so the objective sense of 'should believe' gets identified with truth. The subjective sense, on the other hand, is ordinary epistemic justification. What I now want to suggest, however, is that there is an intermediate sense of 'should believe', that might also be regarded as objective but does not reduce to truth.

It is useful to compare epistemic judgments with moral judgments. Focusing on the latter, let us suppose that a person S subjectively should do A. This will be so *for particular reasons*. There may be relevant facts of which the person is not apprised that bear upon these reasons. It might be the case that even in the face of all the relevant facts, S should still do **A**. That can happen in either of two ways: (1) among the relevant facts may be new reasons for doing A of which S has no knowledge; or (2) the relevant facts may, on sum, leave the original reasons intact. What I have been calling 'the objective sense of "should" ' appeals to both kinds of considerations, but there is also an important kind of moral evaluation that appeals only to considerations of the second kind. This is the notion of the original reasons surviving intact, and it provides us with another variety of objective moral evaluation. We appraise a person and his act simultaneously by saying that he has a moral obligation to perform the act (he subjectively should do it) and his moral obligation derives from what are in fact good reasons (resons withstanding the test of truth). It seems to me that we are often in the position of making such appraisals, although moral language provides us with no simple way of expressing them. The purely objective sense of 'should' pertains more to acts than to agents, and hence does not express moral obligation. Therefore, it should not be confusing if I express appraisals of his third variety artificially by saying that S has an *objective obligation* to do **A** when he has an obligation to do **A** and the obligation derives from what are in fact good reasons (in the face of all the relevant facts).

How might objective obligation be analyzed? It might at first be supposed that S has an objective obligation to do **A** if and only if (1) S subjectively should do **A**, and (2) there is no set of truths **X** such that if these truths were added to S's beliefs (and their negations removed in those cases in which S disbelieves them) then it would not be true that S subjectively should do **A** *for the same reason*. This will not quite do, however. It takes account of the fact that moral reasons are defeasible, but it does not take account of the fact that the defeaters are also defeasible. For example, S might spy a drowning man and be in a position to save him with no risk to himself. Then he subjectively should do so. But suppose that, unbeknownst to S, the man is a terrorist who fell in the lake while he was on his way to blow up a bus station and kill many innocent people. Presumably, if S knew that then he would no longer have a subjective obligation to save the man, and so it follows by the proposed analysis that S does not have an objective obligation to save the man. But suppose it is also the case that what caused the man to fall in the lake was that he underwent a sudden religious conversion that persuaded him to give up his evil ways and devote the rest of his life to good deeds. If S knew this, then he would again have a subjective obligation to save the man, for the same reasons as his original reasons, and so he has an objective obligation to save the man. There is, however, no way to accommodate this on the proposed analysis. On that analysis, if a set of truths defeats an obligation, there is no way to get it undefeated again by appealing to a broader class of truths.

What the analysis of objective obligation should require is that if S were apprised of "enough" truths (all the relevant ones) then he would still be subjectively obligated to do **A**. This can be cashed out as requiring that there is a set of truths such that if S were apprised of them then he would be subjectively obligated in the same way as he originally was, and those are all the relevant truths in the sense that if he were to become apprised of any further truths that would not make any difference. Precisely:

> S has an objective obligation to do **A** if and only if:
> (1) S subjectively should do **A**; and
> (2) there is a set **X** of truths such that, given any more inclusive set **Y** of truths, necessarily, if the truths in **Y** were added to S's beliefs (and their negations removed in those cases in which S disbelieves them) then it would still be true *for the same reason* that S subjectively should do **A**.

Now let us return to epistemology. An important difference between moral judgments and epistemic judgments is that basic moral judgments concern obligation whereas basic epistemic judgments concern permissibility. This reflects an important difference in the way moral and epistemic norms function. In morality, reasons are reasons for obligations. Anything is permissible that is not proscribed. In epistemology, on the other hand, epistemic justification concerns what beliefs you are permitted to hold (not 'obliged to hold'), and reasons are required for permissibility. Thus the analogy between epistemology and morality is not exact. The analogue of objective moral obligation is "objective epistemic permissibility", or as I will say more simply, *objective epistemic justification*. I propose to ignore our earlier concept of objective epistemic justification because it simply reduces to truth. Our new concept of objective epistemic justification can be defined as follows, on analogy to our notion of objective moral obligation:

> S is objectively justified in believing P if and only if:
> (1) S is (subjectively) justified in believing P; and
> (2) there is a set **X** of truths such that, given any more inclusive set **Y** of truths, necessarily, if the truths in **Y** were added to S's beliefs (and

their negations removed in those cases in which S disbelieves them) and S believed P *for the same reason* then he would still be (subjectively) justified in believing P.

Despite the complexity of its definition, the concept of objective epistemic justification is a simple and intuitive one. As is so often the case with technical concepts, the concept is easier to grasp than it is to define. It can be roughly glossed as the concept of getting the right answer while doing everything right. I am construing 'S is justified in believing P' in such a way that it entails that S does believe P, so objective justification entails justified belief. It also entails truth, because if P were false and we added P to Y then S would no longer be justified in believing P. Thus, objective epistemic justification entails justified true belief.

My claim is now that objective epistemic justification is very close to being the same thing as knowledge. We will find in section three that a qualification is required to turn objective justification into knowledge, but in the meantime it can be argued that the Gettier problem can be resolved by taking objective epistemic justification to be a necessary condition for knowledge. This enables us to avoid the familiar Gettier-type examples that create difficulties for other analyses of knowledge. Consider one of Gettier's original examples. Jones believes, correctly, that Brown owns a Ford. He believes this on the grounds that he has frequently seen Brown drive a particular Ford, he has ridden in it, he has seen Brown's auto registration which lists him as owning that Ford, and so forth. But unknown to Jones, Brown sold that Ford yesterday and bought a new one. Under the circumstances, we would not agree that Jones now knows that Brown owns a Ford, despite the fact that he has a justified true belief to that effect. This is explained by noting that Jones is not objectively justified in believing that Brown owns a Ford. This is because there is a truth—namely, that Brown does not own the Ford Jones thinks he owns—such that if Jones became apprised of it then his original reasons would no longer justify him in believing that Jones owns a Ford, and becoming apprised of further truths would not restore those original reasons.

To take a more complicated case, consider Goldman's barn example. Suppose you are driving through the countryside and see what you take to be a barn. You see it in good light and from not too great a distance, and it looks like a barn. Furthermore, it is a barn. You then have justified true belief that it is a barn. But the countryside here is littered with very realistic barn facades that cannot readily be distinguished from the real thing when viewed from the highway. There are many more barn facades than real barns. Under these circumstances we would not agree that you know that what you see is a barn, even though you have justified true belief. This can be explained by noting that if you were aware of the preponderance of barn facades in the vicinity then you would not be justified in believing you see a barn, and your original justification could not be restored by learning other truths (such as that it is really a barn). Consequently, your belief that you see a barn is not objectively justified.[6]

Finally, consider the Grabit example. Here we want to say that I really do know that Tom Grabit stole the book, despite the fact that Mrs. Grabit alleged that Tom was thousands of miles away and his twin brother John was in the library. That she said this is a true defeater, but there is also a true defeater defeater, viz., that Mrs. Grabit is a compulsive and pathological liar and John Grabit is a fiction of her demented mind. If we include *both* of these truths in the set **X** then I remain justified *for my original reason* in believing that Tom stole the book, so in this case my belief is objectively justified despite the existence of a true defeater.

To a certain extent, I think that the claim that knowledge requires objective epistemic justification provides a solution to the Gettier problem. But it might be disqualified as a solution to the Gettier problem on the grounds that the definition of objective justification is vague in one crucial respect. It talks about being justified, *for the same reason,* in believing P. I think that that notion makes pre-theoretic good sense, but to spell out what it involves requires us to construct a complete epistemological theory. That, I think, is why the Gettier problem has proven so intractable. The complexities in the analysis of knowing all have to do with filling out this clause. The important thing to realize, however, is that these complexities have nothing special to do with knowledge per se. What they pertain to is the structure of epistemic justification and the way in which beliefs come to be justified on the basis of other beliefs and nondoxastic states. Thus even if it is deemed that we have not yet solved the Gettier problem, we have at least put the blame where it belongs—not on knowledge but on the structure of epistemic justification and the complexity of our epistemic norms.

Let us turn then to the task of filling in some of the details concerning epistemic justification. In chapter two of *Contemporary Theories of Knowledge*, I proposed an analysis of epistemic justification in terms of ultimately undefeated arguments. That analysis proceeded within the context of a subsequently rejected foundationalist theory, but basically the same analysis can be resurrected within direct realism. For this purpose we must take arguments to proceed from internal states (both doxastic and nondoxastic states) to doxastic states, the links between steps being provided by reasons. Within direct realism, reasons are internal states. They are generally doxastic states, but not invariably. At the very least, perceptual and memory states can also be reasons.

Our epistemic norms permit us to begin reasoning from certain internal states without those states being supported by further reasoning. Such states can be called *basic states*. Paramount among these are perceptual and memory states. Arguments must always begin with basic states and proceed from them to nonbasic doxastic states. What we might call *linear arguments* proceed from basic states to their ultimate conclusions through a sequence of steps each consisting of a belief for which the earlier steps provide reasons. It seems likely, however, that we must allow arguments to have more complicated structures than those permitted in linear arguments. Specifically, we must allow "subsidiary arguments" to occur within the main argument. A subsidiary argument can begin with premises that are merely assumed for the sake of the argument rather than because they have already been justified. For instance, in the forms of conditional proof familiar from elementary logic, in establishing a conditional $(P \supset Q)$, we may begin by taking P as a premise (even though it has not been previously established), deriving Q from it, and then "discharging" the assumption of the antecedent to obtain the conditional $(P \supset Q)$. It seems that something similar occurs in epistemological arguments. We can accommodate this by taking *an argument conditional on a set* X *of propositions* to be an argument beginning not just from basic states but also from doxastic states that consist of believing the members of X. Then an argument that justifies a conclusion for a person may have embedded in it subsidiary arguments that are conditional on propositions the person does not believe. For present purposes we need not pursue all the details of the permissible structures of epistemological arguments, but the general idea of conditional arguments will be useful below.

An argument *supports* a belief if and only if that belief occurs as a step in the argument that does not occur within any subsidiary argument. A person *instantiates* an argument if and only if he is in the basic states from which the argument begins and he believes the conclusion of the argument on the basis of that argument. Typically, in reasoning to a conclusion one will proceed first to some intermediate conclusions from which the final conclusion is obtained. The notion of holding a belief on the basis of an argument is to be understood as requiring that one also believes the intermediate conclusions on the basis of the initial parts of the argument.

Epistemic justification consists of holding a belief on the basis of an ultimately undefeated argument, that is, instantiating an ultimately undefeated argument supporting the belief. To repeat the definition of an ultimately undefeated argument, every argument proceeding from basic states that S is actually in will be *undefeated at level 0* for S. Of course, arguments undefeated at level 0 can embed subsidiary arguments that are conditional on propositions S does not believe. Some arguments will support defeaters for other arguments, so we define an argument to be undefeated at level 1 if and only if it is not defeated by any other arguments undefeated at level 0. Among the arguments defeated at level 0 may be some that supported defeaters for others, so if we take arguments undefeated at level 2 to be arguments undefeated at level 0 that are not defeated by any arguments undefeated at level 1, there may be arguments undefeated at level 2 that were arguments defeated at level 1. In general, we define an argument to be *undefeated at level n + 1* if and only if it is undefeated at level 0 and is not defeated by any arguments undefeated at level n. An argument is *ultimately undefeated* if and only if there is some point beyond which it remains permanently undefeated; that is, for some N, the argument remains undefeated at level n for every n $>$N.

This gives us a picture of the structure of epistemic justification. Many details remain to be filled in, but we can use this picture without further elaboration to clarify the concept of objective epistemic justification. Roughly, a belief is objectively justified if and only if it is held on the basis of some ultimately undefeated argument **A**, and either **A** is not defeated by any argument conditional on true propositions not believed by S, or if it is then there are further true propositions such that the initial defeating arguments will be defeated by arguments conditional on the enlarged set of true propositions. This can be made precise by defining an *argument conditional on* **Y** to be any argument proceeding from basic states S is actually in together with doxastic states consisting of believing members of **Y**. We then say that an argument instantiated by S (not an argument conditional on **Y**) is *undefeated at level n + 1 relative to* **Y** if and only if it is undefeated by any argument undefeated at level n relative to **Y**. An argument is *ultimately undefeated relative to* **Y** if and only if there is an N such that it is undefeated at level n relative to **Y** for every n $>$N. Then the concept of objective epistemic justification can be made more precise as follows:

> S is objectively justified in believing P if and only if S instantiates some argument **A** supporting P which is ultimately undefeated relative to the set of all truths.

I will take this to be my official definition of objective epistemic justification. I claim, then, that the Gettier-style counterexamples to the traditional definition of knowledge can all be met by taking knowledge to require objective epistemic justification. This makes precise the way in which knowledge requires justification that is either undefeated by true defeaters, or if defeated by true defeaters then those defeaters are defeated by true defeater defeaters, and so on.

A common view has been that the reliability of one's cognitive processes is required for knowledge, and thus reliabilism has a place in the analysis of knowledge quite apart from whether it has a place in the analysis of epistemic justification.[7] The observation that knowledge requires objective epistemic justification explains the appeal of the idea that knowledge requires reliability. Nondefeasible reasons logically entail their conclusions, so they are always perfectly reliable, but defeasible reasons can be more or less reliable under various circumstances. Discovering that the present circumstances are of a type in which a defeasible reason is unreliable constitutes a defeater for the use of that reason. Objective justification requires that if a belief is held on the basis of a defeasible reason then there are no true defeaters (or if there are then there are true defeater defeaters, and so on). Thus knowledge automatically requires that one's reasons be reliable under the present circumstances. Reliabilism has a place in knowledge even if it has none in justification. It is worth emphasizing, however, that considerations of reliability are not central to the concept of knowledge. Rather than having to be imposed on the analysis in an *ad hoc* way, they emerge naturally from the observation that knowledge requires objective epistemic justification.

3. Social Aspects of Knowledge

It is tempting to simply identify knowledge with objective epistemic justification. As I have pointed out, objective justification includes justified true belief, and it is immune from Gettier-style counterexamples. It captures the idea underlying defeasibility analyses. The basic idea is that *believed* defeaters can prevent justification, and defeaters that are true but not believed can prevent knowledge while leaving justification intact. However, there are also some examples that differ in important ways from the Gettier-style examples we have discussed so far, and they are not so easily handled in terms of there being true defeaters. These examples seem to have to do with social aspects of knowing. The philosopher most prominently associated with these examples is Gilbert Harman.[8] One of Harman's examples is as follows:

> Suppose that Tom enters a room in which many people are talking excitedly although he cannot understand what they are saying. He sees a copy of the morning paper on a table. The headlines and main story reveal that a famous civil-rights leader has been assassinated. On reading the story he comes to believe it; it is true. . . .
> Suppose that the assassination has been denied, even by eyewitnesses, the point of the denial being to avoid a racial explosion. The assassinated leader is reported in good health; the bullets are said, falsely, to have missed him and hit someone else. The denials occurred too late to prevent the original and true story from appearing in the paper that Tom has seen; but everyone else in the room has heard about the denials. None of them knows what to believe. They all have information that Tom lacks. Would we judge Tom to be the only one who knows that the assassination has actually occurred? . . . I do not think so. ([1968], p. 172)

This example cannot be handled in the same way as the Grabit example. As in the Grabit example there is a true defeater, viz., that the news media have reported that the assassination did not occur. But just as in the Grabit example, there is also a true defeater defeater, viz., that the retraction of the original story was motivated by an attempt to avoid race riots and did not necessarily reflect the actual facts. The appeal to true defeaters and true defeater defeaters should lead us to treat this example just like the Grabit example, but that gives the wrong answer. The Grabit example is one in which the believer has knowledge, whereas the newspaper example is one in which the believer lacks knowledge.

Harman gives a second kind of example in a recent article:

> In case one, Mary comes to know that Norman is in Italy when she calls his office and is told he is spending the summer in Rome. In case two, Norman seeks to give Mary the impression that he is in San Francisco by writing her a letter saying so, a letter he mails to San Francisco where a

friend then mails it on to Mary. This letter is in the pile of unopened mail on Mary's desk before her when she calls Norman's office and is told he is spending the summer in Rome. In this case (case two), Mary does not come to know that Norman is in Italy.

It is important in this case that Mary could obtain the misleading evidence. If the evidence is unobtainable, because Norman forgot to mail the letter after he wrote it, or because the letter was delivered to the wrong building where it will remain unopened, then it does not keep Mary from knowing that Norman is in Italy. ([1981], p. 164)

Again, there is a true defeater, viz., that the letter reports Norman to be in San Francisco. But there is also a true defeater defeater, viz., that the letter was written with the intention to deceive. So Mary's belief is objectively justified. Nevertheless, we want to deny that Mary knows that Norman is in Italy.

Harman ([1981], p. 164) summarizes these examples by writing, "There seem to be two ways in which such misleading evidence can undermine a person's knowledge. The evidence can either be evidence that it would be possible for the person to obtain himself or herself or evidence possessed by others in a relevant social group to which the person in question belongs." We might distinguish between these two examples by saying that in the first example there is a true defeater that is "common knowledge" in Tom's social group, whereas in the second example there is a true defeater that is "readily available" to Mary. I will loosely style these "common knowledge" and "ready availability" defeaters.

It is worth noting that a common knowledge defeater can be defeated by a defeater defeater that is also common knowledge. For example, if it were common knowledge that the news media was disclaiming the assassination, but also common knowledge that the disclaimer was fraudulent, then Tom would retain his knowledge that the assassination occurred even if he were unaware of both the disclaimer and its fraudulence. The same thing is true of ready availability defeaters. If Norman had a change of heart after sending the false letter and sent another letter explaining the trick he played on Mary, and both letters lay unopened on Mary's desk when she called Norman's office, her telephone call would give her knowledge that Norman is in Italy.

What is more surprising is that common knowledge and ready availability defeaters and defeater defeaters can be combined to result in knowledge. For instance, if Norman's trick letter lays unopened on Mary's desk when she makes the call, she will nevertheless acquire knowledge that Norman is in Italy if Norman is an important diplomat and, unbeknownst to her, the news media have been announcing all day that Norman is in Rome but has been trying to fool people about his location by sending out trick letters. This shows that despite the apparent differences between common knowledge and ready availability defeaters, there must be some kind of connection between them.

My suggestion is that these both reflect a more general social aspect of knowledge. We are "socially expected" to be aware of various things. We are expected to know what is announced on television, and we are expected to know what is in our mail. If we fail to know all these things and that makes a difference to whether we are justified in believing some true proposition P, then our objectively justified belief in P does not constitute knowledge. Let us say that a proposition is *socially sensitive for S* if and only if it is of a sort S is expected to believe when true. My claim is that Harman's examples are best handled by taking them to involve cases in which there are true socially sensitive defeaters. This might be doubted on the grounds that not all readily available truths are socially sensitive. For instance, suppose that instead of having his trick letter mailed from San Francisco, Norman had a friend secrete it under Mary's doormat. We are not socially expected to check regularly under our doormats, but nevertheless this is something we can readily do and so information secreted under our doormats counts as readily available. It does not, however, defeat knowledge. If the trick letter were secreted under Mary's doormat, we would regard her as knowing that Norman is in Italy. Suppose, on the other hand, that we lived in a society in which it is common to leave messages under doormats and everyone is expected to check his doormat whenever he comes home. In that case, if the trick letter were under Mary's doormat but she failed to check there before calling Norman's office, we would not regard that call as providing her with knowledge. These examples seem to indicate that it is social sensitivity and not mere ready availability that enables a truth to defeat a knowledge claim.

My suggestion is that we can capture the social aspect of knowledge by requiring a knower to hold his belief on the basis of an argument ultimately undefeated relative not just to the set of all truths, but also to the set of all socially sensitive truths. My proposal is:

S knows P if and only if S instantiates some argument **A** supporting P which is (1) ultimately undefeated relative to the set of all truths, and (2) ultimately undefeated relative to the set of all truths socially sensitive for S.

This proposal avoids both the Gettier problem and the social problems discussed by Harman. At this stage in history it would be rash to be very confident of any analysis of knowledge, but I put this forth tentatively as an analysis that seems to handle all of the known problems.

Notes

1. See, for example, Michael Clark [1963].
2. See, for example, Brian Skyrms [1967].
3. The first defeasibility analysis was that of Keith Lehrer [1965]. That was followed by Lehrer and Thomas Paxson [1969], Peter Klein [1971], [1976], [1979], [1980], Lehrer [1974], [1979], Ernest Sosa [1974], [1980], and Marshall Swain [1981].
4. This is basically the analysis proffered by Klein [1971].
5. A good survey of the literature on the Gettier problem, going into much more detail than space permits here, can be found in Shope [1983].
6. This can be formulated in terms of defeaters and defeater defeaters. 'Most of the things around here that look like barns are not barns' is a true reliability defeater, but there is no true defeater defeater. In particular, 'That really is a barn', although true, does not restore your original justification— instead, it constitutes a new reason for believing that what you see is a barn.
7. See, for example, Alvin Goldman [1981], pp. 28–9.
8. See Harman [1968] and [1980]. Harman credits Ernest Sosa [1964] with the original observation that social considerations play a role in knowledge.

References

Clark, Michael
   1963 Knowledge and grounds: A comment on Mr. Gettier's paper. *Analysis* 24: 46–48.
Gettier, Edmund
   1963 Is justified true belief knowledge? *Analysis* 23: 121–23.
Goldman, Alvin
   1976 Discrimination and perceptual knowledge. *Journal of Philosophy* 73: 771–91.
   1981 The internalist conception of justification. *Midwest Studies in Philosophy*, vol. 5, pp. 27–52. Minneapolis: University of Minnesota Press.
Harman, Gilbert
   1968 Knowledge, inference, and explanation. *American Philosophical Quarterly* 5: 164–73.

1981   Reasoning and evidence one does not possess. *Midwest Studies in Philosophy*, vol. 5, pp. 163–82. Minneapolis: University of Minnesota Press.

Klein, Peter
1971   A proposed definition of propositional knowledge. *Journal of Philosophy* 68: 471–82.
1976   Knowledge, causality, and defeasibility. *Journal of Philosophy* 73: 792–812.
1979   Misleading "misleading defeaters". *Journal of Philosophy* 76: 382–86.
1980   Misleading evidence and the restoration of justification. *Philosophical Studies* 37: 81–89.

Lehrer, Keith
1965   Knowledge, truth, and evidence. *Analysis* 25: 168–75.
1974   *Knowledge*. Oxford: Oxford University Press.
1979   The Gettier problem and the analysis of knowledge. In *Justification and Knowledge: New Studies in Epistemology*, ed. George Pappas, pp. 65–78. Dordrecht: Reidel.

Lehrer, Keith, and Thomas Paxson
1969   Knowledge: Undefeated justified true belief. *Journal of Philosophy* 66: 225–37.

Shope, Robert K.
1983   *The Analysis of Knowing*. Princeton: Princeton University Press.

Skyrms, Brian
1967   The explication of 'X' knows that p'. *Journal of Philosophy* 64: 373–89.

Sosa, Ernest
1964   The analysis of 'knowledge that p'. *Analysis* 25: 1–8.
1974   How do you know? *American Philosophical Quarterly* 11: 113–22.
1980   Epistemic presupposition. In *Justification and Knowledge: New Studies in Epistemology*, ed. George Pappas. Dordrecht: Reidel.

Swain, Marshall
1981   *Reasons and Knowledge*. Ithaca, NY: Cornell University Press.

12

Why Solve the Gettier Problem?

Earl Conee

The value of work on the Gettier Problem has been called into question. Michael Williams concludes a paper on this dark note: "That anything important turns on coming up with a solution to Gettier's problem remains to be shown."[1]

Mark Kaplan argues for a gloomier view: "My message is that it is time to stop and face the unpleasant reality that we simply have no use for a definition of propositional knowledge."[2]

This is a fitting occasion to offer a proof of the philosophical importance of the Gettier problem. Here is the proof:

1. Discovering an analysis of factual knowledge turns on solving the Gettier problem.
2. Discovering an analysis of factual knowledge is philosophically important.
3. Something philosophically important turns on solving the Gettier problem.

Not surprisingly, neither Williams nor Kaplan addresses this argument. They provide nothing that amounts to an objection to the first premise. Williams's paper contains no conspicuous objection to either premise. Williams does contend that the Gettier problem falls outside of a certain tradition in epistemology, and he may intend thereby to cast doubt on the problem's significance. In particular, Williams claims that traditional theories of knowledge are best construed as responses to radical skepticism, where radical skepticism is understood as the view that we can never have any reason to believe anything. Thus, the

Reprinted by permission of Kluwer Academic Publishers from D. F. Austin, ed., *Philosophical Analysis*, pp. 55–58. Copyright 1988 by Kluwer.

radical skeptic denies the possibility of justified belief. Those who accept Gettier's counterexamples assume that justified beliefs exist in the prosaic circumstances described in giving the examples. So, when such examples are simply taken as data for constructing an extensionally adequate analysis of factual knowledge, radical skepticism is assumed false. Thus, this project fails to respond to what Williams takes to be the traditional issue for a theory of knowledge—the threat of radical skepticism.[3]

It is surprising that Williams does not include the goal of explaining what knowledge really is as part of the traditional project. Williams does not argue for the merit of his extraordinary construal of the point of traditional theories of knowledge. The rest of his paper defends his further contention that solving the Gettier problem would also fail to contribute to a theory of reasoning, contrary to claims by Gilbert Harman.[4]

Neither of these negative points about work on the Gettier problem will be disputed here. It does no harm for present purposes simply to grant Williams' conclusions. It can be granted that an analysis of knowledge that solved the Gettier problem would not refute or support radical skepticism, and that would not contribute to a theory of reasoning. This is as far as Williams' paper goes toward establishing his conclusion about the lack of importance of the Gettier problem. It is sufficient for present purposes to note that Williams' paper does not make a decisive case against our premise 2 if, as is argued below, analyzing knowledge can be seen to have some other philosophical importance.[5]

Let us now see what Mark Kaplan's efforts add to a case against premise 2. First he argues that the sort of analysis against which Gettier's objections are effective is not a historically important sort of analysis. He contends that the Platonic accounts of knowledge in the *Theaetetus* and the *Meno* are in fact intended as accounts of knowledge of nonpropositional objects, and thus are not even on Gettier's topic of factual knowledge.[6] Kaplan observes that Descartes' account, though concerned with propositional knowledge, is immune to Gettier's objections. This is because on the Cartesian account false yet justified belief is impossible, while it must be possible for Gettier's counterexamples to succeed.[7]

This is Kaplan's case against Gettier's having criticized a historically important account of knowledge. As in the case of Williams' conclusions, Kaplan's conclusion need not be disputed here. But it seems appropriate to mention that his argument for the conclusion is unfair to Gettier in one respect. The argument neglects that part of history which lies in the twentieth century. This is peculiar because Kaplan mentions the accounts that Gettier's objections are explicitly targeted against Ayer's account. Russell, Ayer, and Chisholm provide a historically important conception of factual knowledge against which Gettier's criticisms apply.

For present purposes it does no harm to waive this objection too. Let it be granted that the analyses that Gettier refutes do not have historical importance. Again, this is innocuous to the above argument for the importance of the Gettier problem, as long as analyzing factual knowledge is shown to be important for some other reason.

Kaplan also argues that a solution to the Gettier problem would neither advance nor clarify the proper conduct of rational inquiry. He indicates that once one has made best use of one's evidence and arrived at a justified belief, it follows that one has made a proper inquiry from the evidence. Gettier has established that knowledge requires something beyond justified true belief. But whatever that is, it is not something the inquirer can simultaneously make a separate check on. If one comes to have further evidence, it might show that one's justification relied on a falsehood, as in Gettier's examples. But at that further point either one lacks justification or one has new evidence that provides a new justification. The fact that justified belief is corrigible over time does not show that some justified true belief that is not known is a result of faulty inquiry. Kaplan contends that since the properties of rational inquiry determine only how to gain justification, solving the Gettier problem would not advance our understanding or practice of rational inquiry.[8]

One fault in this contention is noteworthy here, since it brings to light one philosophical contribution that a solution to the Gettier problem would make. The nature of proper inquiry depends on the goal of the inquiry. The goal of pure rational inquiry is not justified belief, nor is it justified true belief. The goal is knowledge. Thus, a rational inquirer who has merely done all that is required to gain a justified true belief may not have what he or she seeks. It is true that when a person's justified true belief is not knowledge because of the factors that Gettier's examples illustrate, the person does not realize this and may be unable to do anything that would result in knowledge. So any further inquiry will seem superfluous at the outset and may be futile. But success at attaining knowledge to replace mere justified true belief is sometimes available. When, as in Gettier's examples, a person has only a justified true belief in a certain existential generalization or a certain disjunction, further inquiry would sometimes result in the person's perceiving the fact that really does make the generalization or the disjunction true and thus yield knowledge. Additional inquiry may be crucial for achieving the goal of knowledge, though it is unnecessary for securing justification. Since a solution to the Gettier problem would informatively describe exactly what is sufficient for knowledge, it would illuminate what must be accomplished by further rational inquiry for the sake of genuinely knowing.

Once again, however, it is harmless for present purposes to set aside such objections. Let it be granted that solving the Gettier problem is useless for understanding or conducting rational inquiry. This point, together with his historical contention, constitutes Kaplan's case against the utility of analyzing knowledge.

The combined cases of Williams and Kaplan do not refute premise 2. Granting every point, they simply give us a list of four philosophically important things that would not be done by a solution to the Gettier problem: refuting or supporting radical skepticism, contributing to a theory of reasoning, improving some historically important account of knowledge, and improving our understanding or execution of rational inquiry.

An analysis of knowledge that solved the Gettier problem would accomplish something else that is philosophically important. It would provide us with independent conditions which are severally necessary and jointly sufficient for the existence of a case of factual knowledge. This is important because the nature of factual knowledge is itself a philosophical topic. The insightful and dedicated work on the Gettier problem by a diversity of philosophers attests to the philosophical interest of the topic. The primary philosophical goal of this research is an analysis of factual knowledge. There is no good reason to require utility toward some other philosophical objective in order to justify this enterprise as worthwhile philosophy.[9] Learning which conditions constitute a solution to the Gettier problem would greatly enhance our understanding of factual knowledge. That makes solving the problem important.

Notes

1. Michael Williams [1978], 'Inference, Justification and the Analysis of Knowledge', *The Journal of Philosophy* LXXV (May) p. 263.

2. Mark Kaplan [1985], 'It's Not What You Know that Counts,' *The Journal of Philosophy* LXXXII (July), p. 363.

3. Williams [1978], pp. 249–250.

4. *Ibid.*, pp. 250–262. Harman responds in Harman [1978], 'Using Intuitions about Reasoning to Study Reasoning: A Reply to Williams', *The Journal of Philosophy* LXXV (August) pp. 433–483.

5. In fairness to Williams, he may not have intended to make a conclusive case. He may have intended to establish no more than he asserts in the above citation—that the importance had not been shown.

6. Kaplan [1985], pp. 352–353.

7. *Ibid.*, p. 353.

8. *Ibid.*, pp. 354–356.

9. Sometimes the remark is made, as though in objection to such work, that conceptual analysis is a 'sterile' or 'fruitless' undertaking. But it is obvious that the search for an analysis of knowledge has spawned illuminating contributions on such topics as the defeat of evidence, causal and counterfactual conditions on knowledge and justification, and the epistemic role of external factors like social context and the reliability of belief-forming mechanisms. Anyway, suppose that the search for a solution to the Gettier problem had borne no fruit. Suppose that it had lead quickly to an accurate but 'sterile' analysis, useless in other philosophical undertakings. So what? How could this show that the analysis would be philosophically unimportant?

Part III

Skepticism

13

A 'Doxastic Practice' Approach to Epistemology

William P. Alston

I

How can we determine which epistemic principles are correct, valid, or adequate? One way to motivate concern with this issue is to consider controversial principles. What does it take to be justified in perceptual beliefs about the physical environment? Can I be justified in believing that there is a tree in front of me just by virtue of that belief's stemming, in a certain way, from a certain kind of visual experience? Or do I also need reasons, in the form of what I know about my visual experience or about the circumstances of that perception? How do we tell what set of conditions is sufficient for the justification of such beliefs? Another and more usual way to motivate concern with the issue is to raise the specter of skepticism. Why suppose that *any* set of conditions we can realize is sufficient? No matter what experiences and beliefs I have, couldn't they have been produced directly by an omnipotent being that sees to it that there is no physical world at all and that all my perceptual beliefs are false? That being the case, why should we suppose that our sensory experience justifies us in holding any beliefs about the physical world?

As the above paragraph suggests, the epistemic principles I will be thinking of lay down conditions under which one is justified in holding beliefs of a certain kind. I shall be using the justification of perceptual beliefs as my chief example. For a more specific focus, you can take

Reprinted with permission of the author and publisher from M. Clay and K. Lehrer, eds., *Knowledge and Skepticism*, pp. 1–29. Westview, 1989.

your favorite principle of justification for perceptual beliefs. Following my own injunction, I will focus on my favorite, which runs as follows.

> I.—S is *prima facie* justified in perceptually believing that x is P iff S has the kind of sensory experience that would normally be taken as x appearing to S as P, and S's belief that x is P stems from that experience in the normal way.

If we were interested in this principle for its own sake, much more would have to be done by way of elucidation. Here I will just say that the justification is only *prima facie* because it can be overridden by sufficient reason to suppose that x is not P or that the experience in this case is not sufficiently indicative of x's being P. I present this particular principle only to have something fairly definite to work with. Our concerns in this paper lie elsewhere. Nothing will hang on the specific character of I.

What it takes to be justified in accepting a principle of justification depends on what justification is. I have discussed this matter at some length elsewhere.[1] Here I must confine myself to laying it down that epistemic justification is essentially "truth conducive." That is, to be justified in believing that p is to believe that p in such a way that it is at least quite likely that one's belief is true.[2] One way of developing this idea is to say that S is justified in believing that p only if that belief was acquired in a reliable manner. This is not to *identify* justification with reliability; the 'only if' principle leaves room for other necessary conditions. I shall be thinking of justification as subject to a "reliability constraint." If this is distasteful to you, you can take the chapter as having to do with the epistemic status of principles of reliability, and leave justification out of the picture altogether.

So to determine which of the competing principles of the justification of perceptual beliefs is correct, if any, we have to determine, *inter alia,* which of them, if any, specify a reliable mode of belief-formation. And to show, against the skeptic, that perception is a source of justified belief (knowledge), we have to show that some mode of forming perceptual beliefs is reliable. But how to do this? Let's take a particular principle that specifies a mode of perceptual belief-formation, e.g., I, and consider what it would take to show that the mode so specified is reliable. The main difficulty is that there seems to be no otherwise effective way of showing this that does not depend on sense perception for some or most of its premises. Take the popular argument that sense perception proves its veridicality by the fact that when we trust our sense perception and build up systems of belief on that basis we have remarkable success in predicting and controlling the course of events. That sounds like a strong argument until we ask how we know that we have been successful at prediction and control. The answer is, obviously, that we know this only by relying on sense perception. Somebody has to take a look to see whether what we predicted did come to pass and whether our attempts at control were successful. Though I have no time to argue the point here, I suggest that any argument for the reliability of perception that is not otherwise disqualified will at some point(s) rely on perception itself. I shall assume this in what follows.[3]

What I have just been pointing to is a certain kind of circularity, one that consists in assuming the reliability of a source of belief in arguing for the reliability of that source. That assumption does not appear as a premise in the argument, but it is only by making the assumption that we consider ourselves entitled to use some or all of the premises. Let's call this *epistemic circularity*. In a recent essay I argue that, contrary to what one might suppose, epistemic circularity does not render an argument useless for justifying or establishing its conclusion.[4] Provided that I can *be* justified in certain perceptual beliefs without already being *justified* in supposing sense perception to be reliable,[5] I can legitimately use perceptual beliefs in an argument for the reliability of sense perception.

However, this is not the end of the matter. What I take myself to have shown in "Epistemic Circularity" is that epistemic circularity does not prevent one from showing, on the basis of empirical premises that are ultimately based on sense perception, that sense perception is reliable. But whether one actually does succeed in this depends on one's being justified in those perceptual premises, and that in turn, according to our assumptions about justification, depends on sense perception being a reliable source of belief. In other words, *if* (and only if) sense perception is reliable, we can show it to be reliable.[6] But how can we cancel out that *if*?

Here is another way of posing the problem. If we are entitled to use beliefs from a certain source in showing that source to be reliable, then any source can be validated. If all else fails, we can simply use each belief twice over, once as testee and once as tester. Consider crystal ball gazing. Gazing into the crystal ball, the seer makes a series of pronouncements: p, q, r, s. . . Is this a reliable mode of belief-formation? Yes. That can be shown as follows. The gazer forms the belief that p, and, using the same procedure, ascertains that p. By running through a series of beliefs in this way, we discover that the accuracy of this mode of belief-formation is 100%! If some of the beliefs contradict others, that will reduce the accuracy somewhat, but in the absence of massive internal contradiction the percentage of verified beliefs will still be quite high. Thus, if we allow the use of mode of belief-formation M to determine whether the beliefs formed by M are true, M is sure to get a clean bill of health. But a line of argument that will validate any mode of belief-formation, no matter how irresponsible, is not what we are looking for. We want, and need, something much more discriminating. Hence the fact that the reliability of sense perception can be established by relying on sense perception does not solve our problem.[7]

II

This is where the "doxastic practice" approach of the title comes into the picture. For help on the problem of the first section, I am going to look to two philosophers separated by almost two hundred years, Thomas Reid and Ludwig Wittgenstein. Both were centrally concerned with our problem, albeit in somewhat different guises. Since within the limits of this paper I am simply drawing inspiration from these figures, mining their work for ideas that I will develop in my own way, I will not attempt to present their views in anything like an adequate fashion.

First Wittgenstein. In *On Certainty*[8] Wittgenstein is concerned with the epistemic status of propositions of the sort G. E. Moore highlighted in his "Defence of Common Sense" and "Proof of an External World"—such propositions as *This is my hand, The earth has existed for many years,* and *There are people in this room.* The gist of Wittgenstein's position is that the acceptance of such propositions is partially *constitutive* of participation in one or another fundamental "language-game."[9] To doubt or question such a proposition is to question the whole language-game of which it is a keystone. There is no provision within that language-game for raising such doubts. In fact, there is no provision within the language-game for justifying such beliefs, exhibiting evidence for them, or showing that we know such matters, as Moore tried to do. Hence we cannot even say that we know or are certain of such matters. They are too fundamental for that. By accepting these and other "anchors" of the game we are thereby enabled to question, doubt, establish, refute, or justify less fundamental propositions. Nor can we step outside the language-game in which they figure as anchors and critically assess them from some other perspective. They have their meaning only within the game in which they play a foundational role; we cannot give sense to any dealings with them outside this context.

Thus, if we ask why we should suppose that some particular language-game is a reliable source of belief, Wittgenstein responds by denying the meaningfulness of the question. The concept of a trans- or inter-language-game dimension of truth or falsity is ruled out on verificationist grounds. We can address issues of truth and falsity only *within* a language-game, by employing its criteria and procedures to investigate issues that are within its scope. Hence there is no room for raising and answering questions about the reliability of a language-game as a whole. To be sure, language-games are not sacrosanct or fixed in cement. It is conceivable that they should be abandoned and new ones arise in their place. But even if we should have some choice in the matter, something that Wittgenstein seems to deny, the issue would be a practical, not a theoretical, one. It would be a choice as to what sort of activity to engage in, not a choice as to whether some proposition is true or false.[10] The foundation of the language-game is action, not intuition, belief, or reasoning.

Applying this to the problem raised in section I, Wittgenstein's view is that no sensible question can be raised concerning the reliability of the language-game that involves forming beliefs on the basis of sense-perception. There is no perspective from which the question can be intelligibly raised. This is a sphere of activity in which we are deeply involved; "this language-game is played."[11] We could try to opt out, but even if, *per impossible,* we could do so, that would have been a practical decision; and what possible reason could we have for such a decision? If, as is in fact the case, we continue to be a whole-hearted participant, we are simply engaged in (perhaps unconscious) duplicity in pretending to question, doubt, or justify the practice.

Now I do not accept for a moment Wittgenstein's verificationist restrictions on what assertions, questions, and doubts are intelligible. There is no time here for an attack on verificationism. I will simply testify that I can perfectly well understand the propositions that sense perception is (is not) reliable, that physical objects do (do not) exist, and that the earth has (has not) been in existence for more than a year, whether or not I or anyone else has any idea of how to go about determining whether one of these propositions is true. This confidence reflects a realistic concept of truth, on which a proposition's being true is *not* a matter of anyone's actual or possible epistemic position vis-à-vis the proposition. Hence I cannot accept Wittgenstein's solution to skepticism about perception and his answer to the question of the epistemic status of epistemic principles, the solution that seeks to dissolve the problem by undercutting the supposition that it can be meaningfully posed.

But then how can I look to Wittgenstein for inspiration? I shall explain. First a terminological note. Because I am concentrating on ways of forming and critically evaluating beliefs, I shall use the term 'doxastic practice,' instead of 'language-game.' The term 'practice' will be misleading if it is taken to be restricted to voluntary activity; for I do not take belief-formation to be voluntary. I am using 'practice' in such a way that it stretches over, e.g., psychological processes such as perception, thought, fantasy, and belief-formation, as well as voluntary action. A doxastic practice can be thought of as a system or constellation of *dispositions* or habits, or, to use a currently fashionable term, *mechanisms,* each of which yields a belief as output that is related in a certain way to an "input." The sense perceptual doxastic practice (hereinafter SPP) is a constellation of habits of forming beliefs in a certain way on the basis of inputs that consist of sense experiences.

Let me now set out the basic features of the view of doxastic practices I have arrived at, partly inspired by Wittgenstein. Some of these features are not stressed by Wittgenstein and some are only hinted at. But I believe that all of them are in the spirit of his approach.

1. We engage in a plurality of doxastic practices, each with its own sources of belief, its own conditions of justification, its own fundamental beliefs, and, in some cases, its own subject matter, its own conceptual framework, and its own repertoire of possible "overriders." There is no one unique source of justification or knowledge, such as Descartes and many others have dreamed of. However, this point needs to be handled carefully. What it is natural to count as distinct doxastic practices are by no means wholly independent. We have to rely on the output of memory and reasoning for the overriders of perceptual beliefs. Apart from what is stored in memory, and used in reasoning, concerning the physical world and our perceptual interactions therewith, we would have nothing to go on in determining when sensory deliverances are and are not to be trusted. Reasoning is beholden to other belief-forming practices for its premises. We can, of course, reason from the output of previous reasoning, but somewhere back along the line we must have reasoned from beliefs otherwise

obtained.[12] Thus we must avoid any suggestion that these practices can be engaged in separately.

We need to distinguish between what we may call "generational" and "transformational" practices. Generational practices produce beliefs from non-doxastic inputs; transformational practices transform belief inputs into belief outputs.[13] Generational practices *could* be used without reliance on other practices, as in forming perceptual beliefs without any provision for a second, "censor" stage that filters out some beliefs as incompatible with what we already firmly believe. This would be a more primitive kind of practice than we actually have in mature human beings, but it is possible, and may well be actual in very young children and lower animals. Moreover, our mature "introspective" practice is of this independent sort if, as seems likely, beliefs about one's current conscious states do not regularly pass any test of compatibility with what we believe otherwise. Transformational practices, on the other hand, cannot be carried on in any form without dependence on other practices. We have to acquire beliefs from some other source in order to get reasoning started.

Each of the generational practices has its own distinctive subject matter and conceptual scheme. SPP is a practice of forming beliefs about the current physical environment of the subject, using the common sense "physical object" conceptual scheme. Introspective practice is a practice of forming beliefs about the subject's own current conscious states, using the "conscious state" conceptual scheme, whereas beliefs formed by reasoning and by memory can be about anything whatever and can use any concepts whatever.

Then is there anything common to all doxastic practices, other than the fact that each is a regular systematic way of forming beliefs? Yes. In the initial statement I said that each practice has its own "sources of belief" and its own "conditions of justification." These are two sides of the same coin. We may take the former as our fundamental criterion for distinctness of doxastic practices. The practices we have distinguished differ in the kind of belief-forming "mechanism" involved. Such a mechanism consists of a "function" that yields a certain belief as output, given a certain input. This means that belief-forming mechanisms differ as to the sorts of inputs involved and as to the way in which inputs map onto belief outputs. There will be as many (possible) deductive inference belief-forming mechanisms as there are forms of deductive inference.[14] And perceptual belief-forming mechanisms will differ as to the type of sensory experience inputs, and as to the way in which beliefs about environmental states of affairs are extracted from a certain kind of sensory experience. The conditions of justification for each practice simply amount to an epistemic version of the psychological notion of a belief-forming mechanism.[15] Thus the criteria of justification built into SPP have to do with the way a perceptual belief is standardly based on sense experience. The criteria of justification built into an inferential practice have to do with the way a belief is based on the kind of inference that constitutes the basic source for that practice.[16]

Thus we can translate our basic issue concerning the reliability of belief sources, or modes of belief-formation, into an issue concerning the reliability of doxastic practices. A practice is reliable *iff* its distinctive belief-forming mechanisms (modes of belief-formation) are reliable. And we can similarly restate the "reliability constraint" on principles of justification in these terms. A (general enough) principle of justification, e.g., I., will be true (valid, acceptable. . .) only if the doxastic practice in which we form beliefs in the way specified in that principle is reliable. From now on we will be thinking of reliability as attaching to doxastic practices.

We have also spoken of each practice as possessing its own distinctive set of foundational presuppositions. This is an idea that bulks large in *On Certainty*. I feel that Wittgenstein is much too generous in according this status to beliefs. It seems clear to me that *This is my hand* and *The earth has existed for more than a year* are propositions for the truth of which I have a great deal of empirical evidence *within* SPP (or rather within some combination of that with memory and reasoning of various sorts), rather than a basic presupposition of the practice. However, I do recognize this latter category. The existence of physical objects and the general reliability of sense perception are basic presuppositions of SPP; we couldn't engage in it wholeheartedly without at least tacitly accepting those propositions. Similarly, the reality of the past and the reliability of memory are basic presuppositions of the practice of forming memory beliefs.

2. These practices are acquired and engaged in well before one is explicitly aware of them and critically reflects on them. When one arrives at the age of reflection, one finds oneself ineluctably involved in their exercise. Here especially, the owl of Minerva flies only at the gathering of the dusk. Philosophical reflection and criticism build on the *practical* mastery of doxastic practices. Practice precedes theory; and the latter would be impossible without the former. This is a recurrent them in *On Certainty*. If we hadn't learned to *engage* in inference, we could never develop a system of logic; we would have

nothing either to reflect *on* or to reflect with. If we had not learned to form perceptual beliefs, we would have no resources for formulating the philosophical problems of the existence of the external world and of the epistemic status of perceptual beliefs.

3. Practices of *belief-formation,* on which we have been concentrating, are set in the context of wider spheres of practice. We learn to form perceptual beliefs along with, and as a part of, learning to deal with perceived objects in the pursuit of our ends. Our practice of forming beliefs about other persons is intimately connected with interpersonal behavior, treating persons as persons and forming typically interpersonal relations with them.

4. These practices are thoroughly *social:* socially established by socially monitored learning, and socially shared. We learn to form perceptual beliefs about the environment in terms of the conceptual scheme we acquire from our society. This is not to deny that innate mechanisms and tendencies play a role here. We still have much to learn about the relative contributions of innate structures and social learning in the development of doxastic practices. Reid places more stress on the former, Wittgenstein on the latter. But whatever the details, both have a role to play; and the final outcome is socially organized, reinforced, monitored, and shared.

At the beginning of this section I said that I was going to develop an approach to epistemology that was inspired by Reid and Wittgenstein. So far nothing has been said about the former. But only the name has been absent. The conception of doxastic practices just outlined is, in its essentials, the view of Reid, even though the terminology is different.[17] Where I speak of various doxastic practices Reid speaks of various kinds of "evidence": "the evidence of sense, the evidence of memory, the evidence of consciousness, the evidence of testimony, the evidence of axioms, the evidence of reasoning."[18] "We give the name of evidence to whatever is a ground of belief."[19] Alternatively, he speaks of "general principles of the human mind" by which we form beliefs of certain sorts under certain conditions.[20] Reid stresses the plurality of these principles or sorts of evidence, and the impossibility of reducing them to a single supreme principle. ". . .I am not able to find any common nature to which they may all be reduced. They seem to me to agree only in this, that they are all fitted by nature to produce belief in the human mind. . ."[21] Again, Reid often stresses the point that we utilize these principles in practice long before we are explicitly aware of them as such. As mentioned above, he stresses the contribution of innate structure, whereas Wittgenstein stresses social learning, but in both cases there is emphasis on the point that we have them and use them before we reflect on them. Reid, much more than Wittgenstein, goes into the way in which belief-forming dispositions, once established, can be modified by experience.[22] On the other hand, Reid does not stress the way in which cognitive practices are set in the context of practices of overt dealings with the environment. Reid's perspective is that of a purely cognitive, mentalistic psychology. Finally, I should mention the point that one reason my account is closer to Reid's is that Reid had the advantage of philosophizing before the advent of verificationist and other anti-realist philosophies. Reid never suggests that there is anything unintelligible about the idea that, e.g., sense perception is or is not reliable, or that we cannot meaningfully raise the question of whether this is so, however difficult it may be to find a way to answer the question. As we shall see, this leaves Reid, and me, free to look for ways of evaluating basic doxastic practices.[23]

III

But how does my Reidian view of doxastic practices provide us with a solution of our central problem, viz., how we can determine, with respect to a particular practice such as SPP, whether it is reliable? Thus far I have presented my view as what we might call "cognitive social psychology," an account of how it is in fact with our activities of belief-formation. I believe that there can be no doubt that this account is correct, at least in its general outlines. But so far this is just psychology. What bearing does it have on our central epistemological question? How does it help us to determine which practices are reliable ones?

I am not going to tackle this question head on. Instead I am going to shift ground in this section and the next, and consider what resources our approach gives us for determining whether a given practice is *rationally* accepted (engaged in). Having completed that task, I shall turn, in section V, to the question of what bearing all this has on our central issues of the reliability of practices and the assessment of principles of justification.

Our two role models seek to make epistemological hay out of their psychology in different fashions. In a word, Wittgenstein draws linguistic conclusions from the psychology (while not admitting for a moment that it is psychology) and then applies these linguistic points to epistemology, while Reid tries to move more directly from the psychology to the epistemological position, if indeed he does clearly distinguish the two. Wittgenstein's linguistic solution, as already pointed out, is that no meaning can be given to a question as to the truth or justifiability of beliefs that are constitutive of a practice. We can't address such questions in the practice itself, nor can we address them in any other practice. The only meaningful questions are those for the investigation of which a practice makes provision; and no such provision is made for questions as to the fundamental presuppositions of a practice or as to its own reliability. I have already made clear that I do not accept the verificationist assumptions that underlie Wittgenstein's restrictions on meaningfulness, and hence I cannot avail myself of his solution. Reid's response is hazier and more difficult to summarize neatly, at least insofar as it goes beyond reminding the skeptic that he is deeply involved in practices the presuppositions and outputs of which he is questioning; and despite the popular picture of Reid, it is clear that his response does go beyond this, however difficult it may be to say in exactly what way.[24] Since my aims in this paper are not historical, I shall state in my own way what I take to be essentially a Reidian response.

Consider a typical reaction of a contemporary American epistemologist to my suggestion that a study of social cognitive psychology can throw light on our epistemic question about the rationality of a practice. "What does all this have to do with epistemology? The fact that a given practice is socially established cuts no ice whatever epistemologically. The function of the epistemologist is to subject any such practice to critical standards, bring it before the bar of reason, playing no favorites on grounds of familiarity, general acceptance, practical indispensability, irresistibility, innateness, or commonsense plausibility."[25]

Let's term this position "Autonomism." It holds that epistemology is autonomous vis-à-vis psychology and other sciences dealing with cognition. It holds that epistemology is essentially a normative or evaluative enterprise, and that here as elsewhere values are not determined by fact.

But this non-naturalist philippic inevitably provokes a naturalist rejoinder. "You say that the province of epistemology, so far as it is concerned with doxastic practices, is to carry out a rational assessment of such practices. Well and good. But where is the epistemologist to obtain the standards by which that evaluation will be carried out? I doubt that there is any such special epistemological procedure for setting standards. Certainly there is none that is utilized by all or most epistemologists; or if there is, its employment does not yield general agreement. I suggest that when an epistemologist propounds principles of justification, these utterances, no matter how solemn the intonation, are rooted in one or another of the established practices we have been discussing. Does the epistemologist claim to be proceeding on the basis of self-evident principles of evaluation? Well then, he is participating in the well-established practice of forming beliefs on the basis of their appearing to be obviously true just on consideration. Even if this enables him to pass judgment on other practices, these judgments are worth only as much as the credentials of the practice within which they were pronounced. And if his epistemological judgments are made on some other basis, e.g., coherence or argument to the best explanation, he is still *presupposing* the acceptability of that mode of forming beliefs, in passing judgment on other practices. And he can't critically evaluate that mode in the same way without falling into epistemic circularity. Thus the autonomist, however lordly his pretensions, cannot, in the end, avoid reliance on one or more of the doxastic practices from which he was seeking to distance himself. He avoids a wholesale commitment to established doxastic practices only by taking one or more uncritically so as to have a platform from which to judge others. We cannot avoid dependence on the doxastic practices in which we find ourselves engaged when we begin to reflect. At most, we can restrict ourselves to one or two as the only ones we will accept without rational warrant, subjecting the others to the standards of these chosen few. Thus, on closer scrutiny, the autonomist turns out to be a selective heteronomist. And this is arbitrary partiality. It can have no rational justification. What justification can there be for accepting the pretensions of, e.g., rational intuition or introspection without critical scrutiny, while refusing the same privilege to sense perception?[26] If the epistemologist is to escape such arbitrariness, he must content himself with delineating the contours of established doxastic practices, perhaps neatening them up a bit and rendering them more internally coherent and more consonant with each other. He must give up pretensions to an Archimedean point from which he can carry out an impartial rational evaluation of *all* practices.

Let's call the position suggested by the last two sentences of this retort, "Heteronomism." We may think of Autonomism and Heteronomism as constituting an antinomy. Our present task is to resolve this antinomy.

The first step in that resolution is to point out that neither side of the antinomy does full justice to the epistemological enterprise. As for the autonomist, his opponent has already made explicit where he falls short. The autonomist, since he eschews implicit trust in established doxastic practices, needs some other source and warrant of his critical standards, and what could that be? But this criticism can be pushed further by pointing out that the practice of epistemology reveals, at several points, an uncritical reliance on the practices we acquired with our mothers' milk. If we look at attempts to formulate and establish principles of justification, we will find the protagonists engaged in two sorts of activities. First, they put forward various principles as plausible, reasonable, sensible, or evident. Second, they test these principles by confronting them with various examples of justified and unjustified beliefs. Now where do they get these principles, and what is the source of their plausibility? Why is it that I, and its many near relatives seem so reasonable? A plausible answer is that such principles formulate, or come close to formulating, the principles of belief-formation and assessment built into our familiar practice of forming perceptual beliefs. Why else should these principles make a strong claim on our assent? Is it that we have some special access to a realm of being known as "epistemic justification"? That seems unlikely. *Nous n'avons pas besoin de cette hypothese*. When we encounter a formulation of some deeply embedded practice of ours, it naturally makes a strong appeal. As for examples of justified and of unjustified beliefs, why do they evoke such widespread concurrence? Again, the most reasonable hypothesis is that the judgments are being made from within widely shared doxastic practices. Thus Chisholm et al. are, much of the time, doing just what the heteronomist says they should do, viz., making explicit the structure of one or another common doxastic practice.

But the heteronomist doesn't have the whole story either. In seeking to make the delineation and refinement of established practices the whole task of epistemology, she neglects the fact that making judgments on absolutely general questions, and deciding between opposing positions on such questions, is constitutive of the philosophical enterprise, in epistemology as elsewhere. How can the epistemologist fail to ask about the rationality of forming beliefs in one or another way, without violating her Socratic oath? The unexamined practice is not worth engaging in, at least not once it has been dragged into the light and made a possible subject of philosophical criticism. It is absolutely fundamental to the philosophical enterprise to subject all the basic features of our life to rational criticism, and not the least of these is the set of belief-forming tendencies with which we are endowed, or saddled as the case may be. Any "naturalism" that spurns this task is unworthy of the name of philosophy.

So where does this leave us? If epistemology is confined to the delineation of existing doxastic practices, it will thereby renounce its most sacred charge—to carry out a rational criticism of all claims to knowledge and justification. And yet how can it assess any particular doxastic practice without making use of some other in order to do so? And in that case, how can it subject all epistemic claims to rational scrutiny? Even if epistemology had a distinctive epistemic practice all its own, what would give this practice a licence to set itself up in judgment over its fellows? Don't the Reidian charges of arbitrary partiality come back to haunt us?

I think we can find a way out of this thicket by attending to the distinction between a more or less tightly structured *practice* with more or less fixed rules, criteria, and standards on the one hand, and a relatively free, unstructured "improvisational" activity on the other. When we engage in an organized practice, whether it is a doxastic practice, a game, a traditional craft such as carpentry, or speaking a language, our activity is more or less narrowly confined by antecedent rules and procedures, which themselves constitute the substance of the practice. This is not to say that all the details are laid down in advance. There will be room for free variation, and the degree of this will vary. When a carpenter puts together a wall of a room from plans and blueprints, his activity is fairly well predetermined in its gross outlines, though no set of plans specifies exactly how many hammer strokes are to be given to each nail. The rules of a language determine what combinations are acceptable and what ways there are to express a given meaning, but they do not dictate just what one is to say at a given stage of an extended conversation.

In contrast to these highly circumscribed forms of activity, there are others that call for the exercise of "judgment," where no established rules or criteria put tight constraints on what judgment is to be made in a particular situation. Familiar examples are found in aesthetics, religion, and science. When it is a question of the comparative worth of two works of art, or of what makes a particular work of art so striking, there are no formulable canons that the critic can consult to determine what the verdict should be. The critic must use her sensitivity, experience, familiarity with the field, and "intuition" to arrive at a considered judgment. And that judgment can in turn be validated or challenged only by the use of similar resources. From the sphere of religion a similar story is to be told concerning, e.g., the spirituality of a particular person. No generally accepted checklist of observable features will settle the matter. What is required is trained judgment and sensitivity. Finally, in science, although many things are to be done by following definite rules, e.g., the preparation of chemical solutions, competing high-level theories are to be evaluated in terms of their relative fecundity, explanatory power, simplicity, and the like; and for the determination of these matters there is no calculus. Again trained judgment is called for.[27]

Where philosophy is concerned with ultimate questions it falls, I suggest, on the latter side of our contrast. It is distinctive of philosophy, in epistemology and elsewhere, to be operating at a level deeper than those spheres of intellectual activity for which there are established rules. In philosophy *everything* is up for grabs. If anyone suggests a set of rules, methods, or procedures for philosophy, that itself immediately becomes a matter of controversy. Just think of the historically prominent attempts to provide effective decision procedures for philosophical problems, from Descartes' *Rules for the Direction of Mind,* through Locke's *Essay,* Kant's *Critique of Pure Reason,* Russell's *Our Knowledge of the External World as a Field for Scientific Method in Philosophy,* and the Vienna Circle. Each proposed methodology, instead of setting philosophy onto the secure path of science, simply becomes an additional disputed claim. So far from being susceptible of regularization, philosophy is, rather, *inter alia,* the activity of subjecting proposed methodologies to reflective examination. The philosopher must search for the best way of answering questions, as well as search for the answers. The philosopher must arrive at whatever *judgment* best recommends itself after careful reflection, rather than proceed according to rules that are constitutive of the enterprise. That is what makes philosophy so uncomfortable, so unsettling, and at the same time so exciting and challenging. One can never rest secure in the realization that one has a bedrock from which to proceed further. One is always dangling in the air from a rope that isn't tied down anywhere. Or, to pile on still another metaphor, epistemology is, like the rest of philosophy, largely a "seat of the pants" enterprise.

All this suggests that we resolve the antinomy of Autonomism and Heteronomism by distinguishing two aspects of the epistemological enterprise. First, as the heteronomist points out, the epistemologist makes explicit the structure of one or another established doxastic practice. This is an activity with fairly definite criteria and procedures, in which one can be shown to be correct or incorrect, although it is much less firmly regularized than, say, census taking or chemical engineering. The final appeal is always to one's implicit acquaintance with the doxastic practice in question. However, this is sufficiently widespread and sufficiently retrievable to provide a usable touchstone.

But, second, as the autonomist points out, the epistemologist is called on to make judgments on the acceptability of one or another such practice, especially those which are not universally engaged in, such as practices of religious belief-formation as well as judgments on how to settle conflicts between practices. Here, for reasons we have already rehearsed, no set of rules or criteria can be appealed to. If the question is whether SPP (the sense perceptual doxastic practice) is rationally engaged in, we cannot make use of SPP in answering the question; and if we try to use the criteria of some other practice(s), we throw ourselves open to the Reidian charge of arbitrary partiality. Hence, when doing this more ultimate sort of epistemology, one is forced into the wilderness, outside settled territory with its laws and regulations. Thus the epistemologist in her critical function is not, or need not be, arbitrarily setting up one established practice in judgment over another, because she is not operating within any established practice at all. She is improvising, seeking to reach a *judgment* on the basis of what commends itself to her on due reflection as decisive for the question at hand.

This, then, is the resolution of our perplexity. The epistemologist, in seeking to carry out a rational evaluation of one or another doxastic practice, is not working from within a particular such practice. Nor need she be proposing to establish a novel practice, the specifications of which she has drawn up herself in her study. On the other hand, she need not abjure everything, or anything, she has learned from the various practices she has mastered. She makes use of her doxastic skills and tendencies, not by following the relatively fixed rules and procedures of some particular practice, but by using all this in a freer fashion. In that way an epistemologist can carry out the traditional philosophical function of critical evaluation without chauvinistically picking one or more practices to set in judgment over the others.

IV

But how far will this free "exercise of judgment" take us? If we eschew reliance on the criteria and standards of any established practice, how firm a judgment can we rationally pass on a given practice? What considerations will commend themselves to us as relevant when we confront the question of the rationality of SPP? And will those considerations be sufficient to justify a definite answer?

I will address these questions by practicing what I have been preaching, i.e., by engaging in the improvisational activity of which I have been speaking. At least I will give you the meta-theory thereof, reflecting on what considerations are relevantly adduced in such an evaluation, while avoiding both epistemic circularity and chauvinism.

But first I should warn you not to get carried away by the rhetoric of the last few paragraphs into an expectation of more unanimity here than we find on other fundamental philosophical issues. As we have seen, the exercise of judgment is not subject to rules or criteria by appeal to which it can be definitively established that a particular judgment is correct. Hence it is to be expected that different persons, with different backgrounds, orientations, sensitivities, and experiences will sometimes arrive at different judgments. This is not to say that no reasonable argument is possible, but only that we should not expect such argument to always produce agreement. These facts of life are, no doubt, excessively familiar to readers of this essay.

To turn, then, to the principles of my "seat of the pants" critical epistemology, I will begin by reminding you of an earlier point. If we eschew epistemically circular support, we will not be able to establish the reliability of any of our basic practices.

This being the case, what is the most rational attitude to take toward established fundamental practices? Disallowing epistemically circular arguments, none can be shown to be reliable; and if we admit such arguments, an airtight case can be made for each of them, as pointed out in section I. Hence as far as proofs of reliability are concerned, they are all on a par. Is there any other basis for distinguishing between them as candidates for rational acceptance? A number of candidates suggest themselves. Some may be more essential for the conduct of life than others; some may yield greater payoffs in the way of prediction and control; and so on. However, we can't establish any of this either without falling into epistemic circularity. How, for example, can we show that the formation of perceptual beliefs is necessary for the conduct of life without relying on what we learn from perception to do so? To be sure, we might be able to show one or another practice to be unreliable. However, since a large part of our resources for doing this involves using the outputs of one practice to discredit those of another, this enterprise is best considered at a second stage, after we already have a number of *prima facie* acceptable practices to work with. Therefore, apart from reasons for unreliability and at a first stage of evaluation, it would be arbitrary to distinguish between established basic practices; reason dictates that they all be accepted as rational or all be rejected as irrational. Is the latter option a live one? Clearly, abstention from all doxastic practices is not a real possibility; therefore, we could brand them all as irrational only at the price of a severe split between theory and practice, a state of affairs I take to be rationally unacceptable.[28] Hence the only rational alternative open to us is to regard all established doxastic practices as rationally engaged in, pending sufficient reasons to take any of them as unreliable, and pending any other sufficient disqualifying considerations, if any. In other words, the only rational course is to take all established basic practices to be *prima facie* rational. This is a sort of "negative coherentism"[29] with respect to established practices; each such practice is innocent until proved guilty. Note that I am not endorsing negative coherence with respect to *beliefs*; I take a belief to require positive support in the form of adequate grounds on which it is based.[30]

It might be contended that we should restrict our *prima facie* acceptance to practices that are not only socially established but also *universally* engaged in, at least by all normal adult human beings. This would cut out, e.g., such religious belief-forming practices as we find in established religions such as Christianity and Hinduism. But I find this exclusion unwarranted. Why suppose that the outputs of a practice are unworthy of *prima facie* acceptance just because it is engaged in by only a part of the normal adult population? Why this predilection for egalitarianism in the epistemic sphere, where its credentials are much less impressive than in the political sphere? Of course, in taking non-universal established practices to be *prima facie* acceptable, the possibility is left open of taking non-universality as a sufficient disqualifying consideration.[31]

My judgment may also be assailed as not permissive enough. "Why not take *all* practices to be *prima facie* acceptable, not just socially established ones? Why this prejudice against the idiosyncratic? If Cedric has developed a practice of consulting sundried tomatoes to determine the future of the stock market, why not take that as reliable too unless we have something against it?" Now in fact, I think that we will almost always have something decisive against idiosyncratic doxastic practices; and so it would perhaps do no harm to let all of them in as *prima facie* acceptable, knocking them out at a second stage by sufficient reasons to the contrary. Nevertheless, there is a significant reason for doing it my way. When a doxastic practice has persisted over a number of generations, it has earned a right to be considered seriously in a way that Cedric's consultation of sundried tomatoes has not. It is a reasonable supposition that a practice would not have persisted over large segments of the population unless it was putting people into effective touch with some aspect(s) of reality and proving itself as such by its fruits. But there are no such grounds for presumption in the case of idiosyncratic practices. Hence we will proceed more reasonably, as well as more efficiently,[32] by giving initial, ungrounded credence only to the socially established practices. Newcomers will have to prove themselves.

If all established practices are *prima facie* acceptable, the remaining question concerns what sorts of disqualifying considerations, what sorts of "reasons to the contrary," can be identified, from the free swinging, non-rule-bound perspective that is appropriate for radical epistemological criticism. Before presenting my choices, let me emphasize the point that we must avoid holding one practice subject to the special requirements of another, unless we have sufficient reason for taking those requirements to be universally applicable. We must avoid, *pace* Plato and Descartes, supposing that SPP is disqualified by failing to come up to mathematical standards of precision and determinateness; and we must likewise avoid supposing, *pace* Hume, that induction is disqualified for failing to come up to the standards of deduction. In the papers cited in note 31, I have argued in like fashion that it is illegitimate to fault the practice of forming beliefs about God on the basis of religious experience for failing to satisfy constraints of predictability and checkability that are laid down by SPP for beliefs in its domain.

My first suggestion is that a practice is disqualified by persistent and irremediable inconsistency in its output. Consistency is a requirement of unrestricted generality just because its violation frustrates the most basic cognitive aim; to believe what is true and not to believe what is false. Massive internal inconsistency guarantees that a significant proportion of one's beliefs are false. But note that I am taking only a "persistent and irremediable" inconsistency to be disqualifying. Some degree of inconsistency pops up in all practices, and it is undoubtedly healthy that it should. Since it is often not crystal clear which side of a contradiction is true, it is well that different practitioners should be free to explore different sides.[33] If a practice should persistently deliver large numbers of mutually inconsistent beliefs, without any tendency over time to reduce their incidence, that would be a disqualification.

Second, and for the same reason, a massive and persistent inconsistency between the outputs of two different practices is a good reason for disqualifying at least one of them. This principle, of course, does not tell us which of the contenders to eliminate; and I don't see how to lay down any decision procedure for this from the standpoint of radical epistemological criticism. The only principle that suggests itself to me as both unchauvinistic and eminently plausible is the conservative principle that one should give preference to the more firmly established practice. What does being more firmly established amount to? I don't have a precise definition, but it involves such components as (a) being more widely accepted, (b) having a more definite structure, (c) being more important in our lives, (d) having more of an innate basis, (e) being more difficult to abstain from, and (f) its principles seeming more obviously true. But mightn't it be the case in a particular conflict that the less firmly established practice is the more reliable? Of course that is conceivable. Nevertheless, in the absence of anything else to go on, it seems the part of wisdom to go with the more firmly established. It would be absurd to make the opposite choice; that would saddle us with all sorts of bizarre beliefs.

It is easy to suppose that this principle can be illustrated by the way in which religious beliefs have progressively given way to scientific beliefs in the last few centuries. However, we must be careful to distinguish a choice between doxastic practices and a choice between particular beliefs (outputs of doxastic practices). When *some* religious beliefs contradict *some* scientific beliefs, the less firmly established practice can be preserved by sacrificing some of its beliefs and/or modifying its belief-forming procedures. This is what has happened in our culture over the last few centuries. What we may term the "Christian doxastic practice" has been modified by, e.g., changes in Biblical interpretation, so that it no longer generates a belief structure that is massively inconsistent with the belief structure generated by the "scientific doxastic practice." Thus the former is modified, not abandoned. To be sure, as this example indicates, our principle favoring the more firmly established could also be applied to choices between inconsistent beliefs, but that is not our present concern.

My final suggestion for a disqualifying consideration has to do not with a ground for definitive rejection, but with something that will strengthen or weaken the *prima facie* acceptability. The point is this. A practice's claim to acceptance is strengthened by significant "self-support," and the claim is weakened by the absence of such.

But how can self support enter into the picture if we are eschewing epistemically circular considerations? To answer this question we must distinguish different sorts of self-support. The reasons given earlier for not taking just any epistemically circular support to establish the credentials of a practice was that this was too easy, so easy that it would result in validating any practice whatever, no matter how absurd. For any practice can be conclusively self-supported if we allow ourselves to use each doxastic output twice, once as testee and once as tester. But we were too hasty in supposing that any self-support would be equally trivial. Consider the following ways in which SPP supports its own claims. (1) By engaging in SPP and allied memory and inferential practices we are enabled to make predictions, many of which turn out to be correct, and thereby we are able to anticipate and control, to some considerable extent, the course of events. (2) By relying on SPP and associated practices we are able to establish facts about the operation of sense perception that show both that it is a reliable source of belief and why it is reliable. These results are by no means trivial. It cannot be taken for granted that any practice whatever will yield comparable fruits. It is quite conceivable that we should not have attained this kind or degree of success at prediction and control by relying on the output of SPP; and it is equally conceivable that this output should not have put us in a position to acquire sufficient understanding of the workings of perception to see why it can be relied on. To be sure, an argument from these fruits to the reliability of SPP is still infected with epistemic circularity; apart from reliance on SPP we have no way of knowing the outcome of our attempts at prediction and control, and no way of confirming our suppositions about the workings of perception. Nevertheless, this is not the trivial epistemically circular support that necessarily extends to every practice. Many practices cannot show anything analogous; crystal ball gazing and the reading of entrails cannot. Since SPP supports itself in ways it conceivably might not, and in ways other practices do not, its *prima facie* claims to acceptance are thereby strengthened; and if crystal ball gazing lacks any non-trivial self support, its claims suffer by comparison.

We must be careful not to take up another chauvinistic stance, that of supposing that a practice can be non-trivially self supported only in the SPP way. The acceptability of rational intuition or deductive reasoning is not weakened by the fact that reliance on the outputs of these practices does not lead to achievements in prediction and control. The point is that they are, by their very nature, unsuitable for this use; they are not "designed" to give us information that could serve as the basis for such results. Since they do not purport to provide information about the physical environment, it would be unreasonable in the extreme to condemn them for not providing us with an evidential basis for predictive hypotheses. Similarly, I have argued in the articles cited in note 31 that it is equally inappropriate to expect predictive efficacy from the practice of forming beliefs about God on the basis of religious experience, and equally misguided to consider the claims of that practice to be weakened by its failure to contribute to achievements of this ilk. On the other hand, we can consider whether these other practices yield fruits that are appropriate to their character and aims. And it would seem that the combination of rational intuition and deduction yields impressive and fairly stable abstract systems, while the religious experiential practice mentioned earlier provides effective guidance to spiritual development.

Much more could and should be said about the ways in which the *prima facie* acceptability of one or another doxastic practice can be strengthened or weakened. But perhaps the above will suffice to indicate that it is possible, without falling into reprehensible chauvinism, to carry out a rational assessment of doxastic practices. But again we must be careful not to expect too much from this activity. The most we can hope from radical epistemic criticism is that some of the *prima facie* acceptable doxastic practices may be weeded out, and the claims of some strengthened and of others weakened, so that we may have a rank ordering of preferability to use when massive conflicts arise. For the most part, then, the epistemologist's proper and distinctive work will be that delineated by the heteronomist: to make explicit the structure of various established doxastic practices.

V

In the last two sections we have departed from our original issues concerning the epistemic status of principles of justification and concerning the *reliability* of doxastic practices, and have instead been considering how we could determine whether it is *rational* to engage in a certain practice. It is now time to return to the question of reliability and see what our conclusions concerning rationality have to tell us about that. I have suggested that the epistemologist, in her critical function, can make a sound (valid, acceptable. . .) judgment to the effect that it is rational to engage in SPP and other established doxastic practices. What implications does that have for the reliability of SPP? It might seem to have none. First of all, it is clear that the rationality of a practice does not *entail* its reliability. At least this is clear if the notion of a rational doxastic practice works at all as we have been assuming. We have supported the *prima facie* rationality of engaging in established doxastic practices without producing any evidence in support of a reliability claim. Instead, our main point was that all such practices are on a par with respect to the crucial issue of evidence for reliability, so that, as far as that is concerned, they all stand or fall together. Since abstaining from all doxastic practices is not a live option, the only reasonable course is to regard them all as rationally engaged in, pending sufficient disqualifying reasons. It is clear that all this could be the case, and hence that we are rational in engaging in, e.g., SPP, even if it were in fact unreliable. Moreover, the rationality of SPP does not even provide non-deductive but sufficient grounds for supposing it to be reliable. That follows from our initial assumption that there can be no adequate non-epistemically circular argument for the reliability of SPP and other fundamental doxastic practices. But then it looks as if the judgment that the practice is "rationally" engaged in has no bearing on the likelihood that it will yield truths; rationality has no "truth-conductivity" force. And so that judgment will not advance our original aim of determining how principles of justification are to be established, by determining how the reliability of modes of belief-formation is to be ascertained.

Before responding to this challenge, let's digress long enough to note that the above claim that rationality does not entail reliability may seem to conflict with our earlier claim that a principle of justification is correct only if the mode of belief-formation it takes to be sufficient for justification is reliable. This would seem to be true only if justification entails reliability. But now we are denying that the rationality of a practice entails its reliability. There are several ways in which one might argue that there is no real conflict here. (1) There's a difference between justification and rationality. (2) There's a difference between the justification (rationality) of *beliefs* and the justification (rationality) of *practices*. (3) However, I feel that the real explanation for the seeming discrepancy is that I have been speaking of the justification of *beliefs* in an objective, "externalist" sense of 'justification,' while speaking of the rationality of *practices* in a more subjective or "internalist" sense of 'rationality.'[34] In the externalist sense, a belief is justified (rational) only if it was formed in what is in fact a reliable way; and similarly a practice is rational (justified) only if it is in fact reliable. Whereas in the sort of internalist sense in question here, a belief is justified (rational) if it is more reasonable, all things considered, to adopt it than not to do so; and similarly a practice is rational (justified) if, all things considered, it is more reasonable to engage in it than to refrain from doing so. It would be a long story to explain fully my proceeding in this way (an externalist concept of the justification of beliefs and an internalist concept of the rationality of practices), and to defend that choice. The short story is this. I have tried to be objectivist as long as possible. But the difficulties in establishing justification (rationality) for beliefs in an objectivist sense drives us (sooner or later, and why make it any later?) to appeal to an internalist rationality for practices. If one still wonders why we couldn't have used an internalist conception of justification for beliefs in the first place and saved ourselves this extended treatment of practices, that issue will be addressed a few paragraphs down the road.

But, to return to the contention that a judgment of the rationality of a practice has no bearing on the question of the reliability of the practice, this is an illusion, born of a confusion of the most basic way of *establishing* a reliability claim with the *import* of that claim, another manifestation of that many-headed monster, verificationism. The fact that we cannot give a conclusive deductive or inductive argument for the reliability of SPP does *not* show that we are not *judging* that practice to be reliable when, as a critical epistemologist, we judge it to be rational. How could we judge it to be rationally engaged in without judging it to be reliable? To accept some doxastic practice, like SPP, as rational is to judge that it is rational to take it as a way of finding out what (some aspect of) the world is like; it is to judge that to form beliefs in accordance with this practice is to reflect the character of some stretch of reality. That means that to judge SPP to be rational is to judge that it is a reliable mode of belief-formation; for the beliefs thus formed could not be an accurate reflection of the facts without being generally true. Hence in explaining how one can form a sound, defensible judgment to the effect that SPP is rational, we have explained how one can arrive at a sound judgment that SPP is reliable.

It may seem that if a judgment of the rationality of SPP carries with it a judgment of its reliability, this can only be because rationality entails reliability, contrary to what I have said. But entailment of q by p is only one way in which judging that p amounts to judging that q. There is also pragmatic implication, as when in judging that I believe that p I thereby judge that p is true or, perhaps better, commit myself to the truth of p. And, closer to home, when I take myself to be rational or justified in believing that p, or take it that p is adequately supported by all relevant evidence, I thereby take p to be true. And in none of these cases is the truth of p entailed by what I explicitly judge.

It could be the case that I believe that p, that p is justifiably or rationally believed, and that p is adequately supported by all relevant evidence, and still not be the case that p is true. Likewise in our case. In taking SPP to be rationally engaged in, I thereby commit myself to regarding it as reliable; so that I cannot, if I know what I am about, affirm the one and deny the other. But that is because of what I am committing myself to in making the judgment of rationality, not because of an entailment, or other species of logical support, that obtains between the proposition that SPP is rational and the propositions that SPP is reliable.

Let's approach the whole matter from another angle. We have seen (at least I have laid it down) that a basic practice like SPP cannot be shown to be reliable in what we may call a "direct" fashion. That is, we cannot exhibit the proposition *SPP is reliable* as the conclusion of a deductive or inductive argument that meets the epistemic conditions for establishing the truth of the conclusion (non-circularity, premises knowable without already accepting the conclusion, and so on). That means that if we are to have any basis at all for making a judgment as to the reliability of SPP, we will have to take a more roundabout approach. More specifically, we need to consider the higher-level question as to whether the proposition *SPP is reliable* is justifiably or rationally accepted. Blocked from giving direct support to the proposition, we are led to seek support for a proposition about its epistemic status. This will provide indirect support to the proposition itself; for it is certainly a recommendation of a proposition that it is rational to accept it. This is just what we have been doing in indicating how one could make a sound judgment to the effect that it is rational to engage in SPP, and thereby a sound judgment to the effect that SPP is reliable. No doubt, it would be much more satisfying to produce a direct demonstration of the truth of the proposition that *SPP is reliable*. But since that is impossible, the next best thing is to show that it is reasonable to believe that SPP is reliable.[35]

Now for the question: "Is this detour through doxastic practices necessary?" If we are unable to give a straightforward argument for the reliability of ordinary perceptual beliefs, and are thereby unable to provide a straightforward defense of an externalist principle of justification for perceptual beliefs, why shouldn't we simply defend an internalist principle of justification for perceptual beliefs, rather than erect an elaborate framework of doxastic practices, and then defend an internalist principle of justification (rationality) for the *practice* of forming perceptual beliefs in the usual way? If we must retreat from externalist to internalist judgments of rationality, why not execute the maneuver on the field of beliefs rather than move into an entirely different sector?

A good question. But the answer is quite simple. So long as we consider beliefs in isolation, we have no sufficient basis for an internalist judgment of rationality. Take my present perceptual belief that there is a squirrel on the telephone wire outside my window. Recall that I cannot give a non-epistemically-circular argument for its being formed in a reliable manner, and hence am debarred from arguing directly for its being justified in an externalist sense. How can I support the thesis that it is, nevertheless, reasonable for me to believe this in the present situation? If I am unable to ascertain that it is formed in a reliable fashion, what other epistemically relevant recommendation can I give to this particular belief? I am at a loss to say. Hence the switch to something more general: the general mode of forming perceptual beliefs of which this is an instance. By focusing on this general *way* of forming beliefs, we may have a hope of finding some basis for an internalist judgment of (*prima facie*) reasonableness for all beliefs so formed. But how does the move to generality help us? We are still unable to carry out a direct demonstration of reliability. Well, I don't think generalization is a help, so long as we think of the mode of belief-formation abstractly, as a possible input-output device, or as the function that defines the device. Why is it more rational to instantiate one such function than another? We come onto something really helpful only when we take the mode of belief-formation *concretely,* as an aspect of a practice that is socially established and that plays a central role in human life. Then, and only then, do we find reasons for a judgment that it is reasonable to engage in the practice. Those reasons essentially depend on the fact that this practice, and other fundamental practices, are thoroughly entrenched in our lives, to such an extent that we can hardly, if at all, imagine life without them. It is against this background that we find the most reasonable judgment to be that any such way of forming beliefs is rationally employed, in the absence of sufficient disqualifying considerations. Hence the "detour" through practices is essential to the line of argument of this paper.

VI

How is the "doxastic practice approach" to epistemology presented in this paper related to the most prominent issues in epistemology and the most prominent positions on those issues? Is it foundationalist or coherentist? Is it internalist or externalist? What position does it take, or dictate, on the nature of knowledge, the nature of justification, and their interrelation? For that matter, how is this "approach" related to positions I have taken in recent publications?[36] Is this an about-face on my part, or does it simply continue and extend my recent work?

The key to answering these questions lies in the distinction between meta-epistemology and substantive epistemology, parallel to the more familiar distinction between meta-ethics and normative ethics. My doxastic practice approach is, most centrally, a position on the nature of epistemology: what its central problems and tasks are; how to go about those tasks; what sort of success it is reasonable to expect in this enterprise. This is meta-epistemology—a view about epistemology, its nature, conduct, methodology, and prospects—rather than a position developed in the prosecution of the discipline itself. Therefore, we should not expect this paper to present any position on those central issues mentioned in the last paragraph. And that is what we find. The *approach* developed here tells us that epistemology is primarily a reflection, both descriptive and critical, on established doxastic practices. But it doesn't tell us what that reflection will uncover. It doesn't tell us what practices there are, how they are structured, what criteria of justification or rationality are built into them, or how resistant to criticism those criteria are. We need to go about doing epistemology along these lines in order to resolve those issues. The meta-epistemology won't do that for us.

Nevertheless, just as with the relation between meta-ethics and normative ethics, the compartments are not hermetically sealed. Being an intuitionist or a naturalist in meta-ethics is going to put some constraints on the positions that can reasonably be taken in normative ethics. And so it is here. Thus, in conclusion, I will give a sketchy presentation of the major respects in which my doxastic practice approach does and does not carry commitments in substantive epistemology.

First, in taking epistemology to be centrally concerned with doxastic practices we commit ourselves to the view that the source of a belief is crucial for its epistemic status. Thus this approach rules out those views that take rationality or justification to be determined by, e.g., the evidence a person *has* for a belief, regardless of what the belief was based on; and it definitely aligns itself with views that see a close connection between the psychology of belief-formation and epistemology. Second, the approach rules out what we might call "universalism" in epistemology, any view that, like classical forms of both foundationalism and coherentism, suppose that one's knowledge or justified belief forms a single unified structure, so that the epistemic credentials of any belief is to be determined by locating it within that unique structure. On this approach, by contrast, pluralism reigns; there is no common measure for all beliefs. The epistemic status of a particular belief depends on the doxastic practice(s) from which it sprang; it depends on whether the belief conforms to the requirement of that practice, and, of course, on whether that practice itself is acceptable. There is no single, all-inclusive system by reference to which the credentials of any belief is to be assessed.[37] Third, the approach carries a strong prejudice toward an emphasis on reliability. This is largely because of the first commitment, to the epistemic relevance of the psychological source of beliefs. Once we take that step, it is almost inevitable that the reliability of the source will be a major factor in epistemic evaluation. If the most important thing to determine about a belief, from an epistemic point of view, is its mode of origin, then, given the acquisition of truth and the avoidance of error as the defining aim of the epistemic point of view, it can hardly be denied that the epistemically most important feature of a source is the reliability with which it yields true beliefs.[38]

Finally, there are more complicated relations with coherentism. There is a definite antipathy to coherentism, but it is multifaceted and qualified. For one thing, coherentism is typically universalist in character and so falls under the ban on universalist theories. But this doesn't show the approach to be more antithetical to coherentism than to classical foundationalism. For another thing, coherentism in its more plausible forms rejects source relevance. A source-relevant form of coherentism would have it that a belief is justified only if it is formed on the basis of its coherence with the total system of one's beliefs; and since beliefs are rarely, if ever, so formed, it would turn out that practically no beliefs are justified. Finally, the argument that attempts to establish the reliability of basic doxastic practices are infected with epistemic circularity, and therefore don't do the desired job, an argument that is basic to this whole approach, presupposes the inadequacy of coherentism. For a classical coherentism wouldn't worry about that circularity. Reliability claims for basic sources can, for such a theorist, be justified in the way any other belief can, by its coherence with one's total system of beliefs.

These, then, are the major respects in which this approach makes commitments, positive and negative, in substantive epistemology.

Now, finally, we turn to some of the major issues it leaves open. First, it carries no implications for the complete analysis of the major epistemological concepts, such as knowledge, justification, rationality; though, as we have seen, it does provide some constraints on such analyses, particularly with respect to justification and rationality. Second, the approach itself does not prejudge the fundamental questions of the structure of one or another doxastic practice. In keeping with its pluralist thrust, it does not assume that they all have the same structure. With respect to any one, it does not assume that the criteria of justification or rationality involved are internalist or externalist, in any of the many understandings of those notions, nor does it prejudge the question of the relative mix of foundationalist and coherentist elements in that structure. And, of course, it leaves wide open the question of what it takes for a ground of a belief to be *adequate*. Thus it is firmly committed to full employment for substantive epistemologists.[39]

Notes

1. "Concepts of Epistemic Justification," *The Monist,* Vol. 68, No. 1 (January, 1985); "An Internalist Externalism," *Synthese,* Vol. 74 (1988), pp. 265–283.

2. For some defense of this position see, e.g., my "Concepts of Epistemic Justification" and Laurence BonJour, *The Structure of Empirical Knowledge* (Cambridge, Mass.: Harvard U. Press, 1985), Ch. I.

3. Note that this thesis is plausible only for the whole of one of our major sources of belief, sense perception, memory, introspection, inductive reasoning, etc. If we were asking about some partial source, e.g., vision, or, still more partial, thermometers, we could easily find non-circular support by using other sense modalities or other instruments.

4. "Epistemic Circularity," *Philosophy and Phenomenological Research,* Vol. 47, No. 1, September, 1986.

5. Note that I., and many other widely accepted principles of the justification of perceptual beliefs, implies that this is possible. For I. lays down sufficient conditions of the justification of perceptual beliefs that do not require S to be justified in believing sense perception to be reliable.

6. It may be that 'show' is most properly used as a "success" word, so that one shows that p only if it is the case that p. If so, then the previous sentence simply makes explicit a conceptual truth about 'show' and has no special bearing on the problem of showing a source of belief to be reliable. But I mean to be using 'show' in a weaker sense in which it is synonymous with 'present adequate reasons in support of.' (If the English word 'show' cannot

properly be used in this way, just take it that I am using 'show' as an abbreviation for the phrase just quoted.) Given that use of 'show,' it will still be true that we can show sense perception to be reliable only if it is reliable, even though it is not generally true that one can show that p only if p.

7. Later we shall see that some forms of self-support are of greater value than others.

8. *On Certainty,* ed., G. E. M. Anscombe and G. H. von Wright, tr., D. Paul and G. E. M. Anscombe (Oxford: Basil Blackwell, 1969).

9. What Wittgenstein called a "language-game" is something much more inclusive than the term would suggest. It involves modes of belief-formation and assessment (the aspect we shall be concentrating on under the rubric "doxastic practice"), characteristic attitudes and feelings and modes of behavior toward certain sorts of things, as well as ways of talking. The Wittgensteinian term, "form of life," is better suited to suggest the richness of the concept.

10. There is a striking similarity here to Rudolf Carnap's distinction between questions that are internal to a conceptual framework and hence theoretical, and questions that are external to a conceptual framework and hence practical. See his "Empiricism, Semantics, and Ontology," *Revue Internationale de Philosophie,* 11 (1950).

11. *Philosophical Investigations,* tr., G. E. M. Anscombe (Oxford: Basil Blackwell, 1953) I, 654.

12. This is a psychological rather than an epistemic regress argument. See Robert Audi, "Psychological Foundationalism," *The Monist,* Vol. 62, No. 4 (1978).

13. See Alvin I. Goldman, "What Is Justified Belief?" in G. S. Pappas, ed., *Justification and Knowledge* (Dordrecht: D. Reidel Pub. Co., 1979). It is not clear just where to put memory in this classification. If it is possible to form beliefs for the first time about a remembered scene, then memory can sometimes be generational, though it will usually operate on previously acquired beliefs. When it does, it should perhaps be termed "preservative" rather than "transformational," since it is a way of storing and retrieving the same belief that constitutes the input, unless we wish to rule that whereas the previously held belief was, e.g., that one *does* have a stomachache, the memory belief is that one *did* have a stomachache.

14. We might think of the input to an inferential belief-forming mechanism as the beliefs that constitute the premises, leaving the inferential part to the operation of the mechanism. Or we might think of the input as the "realization" (conscious or unconscious) that the conclusion follows (deductively or in some other way) from the premises, thereby, in effect, absorbing the inferential link into the input. In either case, the output is the belief in the conclusion. In this paper I will employ the second alternative.

15. At least this is the case on the "source-relevant" and "truth-conducive" concept of justification we are employing in this chapter. On a "source-

irrelevant'' conception of justification, according to which it is enough for justification that the subject *have* adequate grounds for the belief, whether the belief was based on them or not, a different verdict would be forthcoming.

16. Thus far we are simply talking about what criteria of justification are inherent in a given practice, what a practitioner *takes* to be sufficient for justification. Whether this amounts to ''real justification'' is something we will be taking up later.

17. For a discussion of Reid that stresses this point, see Nicholas Wolterstorff, ''Thomas Reid on Rationality,'' in *Rationality in the Calvinian Tradition,* ed., H. Hart, J. Vanderhoeven, and N. Wolterstorff (Lanham, Md.: University Press of America, 1983).

18. *Essays on the Intellectual Powers of Man* (Cambridge, Mass.: MIT Press, 1969), Essay II, Ch. 20, p. 291.

19. *Ibid.*

20. See, e.g., *An Inquiry into the Human Mind* (Chicago: U. of Chicago Press, 1970), Ch. 6, section xxiv.

21. *Essays,* II, 20, pp. 291–292.

22. See *Inquiry,* Ch. 6, sec. xxiv.

23. As should be clear from the above, my theory of doxastic practices differs radically from the various non-realist, verificationist, and relativistic versions of Sprachspielism now current, represented by such writers as D. Z. Phillips and Richard Rorty, and by such rumblings as deconstructionism that emanate from the continent of Europe. I am far from supposing, with many of these writers, that each ''language-game,'' ''conceptual scheme,'' ''discourse,'' or what have you, carries with it its own special concept of truth and reality, that each defines a distinct ''world,'' or that truth is to be construed as ''what one's linguistic peers will let one get away with'' (Rorty). My theory of doxastic practices is firmly realistic, recognizing a single reality that is what it is, regardless of how we think or talk about it. The doxastic practice is a source of criteria of justification and rationality; it does not determine truth or reality. Another way of putting this is to say that for me doxastic practices are crucial epistemologically, not metaphysically.

24. For one attempt to formulate Reid's response, see my ''Thomas Reid on Epistemic Principles,'' *History of Philosophy Quarterly,* Vol. 2, No. 4 (October, 1985).

25. These are all considerations Reid adduces with respect to what he calls ''first principles,'' which include claims to the reliability of our basic doxastic practices. See my article listed in the last note.

26. This is a point stressed by Reid. See my ''Thomas Reid on Epistemic Principles'' for details.

27. This last contrast is well exemplified by the Kuhnian distinction between ''normal science,'' subject to fairly definite standards and procedures, and ''scientific revolutions,'' in which the previously accepted standards and procedures are thrown into question, and it is everyone for himself.

28. A defence of this judgment would be difficult and tortuous, and this is not the place for it. I will only point to the great importance of this issue and its pervasive neglect. Note that even if universal abstention were a live possibility, it would still be the case that we are firmly entrenched in these practices; hence the burden of proof would seem to fall on the proponent of abandonment rather than on the proponent of continuance.

29. For this term see John Pollock, "A Plethora of Epistemological Theories," in G. S. Pappas, ed., *Justification and Knowledge* (Dordrecht: D. Reidel Pub. Co., 1979), p. 101.

30. For a development of this view about the justification of belief see my "An Internalist Externalism."

31. I have argued against taking non-universality to be disqualifying in "Religious Experience and Religious Belief," *Nous*, Vol. 16 (1982), and in "Christian Experience and Christian Belief," in *Faith and Rationality*, ed., A. Plantinga and N. Wolterstorff (Notre Dame, Ind.: U. of Notre Dame Press, 1983).

32. As for efficiency, contrast the procedure of accepting all applicants to a graduate program and then flunking out the unqualified ones, with the procedure of carefully screening the applicants and accepting only those whose record grounds a *prima facie* presumption of success.

33. For other reasons for allowing some degree of inconsistency in a set of beliefs, see Peter Klein, "The Virtues of Inconsistency," *The Monist*, Vol. 68, No. 1 (January, 1985).

34. For an explanation of the internalist-externalist distinction(s), see my "Internalism and Externalism in Epistemology," *Philosophical Topics*, 14, No. 1 (1986).

35. This is analogous to the "fideist" move in religion. Pessimistic about the chances of directly establishing the truth of the existence of God, one seeks to show that it is rational for one to believe in God, as a postulate of pure practical reason, as a requirement for fullness of life, or whatever.

36. Particularly "Concepts of Epistemic Justification" and "An Internalist Externalism."

37. At least this is the story with respect to beliefs. Our approach does countenance a sort of universalism with respect to practices. We need, and have, universally applicable principles for the evaluation of doxastic practices. But even here, as we saw, there are severe limits on what can be expected from the application of these principles. This is, at most, a very weak universalism.

38. Note that this point does not strictly rule out a deontological conception of justification in terms of obligations, duties, blameworthiness, etc. For the stress on reliability can be preserved in the view that the justification of a particular belief is a matter of that belief's being formed with due regard to one's obligation to see to it that one forms beliefs only by procedures that, so far as one can tell, are reliable.

39. Ancestors of this paper have been presented at the Institute of Philoso-

phy in Moscow, Wayne State University, and the University of Utah. I am indebted to my discussants on those occasions for many useful reactions. I am especially grateful to Robert Audi and Jonathan Bennett for comments that greatly improved the paper.

14

Philosophical Scepticism and Epistemic Circularity

Ernest Sosa

Epistemic circularity has dogged epistemology from the time of the Greek sceptics, through Descartes's circle and Hegel's serpent biting its tail, to serve finally as a source of today's relativism and scepticism—an important source, though of course only one of several. 'Since there is no way to justify one's overall practical or theoretical stance without clarity,' we are told, 'all justification must be ultimately relative to one's basic commitments, conceived perhaps as arbitrary creatures of the will. In comparing overall systems, anyhow, especially when these are equally coherent and self-supportive, there is no way to privilege one's own except arbitrarily, irrationally or arationally, perhaps by adopting a frank and honest ethnocentrism.' That is today a widespread attitude. This paper aims to expose questionable assumptions on which it rests.

We shall consider the following thesis and its supporting argument.

Philosophical Scepticism. There is no way to attain full philosophical understanding of our knowledge. A fully general theory of knowledge is impossible.

The Radical Argument (RA)
A1. Any theory of knowledge must be internalist or externalist.
A2. A fully general internalist theory is impossible.
A3. A fully general externalist theory is impossible.
C. From A1–A3, *philosophical scepticism* follows.

Reprinted by courtesy of the Editor of the Aristotelian Society from *Proceedings of the Aristotelian Society*, Supplementary Volume 68 (1994), pp. 263–290. © 1994.

In discussing these, first it will be convenient to define some terminology. 'Formal internalism'—or 'internalism' for short—shall stand for the doctrine that a belief can be justified and amount to knowledge only through the backing of reasons or arguments. This is of course a special sense of the word, but internalism in this sense today enjoys substantial support. Here are some representative passages, drawn from the writings of Donald Davidson, Richard Rorty, Laurence BonJour, and Michael Williams.

[Nothing] . . . can count as a reason for holding a belief except another belief . . . [And it] . . . will promote matters at this point to review very hastily some of the reasons for abandoning the search for a basis of knowledge outside the scope of our beliefs. By 'basis' I mean here specifically an epistemological basis, a source of justification.[1]

[It] . . . is absurd to look for . . . something outside [our beliefs] . . . which we can use to test or compare with our beliefs.[2]

[Nothing] . . . counts as justification unless by reference to what we already accept, and there is no way to get outside our beliefs and our language so as to find some test other than coherence.[3]

[We] can think of knowledge as a relation to propositions, and thus of justification as a relation between the propositions in question and other propositions from which the former may be inferred. Or we may think of both knowledge and justification as privileged relations to the objects those propositions are about. If we think in the first way, we will see no need to end the potentially infinite regress of propositions-brought-for-ward-in-defense-of-other-propositions. It would be foolish to keep conversation going on the subject once everyone, or the majority, or the wise, are satisfied, but of course we *can*. If we think of knowledge in the second way, we will want to get behind reasons to causes, beyond argument to compulsion from the object known, to a situation in which argument would be not just silly but impossible . . . To reach that point is to reach the foundations of knowledge.[4]

To accept the claim that there is no standpoint outside the particular historically conditioned and temporary vocabulary we are presently using from which to judge this vocabulary is to give up on the idea that there can be reasons for using languages as well as reasons within languages for believing statements. This amounts to giving up the idea that intellectual or political progress is rational, in any sense of 'rational' which is neutral between vocabularies.[5]

[The] notion of a [foundational] 'theory of knowledge' will not make sense unless we have confused causation and justification in the manner of Locke.[6]

If we let φ represent the feature or characteristic, whatever it may be, which distinguishes basic empirical beliefs from other empirical beliefs, then in an acceptable foundationalist account a particular empirical belief could qualify as basic only if the premises of the following justificatory argument were adequately justified:

(1) B has feature φ.

(2) Beliefs with feature φ are highly likely to be true.

Therefore, B is highly likely to be true.

. . . But if all this is correct, we get the disturbing result that B is not basic after all, since its justification depends on that of at least one other empirical belief.[7]

Only a legitimating account of our beliefs about the world will give an understanding of our knowledge of the world. This means that an account of our knowledge of the world must trace it to something that is *ours,* and that is *knowledge,* but that is not *knowledge of the world.*[8]

'Formal externalism' shall stand for the denial of formal internalism. And, again, for short we shall drop the qualifier, and speak simply of 'externalism'.

A very wide and powerful current of thinking would sweep away externalism root and branch. This torrent of thought in one way or another encompasses much of contemporary philosophy, both on the Continent and in the Anglophone sphere, as may be seen in the Continental rejection of presence to the mind as well as in the analytic rejection of the given. The Continentals have been led by Heidegger, Gadamer, Habermas, Foucault, and Derrida to a great variety of anti-foundationalisms, ranging from consensualism and hermeneutics to relativism and contextualism. The tide against the given on this side of the Channel is no less powerful and is illustrated by the passages already cited. Having also rejected the given and presence to the mind, others settle into an irresolvable frustration that recognizes the problems but denies the possibility of any satisfactory solution.[9] Many who now object to externalism in such terms offer little by way of support. Barry Stroud and William Alston are exceptional in spelling out the deep reasons why, in their view, externalism will leave us ultimately dissatisfied.[10] They have made as persuasive a case as can be made for the unacceptability in principle of any externalist circles in epistemology, and have done so on a very simple *a priori* basis grounded in what seemed to be demands inherent in the traditional epistemological project itself. What follows will focus on their case against such externalism, but much of it applies *mutatis mutandis* to the reasoning, such as it is, offered by other thinkers as well.[11] Though the issue before us is phrased in the terms of analytic epistemology, it is a wellspring of main currents of thought that reach beyond analysis and epistemology. Yet the issue and its options, rarely faced directly, are very ill-understood.

One thing is already clear. Given our definition of externalism as simply the denial of internalism, premise A1 is trivially true and amounts to *p or not-p*.

Note further that an acceptable *internalist* epistemological account of all one's knowledge in some domain D would be, in the following sense, a 'legitimating' account of such knowledge.

> *A is a legitimating account of one's knowledge in domain D* IFF D is a domain of one's beliefs that constitute knowledge and are hence justified (and more), and A specifies the sorts of inferences that justify one's beliefs in D, without circularity or endless regress.

But such an account cannot be attained for all one's knowledge:

> *The impossibility of general, legitimating, philosophical understanding of all one's knowledge:* It is impossible to attain a legitimating account of absolutely all one's own knowledge; such an account admits only justification provided by inference or argument and, since it rules out circular or endlessly regressive inferences, such an account must stop with premises that it supposes or 'presupposes' that one is justified in accepting, without explaining how one is justified in accepting them in turn.

Accordingly, premise A2 of argument RA seems clearly right. And it all comes down to premise A3. If we are to resist philosophical scepticism we cannot accept that premise. What then are the prospects for a formal externalist epistemology?

The formal externalist has, it seems to me, three main choices today, concerning how a belief attains the status of knowledge, how it acquires the sort of epistemic justification (or aptness or warrant, or anyhow the positive epistemic status) required if it is to amount to knowledge. These three choices are:

> E1. *Coherentism.* When a belief is epistemically justified, it is so in virtue of its being part of a coherent body of beliefs (or at least of one that is sufficiently coherent and appropriately comprehensive).

E2. *Foundationalism of the given.* When a belief is epistemically justified, it is so in virtue of being either the taking of the given, the mere recording of what is present to the mind of the believer, or else by being inferred, appropriately from such foundations.

E3.*Reliabilism.* When a belief is epistemically justified, it is so in virtue of deriving from an epistemically, truth-conducively reliable process or faculty or intellectual virtue of belief acquisition.

E1. There is a lot to be said about coherentism, but I lack the space to say much of it here. Suffice it to say that the most comprehensive coherence accompanied by the truth of what one believes will not yet amount to knowledge. The New Evil Demon problem establishes this as follows. Consider the victim of Descartes' evil demon. In fact, suppose we are now such victims. Could that affect whether or not we are *epistemically justified* in believing what we believe? If we are justified as we are, we would *seem* equally justified, in some appropriate sense, so long as nothing changed within our whole framework of experiences and beliefs. However, if by sheer luck one happened to be right in the belief that one faces a fire, one's being *both* thus justified *and* right still would fall short of one's knowing about the fire. So whatever is to be said for coherence, or even for comprehensive coherence, one thing seems clear: none of that will be enough just on its own to explain fully what a true belief needs in order to be knowledge. One's beliefs can be comprehensively coherent without amounting to knowledge, and the same goes for one's beliefs and experiences together. So the sense of 'epistemic justification' in play here is one that will not capture fully the epistemic status required in a true belief if it is to constitute knowledge.

E2. What of foundationalism of the given? *Cogito ergo sum* exclaimed Descartes, as he at last found a good apple off the tree of knowledge. By that time many other apples had already been judged defective, or at least not clearly enough undefective. Our perceptual beliefs had not qualified, since we could so easily be fooled into believing something false on the basis of sensory experience. For example, one could fall victim to illusion or hallucination, and, more dramatically, to an evil demon or a mad scientist who manipulated one's soul or one's brain directly, thus creating systematically the sorts of experiences that one would normally take to be indicative of a normal environment. None of this will affect the *cogito*, however, since even while hallucinating or while manipulated by evil demon or mad scientist, we must still exist and we must still be thinking, if we are to be fooled into thinking something incorrectly. One thought that could never be incorrect is the thought that one exists, and another is the thought that one is thinking.

What is the feature of the *cogito* that explains its special assurance? Consider the proposition (a) that I am now standing. This proposition is true but only contingently so, since I might have been sitting now. In contrast, it is not only true but necessarily true (b) that either I am standing or I am not standing. Is it the necessity of (b) that accounts for its special certainty as compared with (a)? Not entirely. For much is necessary without being certain, and much is certain without being necessary. And, in any case, it cannot be the necessity of 'I think' or 'I exist' that gives such propositions their special epistemic status. For in itself the *cogito,* the proposition that I am thinking, is only true and not necessarily true: I might have been unconscious, or even dead, in which case I would not have been thinking. What is not just contingently true, what is necessarily true, is the fact that *if* I am thinking that I think, *then* I am right: no-one can think that they think without being right. Is it *this,* then, that distinguishes the *cogito* and makes it a legitimately known contingent truth, of which we can properly be assured?

No, that one must be right in believing something does not entail that one is justified in doing so. Take the proposition that there is no largest prime. Since that proposition is necessarily true, we could not possibly go wrong in believing it. Nevertheless, we are not justifiably assured in believing it if we are just guessing the right and have seen no proof. That a belief could not be wrong is hence not enough to make it apt, nor is a belief necessarily apt just because even the Cartesian demon could not fool one into holding it incorrectly. A groundless belief is one that we hold in the absence of supporting reasons or arguments. Some such beliefs seem far superior to others: some amount to knowledge of the obvious, while others are no better than superstition or dogma. We are now after distinguishing properties or features that will help explain which groundless beliefs might qualify as knowledge and which could never do so, and for some account of why these properties or features can make such a difference.

A second main source of apt, groundless beliefs, according to the epistemological tradition, is presence to the mind, or what is given in sensory and other experience. What is involved in one's aptly believing something about the character of one's present sensory experience? It is required by the tradition that one be reporting simply how it is in one's experience itself. One must be reporting on the intrinsic, qualitative character of some experience.

But here again a similar problem arises. Suppose one eyes a well-lit surface with a medium-sized white triangle against a black background. In that case, assuming one is normally sighted, one would have visual experience of a certain distinctive sort, as if one saw a white triangle against a black background. Introspectively, then, one could easily come to know that one was then having experience of that sort: viz, that one was presented with a white triangular image, or the like. What now is the relevant feature of one's introspective belief, what is the feature that makes one's belief apt, makes it indeed a bit of knowledge? Is it simply that one is just reporting what is directly present to one's mind, what is given in one's experience?

No, that something is thus present to one's mind or given in one's experience is not enough to make it something of which one can be legitimately assured. Take that same situation and change the white image projected on the black surface from a triangle to a dodecagon. And suppose you believe yourself to be presented with a white dodecagon on a black surface, all other conditions remaining as before. Are you then properly assured about the character of your experience so that your introspective belief can then count as apt belief, and indeed as knowledge? What of someone poor at reporting dodecagons in visual experience, who often confuses them with decagons, but who now happens by luck to be right? Such a belief could hardly count as knowledge or even as apt belief.

What Descartes needs in order to explain the special status of the *cogito* is not just that one cannot incorrectly believe that one thinks, but rather that one could not possibly answer incorrectly the question whether one thinks (at least not sincerely and *in foro interno*). And how can one explain this special status enjoyed by that proposition? Descartes's explanation is of course that even a powerful evil demon could not fool one into thinking incorrectly that one thinks. For if the demon gets one to *think* that one thinks—and how else could he fool one into *thinking* incorrectly that one thinks?—then of course inevitably one *does* think and one is bound to be right.

However, that does only half the job. It explains only how one must be right if one thinks that one thinks. It does not explain why it is that one would never think that one does *not* think. Of course Descartes does *claim* that the proposition that one thinks is not only one with regard to which one is infallible, such that if one accepts it one must be right. He also thinks that it is an *indubitable* proposition. But whereas he explains incontestably why one must be right in thinking that one thinks, he does little or nothing to explain why it is that the *cogito* and other similarly simple, clear, distinct propositions are for us indubitable.

What of the doctrine of the given or of presence to the mind? Here the proposal would be that one aptly introspects P iff P describes a present state of one's own consciousness and while considering attentively and with a clear mind the question whether P is the case, one believes P. It is held to be very unlikely that one would ever opt wrong on such a proposition when in such circumstances.

By reflecting on how the doctrine of the given must be formulated in order to meet certain objections, we have arrived at a reliabilist version of foundationalism. What matters is not that one attend to the contents of one's mind, to one's experiences or beliefs or other states of mind, nor is what matters that one attend to simple necessary truths. For simplicity is a relative matter: what is simple for an experienced mathematician is far from it to the schoolchild learning arithmetic. It is important rather that the subject be reliable on the object of knowledge, and unlikely to go wrong on such subject matter.

E3. So we are down to the third and last of the options open to the formal externalist. But I view generic reliabilism as a *very* broad category indeed, one capacious enough to include thinkers as diverse as Descartes and Alvin Goldman. If we are to resist philosophical scepticism it would appear that here we must make a stand. For, remember, if A3 cannot be defeated, then philosophical scepticism seems the inevitable consequence. So let us consider some objections to generic reliabilism. Here we turn to the promised arguments by Barry Stroud and William Alston.

According to Stroud, 'we need some reason to accept a theory of knowledge if we are going to rely on that theory to understand how our knowledge is possible. That is what . . . no form of 'externalism' can give a satisfactory account of.'[12] Against Descartes, for example, and against the 'externalist' in general he objects on the basis of the following *metaepistemic requirement:*

MR In order to understand one's knowledge satisfactorily one must see oneself as having some reason to accept a theory that one can recognize would explain one's knowledge if it were true.

And how is MR to be defended? From the assumptions: (a) that understanding something is a matter of having good reason to accept something that would be an explanation if it were true, and (b) that, as a generality-thirsty theorist of knowledge, one wants to understand how one knows the things one thinks one knows.[13] But MR does not follow from these assumptions. From these assumptions it follows only that in order to understand one's knowledge one must in fact *have* good reason or at least justification to accept some appropriate explanation. Why must one also *see oneself as having* such reason?

Far from being just an isolated slip, MR represents rather a deeply held intuition that underlies a certain way of thinking about epistemology. We have seen already several passages that fit this intuition. According to such 'anti-externalism,' as Stroud might label it, what is important in epistemology is justification; and the justification of any given belief requires appeal to *other* beliefs that constitute one's reasons for holding the given belief. Of course, when one combines this with rejection of circularity, the case for scepticism is very strong, assuming that for limited humans an infinite regress of reasons or justifications is out of the question.

The 'externalist' therefore wants to allow some *other* way for a belief to acquire the epistemic status required for it to be knowledge, some way *other* than the belief's being based on some justification, argument, or reason. Note, moreover, how very broad this sense of 'externalism' is. Even arch-internalist Descartes is an 'externalist' in our present sense. We distinguish our present externalism as '*formal externalism*,' it will be recalled, which induces a corresponding type of internalism, 'formal internalism.' Formal internalism holds that there is only one way a belief can have the positive epistemic status required for knowledge, namely by having the backing of reasons or arguments. Note the connection with the requirement that a philosophically satisfactory account of how one knows must be a *legitimating* account, one that specifies the reasons favouring one's belief. Obviously, a formal internalist will believe that for *every* belief that amounts to knowledge there must be such a legitimating account, and that only once we have such an account can we understand what makes that belief knowledge.[14]

Consider now the naturalist, externalist epistemologist. Will he be able to understand how people know the things they do? He will only if he knows or has some reason to believe his scientific account of the world around him. According to Stroud, this dooms our epistemologist:

> If his goal was, among other things, to explain our scientific knowledge of the world around us, he will have an explanation of such knowledge

only if he can see himself as possessing some knowledge in that domain. In studying other people, that presents no difficulty. It is precisely by knowing what he does about the world that he explains how others know what they do about the world. But if he had started out asking how anyone knows anything at all about the world, he would be no further along towards understanding how any of it is possible if he had not understood how he himself knows what he has to know about the world in order to have any explanation at all. He must understand himself as knowing or having some reason to believe that his theory is true.[15]

But it is again unclear why the epistemologist needs to *see himself as having* justification for his theory, or as knowing his theory, in order for it to give him understanding of how he and others know the things they know, either in general or in the domain in question. Why is it not enough that he in fact *have good reason to accept his theory* or perhaps even *know his theory to be true*? This is different from his knowing that he has good reason to believe his epistemologically explanatory theory, or even knowing that he knows his theory to be true. To this the response is as follows.

> [The externalist epistemologist] . . . is at best in the position of someone who has good reason to believe his theory if that theory is in fact true, but has no such reason to believe it if some other theory is true instead. He can see what he *would* have good reason to believe if the theory he believes were true, but he cannot see or understand himself as knowing or having good reason to believe what his theory says.[16]

> [Even] . . . if it is true that you can know something without knowing that you know it, the philosophical theorist of knowledge cannot simply insist on the point and expect to find acceptance of an 'externalist' account of knowledge fully satisfactory. If he could, he would be in the position of someone who says: 'I don't know whether I understand human knowledge or not. If what I believe about it is true and my beliefs about it are produced in what my theory says is the right way, I do know how human knowledge comes to be, so in that sense I do understand. But if my beliefs are not true, or not arrived at in that way, I do not. I wonder which it is. I wonder whether I understand human knowledge or not.' That is not a satisfactory position to arrive at in one's study of human knowledge—or of anything else.[17]

But again it is hard to see why the externalist theorist of knowledge must be in that position. Suppose that, as suggested earlier, he does *not* have to say or believe that he *does know his theory of knowledge*. *Suppose he does not after all need to satisfy MR. Must he then say or believe that he does not* know his theory of knowledge? Must he begin to wonder *whether* his theory of knowledge is true, or whether he does really understand human knowledge or not?

Here the dialectic is given a further twist. It is replied that the sort of understanding of our knowledge of the external that we want in philosophy is not just understanding by dumb luck. What we want is rather *knowledgeable* understanding. And this we will never have until we are in a good position to accept our view of our own faculties (of perception or memory, for example), a view which properly underlies our trust in their reliability. But this view we will never be able to justify without relying in turn on already attained knowledge of the external. And this precludes our ever attaining a philosophically satis-factory understanding of all our knowledge in that domain.[18]

The demands introduced by this drive for *knowledgeable* philosophi-cal understanding are different from those deriving from the twofold assumption that (a) epistemic justification is required for knowledge, and (b) reasons and arguments are universally required for epistemic justification. This twofold assumption—formal internalism—leads, as we have seen, to the impossibility of any fully general, legitimating, philosophical understanding of one's knowledge (and indeed to the impossibility of one's knowledge altogether). The new demands do not derive simply from such formal internalism. They derive rather from a distinctively epistemic circularity that came to philosophical con-sciousness long ago.

The dialectic of the diallelus is about as ancient as philosophy itself. Nor is Stroud the *only* philosopher today who argues extensively on the basis of epistemic circularity. Recent books by William Alston, for example, contain extensive discussion of these issues, and feature the following main theme:

if sense-perception is reliable, a track-record argument will suffice to show that it is. Epistemic circularity does not in and of itself disqualify the argument. But even granting that point, the argument will not do its job unless we *are* justified in accepting its premises; and that is the case only if sense perception is in fact reliable. And this is to offer a stone instead of bread. We can say the same of any belief-forming practice whatever, no matter how disreputable. We can just as well say of crystal-ball gazing that if it *is* reliable, we can use a track record argument to show that it is reliable. But when we ask whether one or another source of belief is reliable, we are interested in *discriminating* those that can reasonably be trusted from those that cannot. Hence merely showing that *if* a given source is reliable it can be shown by its record to be reliable, does nothing to indicate that the source belongs with the sheep rather than with the goats. I have removed an allegedly crippling disability, but I have not given the argument a clean bill of health.[19]

Both in that book and in more recent work[20] Alston is forthright in his statement of the problem of circularity that he sees, and in his response to that perceived problem:

> Hence I shall disqualify epistemically circular arguments on the grounds that they do not serve to discriminate between reliable and unreliable doxastic practices.[21]

> Hence, when we reflect on our epistemic situation, we can hardly turn our backs *on our inability to give a satisfactory demonstration of SP and other doxastic practices . . .*[22]

In response to this, Alston argues instead that it is 'practically rational' for us to engage in our firmly rooted doxastic practices,[23] such as our 'sense perceptual practice,' SP, 'our customary ways of forming beliefs about the external environment on the basis of sense perception.'[24] And he believes that 'in showing it to be rational to engage in SP,' he has thereby, 'not shown SP to be reliable, but shown it to be rational to suppose SP to be reliable.'[25] This is so in the sense that it would be irrational for one to judge that SP is rational and deny that SP is reliable, or even to abstain from judging that SP is reliable if the question arises. So in accepting that SP is rational one 'pragmatically implies' and thereby 'commits oneself' to its being the case that SP is reliable.

Just how is it shown that it is 'rational' (or 'reasonable') to engage in SP? Here the argument begins by drawing from Thomas Reid the following claim:

1. The 'only (noncircular) basis we have for trusting rational intuition and introspection is that they are firmly established doxastic practices, so firmly established that we cannot help [doing so] . . . ; and we have exactly the same basis for trusting sense perception, memory, nondeductive reasoning, and other sources of belief for which Descartes and Hume were demanding an external validation.'[26]

And it continues as follows:

2. [Even if] we could adopt some basic way of forming beliefs about the physical environment other than SP, or some basic way of forming beliefs about the past other than memory, . . . why should we?'[27]

3. 'The same factors that prevents us from establishing the reliability of SP, memory, and so on without epistemic circularity would operate with the same force in these other cases.'[28]
4. 'These considerations seem to me to indicate that it is eminently *reasonable* for us to form beliefs in the ways we standardly do,'[29] such as SP.

This is presented as an argument for the practical rationality (or reasonableness) of using SP, one which avoids the 'epistemic circularity' that cripples track-record and other arguments for the *reliability* of SP. Where exactly is the circularity, and just how does it do its damage? The answer considers the use of a track-record argument, an argument that appeals to our past cognitive success through using SP:

[If] I were to ask myself why I should accept the premises, I would, if I pushed the reflection far enough, have to make the claim that sense perception is reliable. For if I weren't prepared to make that claim on reflection, why should I, as a rational subject, countenance perceptual beliefs? Since this kind of circularity involves a commitment to the conclusion as a presupposition of our supposing ourselves to be *justified* in holding the premises, we can properly term it 'epistemic circularity'.[30]

However, consider again the earlier argument in favour of the conclusion that it is *rational* (or reasonable) to use SP, the argument presented above as 1–4. If we push reflection far enough with regard to why we should accept the premises of *this* argument, don't we find ourselves appealing precisely to *its* conclusion? And, if so, then is not this argument just as circular, and in a similar way, as the track-record argument in favour of the reliability of SP?

Epistemological reflection therefore leads to a situation that does seem 'fairly desperate' after all. We wonder whether we really know what we take ourselves to know. We wonder how we know whatever it is that we know. We hope that our way of forming beliefs—with its characteristic elements of memory, introspection, perception, and reason—does give us knowledge and explains how we know. But how can we be sure?

Suppose W is our total way of forming beliefs. If we believe that W is reliable, R(W), our belief B:R(W) is itself formed by W. And if a belief is justified iff formed in a reliable way, then our B:R(W) is justified iff W is reliable (given that it is formed by W). B:R(W) is justified, therefore, iff W *is* reliable.

Yet we must sympathize with the critics of 'externalism,' who argue that this is to 'give us a stone instead of bread,' and that the externalist 'is at best in the position of someone who . . . can see what he *would* have good reason to believe if the theory he believes were true.' Let us consider carefully what they have to say.

Alston, in his recent book, argues as follows.

> Consider our sense-perceptual doxastic practice SP, (our total way of forming beliefs based on sense perception). The reliability of SP can be inferred, let us suppose, by relying on the deliverances of SP itself. Hence, assuming our reasoning is otherwise unobjectionable, belief B:R(SP) is justified if SP is reliable. But using the deliverances of SP to argue for B:R(SP) would be unacceptably circular.

Here, again, is how he puts it.

> [When] we ask whether one or another source of belief is reliable, we are interested in *discriminating* those that can reasonably be trusted from those that cannot. Hence merely showing that *if* a given source is reliable it can be shown by its record to be reliable, does nothing to indicate that the source belongs with the sheep rather than with the goats. I have removed an allegedly crippling disability, but I have not given the argument a clean bill of health. Hence I shall disqualify epistemically circular arguments on the grounds that they do not serve to discriminate between reliable and unreliable doxastic practices.[31]

But what exactly is the problem for the justification of B:R(SP)? And, even more generally, what exactly is the problem for the justification of B:R(W), where W is our total way of forming beliefs (of which SP would be only one among several components)?

Justification can be either a matter of one's internal rationality and coherence, or it can go beyond that to encompass some broader (or just different) state pertinent to whether one knows. Thus the victim of Descartes's evil demon may have internal justification for believing that there is a fire before him, but would still lack knowledge even if *by accident* he is right. Similarly, the hopelessly myopic Mr. Magoo may have internal justification for believing that it is safe to step ahead, but would still lack knowledge even if the board over the precipice does by accident still lie ahead.

For now let us focus just on internal justification or rational coherence. Are we bound to fall short of rational coherence if we form our belief that W is reliable—B:R(W)—through W itself? Alston suggests that we do fall short, in *some* way, since in asking whether one or another source of belief is reliable, we wish to *discriminate* sources that we can trust with good reason. Therefore, to show that *if* a given source is reliable it can be shown by its own use to be reliable does nothing to discriminate it from the many other possible sources equally able to pass that test.

We are thus offered the following view of the matter. We have before us a menu of sources, of ways of forming beliefs: W1, . . . , Wn. And we would like to discriminate the reliable from the unreliable. About Wi we discover that it has this much to be said for it: if one uses Wi to form beliefs, then by Wi one can form the belief B:R (Wi), the belief that Wi is reliable. And *if* Wi *is* reliable, then B:R(Wi) will itself of course be justified. When a way of forming beliefs, Wi, has this feature relative to a subject S in circumstances C, let us say that Wi is self-supportive for S in C: i.e., for S in C, Wi will deliver the belief on the part of S *that* Wi is itself reliable—B:R(Wi).

Here then is Alston's point about the feature of being self-supportive relative to oneself and one's circumstances: *several* (indefinitely many) ways of forming beliefs might well have this feature relative to oneself and one's circumstances, but many of these are palpably unacceptable. Indeed they might well be inconsistent in such a way that most by far are bound to be *unreliable*. Therefore, even once we reach the conclusion that Wi is self-supportive relative to us and our circumstances, that by itself does *not* enable us to conclude that it is acceptable, that it is a sheep, not a goat.

That much is surely right. But there is more. There is also the further proposal that if a way of forming beliefs W (a doxastic practice) is 'firmly established' for us, then we *can* conclude that it is *practically* acceptable, that we are practically rational in accepting it.[32] Presumably this feature of a doxastic practice of its being FE (firmly established) is thought to have an advantage over the feature of a doxastic practice of its being R (reliable), with regard to the dialectic above. But it is hard to see how it can possibly enjoy any such advantage. For in order to reach the belief that our total way of forming beliefs W is firmly established—B:FE(W)—we could hardly avoid using W itself. And it is not hard to see that indefinitely many crazy ways W* of forming beliefs might (conceivably) be equally effective, if used by one in one's circumstances, in leading to the belief—B:FE(W*)—that W* is firmly established for us, even though W* is still clearly unacceptable. What is more, it is also conceivable that there be a way W* that might *in fact* become firmly established, even though W* remained unacceptable (by our present lights, of course). Conclusion: It is hard to see the advantage in moving from reliability to firm establishment and practical reasonableness. True, even if using W to settle whether W is reliable would yield a positive verdict, that is not enough to lift W above its many competitors with an analogous feature. But, similarly, even if using W to settle whether W is firmly established would yield a positive verdict, that is not enough to lift W above its similar competitors *either*. It might be answered that we needn't *see W as firmly established* in order for its firm establishment to lend us practical justification for using it. But then why need we *see W as reliable* in order for its reliability to lend us epistemic justification for using it?

Again, suppose we use way W, and that the use of W assures us that W itself *is* reliable. Indeed, consider our situation in the very *best conceivable outcome*. Suppose:

(a) W *is* reliable (and suppose even that, given our circumstances and fundamental nature, it is the *most* reliable overall way we could have).

(b) We are *right* in our description of W: it *is* exactly W that we use in forming beliefs, and it is of course (therefore) W that we use in forming the belief that W is our way of forming beliefs.

(c) We *believe* that W *is* reliable (correctly so, given *a* above), and this belief, too, is formed by means of W.

Now what? Are we still in a 'desperate situation'? What could possibly be missing? How could we possibly improve our epistemic situation?

It might be suggested that perhaps we could still search for some argument that would not be flawed by epistemic circularity. But is such circularity necessarily vicious? After all, what does an argument *ever* accomplish? Suppose you are given argument A with premises P and conclusion C and you correctly accept it as evidently valid. What this gives you in the first instance is the conviction that P entails C. And, unless you go back on this conviction, you are now *restricted* in the combinations of coherent attitudes that are open to you. But that is all that the argument by itself does: i.e., that is all you can derive from its validity. As far as the argument goes, its relevant deliverance is your belief that P entails C, and this justifies your believing C, given that you believe P, only by contrast with believing P and either disbelieving or consciously withholding on C. But it does not justify your believing C, given that you believe P, by contrast with many other optional attitudes: e.g., disbelieving C and disbelieving P. N.B.: it is a kind of intrinsic coherence that lifts the preferable attitudes over the lesser ones: once we have **(a) B:[P entails C]**, we need to avoid **(b) B:P and D:C**, and (c) **B:P and Wh:C**—where D:P means B: ~P, and Wh:P means consciously or deliberately withholding on whether P or ~P. Many other combinations of attitudes remain open options, of course, but so long as we retain (a), both (b) and (c) are to be avoided. Why so? Because they do not cohere well. There is some evident lack of fittingness or harmony in each of them. Here I won't try to spell out the exact nature of the incoherence that attaches not only to (a) & (b) but also to (a) & (c). I'll assume we can agree that it is here, whatever its nature. In fact, it is not really necessary to say anything that strong. A comparative judgment is enough. Consider: **(d) B:P and B:C.** All we need is the judgment that (a) & (d) is more coherent than either of (a) & (b) or (a) & (c). Given (a), which results from our supposed argument above, (d) is lifted over each of (b) and (c) in respect of coherence.

The upshot: all that (the validity of) an argument ever does is to raise some combinations of attitudes (to premises and conclusion respectively) above others in respect of coherence.

But now suppose that by using way W of forming beliefs (which may and probably will include the use of argument) we arrive at the conviction that W is our way of forming beliefs. Now, so long as we do not go back on that conviction, does that not restrict our coherent combinations of attitudes? Take: **(e) B:[W is my overall way of forming beliefs].** And compare **(f) B:[W is reliable], (g) D:[W is reliable]** and **(h) Wh:[W is reliable].** Is it not evident that (e) & (f) would be more satisfyingly coherent than either of (e) & (g) or (e) & (h)?

If so, the question arises: Just how would any further argument provide a fundamentally different and superior source of justification or rationality for our accepting the reliability of our overall way W of forming beliefs, as compared with what we are provided already by our conviction that W is indeed that overall way of ours?

The answer might come back: 'But once we had an argument A for W being reliable from premises already accepted, we would embed our faith in W's reliability within a more comprehensively coherent whole that would include the premises of our argument A.' And it must be granted that such an argument *would* bring that benefit. However: we know that such an argument would *have* to be epistemically circular, since its premises can only qualify as beliefs of ours through the use of way W. That is to say, a correct and full response to rational pressure for disclosure of what justifies one in upholding the premises must circle back down to the truth of the conclusion. *Necessarily* such an argument must be epistemically circular—that much seems clear enough. To rue that fact at this stage is hence like pining for a patron saint of modesty (who blesses all and only those who do not bless themselves), once we have seen that there could not possibly be such a saint.

Perhaps the dissatisfaction emphasized by Alston and Stroud, and many others, has a different source than any we have considered. Perhaps it arises from the following reasoning:

> If we justify our belief in the reliability of our W—B:R(W)—by noting that W itself yields B:R(W), then anyone with a rival but self-supporting method W* would be able to attain an equal measure of justification through parallel reasoning. They would justify their belief B:R(W*) by noting that W* itself yields B:R(W*). So are we not forced to conclude that someone clever enough could attain a measure of rational justification equal to ours so long as their way of forming beliefs, W*, turned out to be, to the same extent, coherently and comprehensively self-supporting?

If *this* is the source of the discomfort, then it is discomfort we must learn to tolerate—though in time reason should be able to dispel it, just as it would dispel any desire to meet the saint who blesses all and only the nonselfblessed. After all, discursive, inferential reason is not our only faculty; and logical brilliance does not even ensure sanity. In light of this, I see no sufficient argument why we must settle, at the end of the day, for any irresolvable theoretical frustration.

We need to distinguish the internal justification that amounts to rational coherence, or even to rational coherence plus rational intuition, from the broader intellectual virtue required for knowledge. In order to know that p, one's belief must not fail the test of rational, internal coherence. But it must be tested in other ways as well: it must be true, for one thing. And, more than that, it must be *apt:* it must be a belief that manifests overall intellectual virtue, and is not flawed essentially by vice. (Mr. Magoo can infer brilliantly and a belief of his can manifest *that* virtue, while it is still flawed by epistemic vice and fails to manifest overall virtue.) Finally, if it is to amount to knowledge a belief must be such that, in the circumstances it *would* be held by that subject iff it were true, and this in virtue of its being apt in the way that it is apt, in virtue of deriving from the complex of virtues that form it and sustain it.

Suppose we are rationally justified in accepting the reliability of our way of forming beliefs W, and suppose our justification derives from the way that very belief coheres within our overall body of beliefs.

Then we do of course commit ourselves to the consequence that anyone intelligent enough to secure an equal measure of coherence for their body of beliefs would attain thereby a comparable degree of rational justification for their belief in the reliability of their way of forming beliefs (a belief we may assume to be already part of their corpus). And this remains so even if their way amounts on the whole to madness! For in granting them logical coherence we need not grant thereby that there is *no* epistemically pertinent distinction between them and us. There are faculties other than reason, surely, and there is plenty of scope for madness and other vices beyond the ability to spin a coherent story.

To sum up: We can legitimately and with rational justification arrive at a belief that a certain set of faculties or doxastic practices are those that we employ *and* are reliable. That remains so, even though someone mad can weave a system of comparable internal coherence and can thereby attain a comparable degree of internal justification. But in granting this we must not grant that such coherently rational belief need only be true in order to be knowledge. A coherently rational belief can fail to be apt, surely, and can even be mad if formed by a mind that is brilliantly logical though deranged in its social and physical perception and perhaps also in its memory. (A rationally coherent belief *can* also be apt, of course, and can thereby amount to knowledge as well.) Anyhow, the point remains: there is no obstacle in principle to our conceivably attaining rationally coherent belief in some general account of our own epistemic faculties and their reliability. This would be bread, not a stone (or a sheep, not a goat). Why could we not conceivably attain thereby a general understanding of how we know whatever we do know?

We have also felt the attraction of Stroud's reasoning, however: his brief for a very general and fundamental doubt against our ever conceivably attaining any such general understanding.

Stroud's reasoning, and that of many others along the historical length and contemporary breadth of philosophy, may perhaps return us to an assumption that seems questionable: the questionable assumption that a satisfyingly general philosophical account of human knowledge would have to be a legitimating account that would reveal how all such knowledge can be traced back to some epistemically prior knowledge from which it can be shown to be derived (without logical or epistemic circularity).[33] There is no good reason to make this assumption, especially when it is evident that no such general account of all our knowledge could conceivably be obtained.

The desire for a fully general, legitimating, philosophical understanding of all our knowledge is unfulfillable. It is unfulfillable for simple, demonstrable logical reasons. In this respect it is like the desire to find the saint who blesses all and only the nonselfblessed. A trek through the Himalayas may turn up likely prospects each of whom eventually is seen to fall short, until someone in the expedition reflects that there could not possibly be such a saint, and this for evident, logical reasons. How should they all respond to this result? They may of course be very unhappy to have been taken in by a project now clearly defective, and this may leave them frustrated and dissatisfied. But is it reasonable for them to insist that somehow the objective is still worthy, even if unfortunately it turns out to be incoherent? Is this a sensible response? How would we respond if we found ourselves in that situation? Would it not be a requirement of good sense or even of sanity to put that obviously incoherent project behind us, to just forget about it and to put our time to better use? And is that not what we must do with regard to the search for fully general, legitimating, philosophical accounts of our knowledge?

If it does not just return us to that questionable assumption, however, then what can be the basis for the objection to a general theory of knowledge, indeed to one so general that it encompasses not only all our knowledge of the external but all of our knowledge in general? Suppose one's belief in one's theory takes the following form:

T   A belief X amounts to knowledge if and only if it satisfies conditions C.

It would not be long before a philosopher would wonder in virtue of what T itself is a piece of knowledge, and if T is held as an explanatory theory for all of our knowledge, then the answer would not be far to seek: T is a piece of knowledge because T itself meets conditions C. And how do we know that T meets conditions C? Well, of course, *that* belief itself must meet conditions C in turn. And so on, without end. Is there any unacceptability in principle here, is there any unavoidable viciousness? Compare the following three things.

E   A belief B in a general epistemological account of when beliefs are justified (or apt) that applies to B itself and explains in virtue of what it, too, is justified (apt).

G   A statement S of a general account of when statements are grammatical (or a sentence S stating when sentences are grammatical) that applies to S itself and explains in virtue of what it, too, is grammatical.

P A belief B in a general psychological account of how one acquires the beliefs one holds, an account that applies to B itself and explains why it, too, is held.

Why should E be any more problematic than G or P? Why should there be any more of a problem for a general epistemology than there would be for a general grammar the grammaticality of whose statement is explained in turn by itself, or for a general psychology belief in which is explained by that very psychology?

It must be granted that what we want is a sort of explanation that would in principle enable us to understand how we have any knowledge at all. Question: "Why are there chickens?" Answer: "They come from eggs." "And why are there eggs?" "They come from chickens." This exchange could not provide a complete answer to a child's question, if the question is, more fully, that of why there are chickens *at all, ever.* To answer this question we need appeal to divine creation, or evolution, or anyhow to something entirely other than chickens. Consider now the analogous question about knowledge, about the sources of the epistemic status of our knowledgeable beliefs (and not now about the causal sources of their existence). A complete answer for this question must appeal to something other than beliefs claimed already to enjoy the status of knowledge. For we want an explanation of how beliefs *ever* attain that status *at all.*

It is important to avert a confusion. We shall never be able really to *have* an explanation of anything without our *having* some knowledge, the knowledge that constitutes our having the explanation, knowledge like

K X is the case in virtue of such and such.

Though we must have such knowledge if we are to understand why X is the case, however, there is no need to include any attribution of knowledge in the explanans of K, in the 'such and such.' The concept of knowledge need not be part of that explanans. Compare again our general theory of knowledge schema:

T A belief X amounts to knowledge if and only if it satisfies conditions C.

T is something we must *know* if it is to give us real understanding, and in offering it we are perhaps, in some sense, 'presupposing' that we know it. This does not mean that our theory must be less than fully general. Our theory T may still be fully general so long as no epistemic status—e.g., knowledge, or justification—plays any role in the 'conditions C' that constitute the explanans of T.

It is true that in epistemology we want *knowledgeable* understanding, and not just 'understanding by dumb luck' (which, in the relevant sense, is incoherent anyhow, and hence not to be had). But there is no apparent reason why we cannot have it with a theory such as T, without compromising the full generality of our account. Of course in explaining how we know theory T, whether to the sceptic or to ourselves, we have to appeal to theory T itself, given the assumptions of correctness and full generality that we are making concerning T. Given those assumptions there seems no way of correctly answering such a sceptic except by 'begging the question' and 'arguing circularly' against him. But, once we understand this, what option is left to us except to go ahead and 'beg' that question against *such* a sceptic (though 'begging the question' and 'arguing circularly' may now be misnomers for what we do, since it is surely no fallacy, not if it constitutes correct and legitimate intellectual procedure). Nor are we, in proceeding thus, by means of a self-supporting argument, assuming that *all self-supporting arguments are on a par*. This would be a serious mistake. It is not just *in virtue of being self-supporting* that our belief in T would acquire its epistemic status required for knowledge. Rather it would be in virtue of meeting conditions C.[34] And conditions C must not yield that a belief for a system of beliefs has the appropriate positive epistemic status provided simply that it is self-supporting. For this would obviously be inadequate. Therefore, our belief in T *would* be self-supporting, as had better be any successful and general theory of knowledge, but it would not amount to knowledge or even to a belief with the appropriate epistemic status, *simply in virtue of being self-supporting*.

In all our reflection and in all our discussion of objections to externalism we have found no good argument for the view that epistemically circular arguments must be disqualified globally as ineffectual in making discriminations between reliable and unreliable doxastic practices. Nor have we been able to find any good reason to yield to the sceptic or to reject externalist theories of knowledge globally and antecedently as theories that could not possibly give us the kind of understanding of human knowledge in general that is a goal of epistemology. And so we have found no good reason to accept *philosophical scepticism,* the main target thesis of this paper. As for any desperate retreat to relativism or ethnocentrism, finally, that now seems ill-conceived and imperceptive.[35] I mean the retreat into relativism that sees no way of adjudicating through reason among clashing, equally coherent systems. The recoil to ethnocentrism (or the like) betrays a rationalist *malgré lui* with no objective way to adjudicate except reason.[36] Who but a philosopher could expect so much from reason?[37] What privileges our positions, if anything does, cannot be that it is self-supportive, as we have seen; *but nor can it possibly be just that it is ours.* Our position would be privileged rather by deriving from cognitive virtues, from the likes of perception and cogent thought, and not from derangement or superstition or their ilk.[38]

Notes

1. Donald Davidson, 'A Coherence Theory of Truth and Knowledge,' in *Kant oder Hegel,* ed. Dieter Henrich (Stuttgart: Klett-Cotta, 1983), pp. 423–438; p. 426.

2. *Ibid.,* p. 431.

3. Richard Rorty, *Philosophy and the Mirror of Nature* (Princeton, N.J.: Princeton University Press, 1979), p. 178.

4. *Ibid.,* p. 159.

5. Richard Rorty, *Contingency, Irony, and Solidarity* (Cambridge, UK: Cambridge University Press, 1989), p. 48. Note the ambiguity between 'reasons for using languages' that one *has* and adduces, versus reasons that there are whether or not one has them or adduces them. And note also the assumption that only what is based on reasonings from adduced reasons can be assessed as 'rational.' (One might of course yield the vocabulary of the 'rational' in the face of such uninhibited assumptions, for the sake of the conversation, so long as one could still distinguish among beliefs, and even among 'choices of vocabulary,' those that are 'apt,' in some apt sense, from those that are not.)

6. Rorty, *Philosophy and the Mirror of Nature,* p. 152.

7. Laurence BonJour, *The Structure of Empirical Knowledge* (Cambridge, MA: Harvard Univ. Press, 1985), p. 31.

8. Michael Williams, 'Epistemological Realism and the Basis of Scepticism,' *Mind* 97 (1988), p. 246. (This paper sketches a view developed and defended in his *Unnatural Doubts* (Oxford: Blackwell Publishers, 1992).) Here Williams is attributing a view to Stroud. But in his paper (and in his book) he evidently agrees that if there were a way of attaining a general philosophical understanding of our knowledge of the world, it would have to be in terms of a legitimating account; and he does not take seriously the possibility of a substantially externalist account.

9. Such overreaction against objective foundations may drive even some-one brilliant to unfortunate excesses. Compare the writings of Paul Feyera-bend. Moreover, the sort of internalism that enforces capitulation to 'circular-ity'-wielding relativists is not confined to the *avant-garde* we have already consulted. For just one example, earlier in the century, in an otherwise most illuminating paper, Alan Gewirth had this to say: 'Consequently, it is circular to say that the basic principles of science are themselves cognitive; for it is these principles or norms which determine whether anything else is to be called cognitive. Moreover, these principles are a selection from among other possible principles—possible, that is, in the sense that they are espoused by people who claim to have "science" or "knowledge" by methods which are in important respects different from those grounded in inductive and deductive logic. These other methods include those of Christian Science, astrology, phrenology, tribal medicine-men, and many others. Each of these other methods has its own way of defining what is to be meant by "fact," "knowl-edge," and so forth. Hence, if any of these latter is to be called "noncogni-tive," it will be by reference not to *its* norms or principles but to those of *some* other way of viewing "science" or "knowledge." To claim that any of those is "absolutely" non-cognitive is to ignore the relativity of all claims of cognitiveness to norms or principles which define what is to be meant by "cognitive." . . . Hence, strictly speaking, the choice among different concep-tions of "knowledge" or "science" cannot itself be said to be made by cognitive means.' (From A. Gewirth, 'Positive "Ethics" and Normative "Sci-ence",' *The Philosophical Review* LXIX (1960); the passage quoted comes from the thirteenth paragraph.) Here again, we might well yield the vocabulary of 'choices made by cognitive means,' so long as we could keep a distinction between choices or commitments that are 'apt' and those that are not, where this is not just something 'relative' to raw or brute or 'arbitrary' commitments.

10. Alston is among those who settle into irresolvable frustration, insofar as he accepts externalism at the cost of a freely avowed dissatisfaction, which, as we shall see, he takes to be inherent in the human *theoretical* condition. Insofar as he tries to struggle against this, it is by conceding the theoretical frustration, and turning to a kind of practical reasonability, in a way we shall consider.

11. The position on these issues of my *Knowledge in Perspective* (Cam-bridge, UK: Cambridge University Press, 1991), has repeatedly drawn an objection (as detailed in notes 19 and 33 below) that we shall consider in what follows.

12. Barry Stroud, 'Understanding Human Knowledge in General,' in *Knowledge and Scepticism*, ed. by Marjorie Clay and Keith Lehrer ((Boulder: Westview, 1989), p. 43.

13. Compare p. 44, *ibid.:* '[Descartes is] . . . a theorist of knowledge. He wants to understand how he knows the things he thinks he knows. And he cannot satisfy himself on that score unless he can see himself as having some

reason to accept the theory that he (and all the rest of us) can recognize would explain his knowledge if it were true. That is not because knowing implies knowing that you know. It is because having an explanation of something in the sense of understanding it is a matter of having good reason to accept something that would be an explanation if it were true.'

14. Compare here again the passages from Davidson and Rorty cited earlier, and the consequences drawn by Rorty not only for theory but also for praxis.

15. *Ibid.*, p. 45.

16. *Ibid.*, p. 46.

17. *Ibid.*, p. 47.

18. Compare Stroud on this: 'We want witting, not unwitting, understanding. That requires knowing or having some reason to accept the scientific story you believe about how people know the things they know. And in the case of knowledge of the world around us, that would involve already knowing or having some reason to believe something in the domain in question. Not all the knowledge in that domain would thereby be explained.' (*Ibid.*, p. 48.) Also: 'The demand for completely general understanding of knowledge in a certain domain requires that we see ourselves at the outset as not knowing anything in that domain and then coming to have such knowledge on the basis of some independent and in that sense prior knowledge or experience . . . [When] we try to explain how we know . . . things [in a domain we are interested in] we find we can understand it only by assuming that we have got some knowledge in the domain in question. And that is not philosophically satisfying. We have lost the prospect of explaining and therefore understanding all of our knowledge with complete generality.' (*Ibid.*, pp. 48–9.)

19. W.P. Alston, *Perceiving God: The Epistemology of Religious Experience* (Ithaca: Cornell University Press, 1991), p. 148. In a review of my *Knowledge in Perspective,* in *Mind* 102 (1993): 199–203, Alston adds that 'it is plausible to suppose that we cannot give an impressive argument for the reliability of sense perception without making use of what we have learned from sense perception. This problem affects Sosa's view as much as it does any other form of externalism that requires for justification or knowledge that the source of a belief be truth-conducive. To apply Sosa's view we would have to determine which belief forming habits are intellectual virtues, i.e., which can be depended on to yield mostly true beliefs. Doesn't epistemic circularity attach to these enterprises, by his own showing? What does he have to say about that?'

20. W.P. Alston, *The Reliability of Sense Perception* (Ithaca: Cornell University Press, 1993).

21. *Ibid.*, p. 17. The problem is supposed to arise from the fact that the data on the basis of which a track-record argument reaches the conclusion that a certain doxastic practice DP is reliable, are data that derive (at some remove, if not immediately) from the use of that very practice DP.

22. *Ibid.*, p. 120. My emphasis.

23. *Ibid.*, p. 130.

24. *Ibid.*, p. 7.
25. *Ibid.*, p. 131.
26. *Ibid.*, p. 127.
27. *Ibid.*, p. 125.
28. *Ibid.*
29. *Ibid.*, p. 126.
30. *Ibid.*, p. 15.
31. *Ibid.*, p. 17.

32. I will use 'firmly established' here as short for 'firmly established in the way described more fully by Alston and proposed by him as sufficient for practical reasonableness or rationalisty'.

33. See p. 6 above. And compare Paul Moser's statement of the difficulty as he sees it (Paul Moser, Review of *Knowledge in Perspective*, in *Canadian Philosophical Reviews* XI (1991): 425–7): 'What . . . can effectively justify one's meta-belief in the virtue of memory? What can effectively justify the claim that "the products of such faculties are likely to be true"? These questions . . . ask what, if anything, can provide a cogent defense of the alleged reliability of memory against familiar sceptical queries . . . The . . . questions ask not for absolute proof, but for a non-questionbegging reason supporting the alleged reliability of memory, a reason that does not beg a key question against the sceptic. It is doubtful that we can deliver such a reason; coherence of mere beliefs will surely not do the job.' T.E. Wilkerson also joins the broad consensus against the supposed 'circularity' in externalism: 'How can I know that I am intellectually virtuous, that I have a settled ability or disposition to arrive at the truth? Indeed, how do I know that I have arrived at the truth? As Sosa points out, it is no good to answer that my beliefs are true in so far as they are justified by other beliefs: that way lies *either* old-fashioned foundationalism *or* coherentism. Nor presumably is it any good to say that they are justified because they have been acquired in an intellectually virtuous way: the circle seems swift and unbreakable.' (Review of *Knowledge in Perspective*, in *Philosophical Books* 33(1992): 159–61.)

34. This seems the key to an answer for Alston's charge that epistemically circular arguments 'do not serve to discriminate between reliable and unreliable doxastic practices,' cited earlier. One can make such discriminations with epistemically circular arguments (ones with premises that are in fact true and justified, etc.) even if it is not the circular character of the reasoning that by itself effects the discrimination.

35. And a similar objection can be lodged, based on similar reflections, on the analogous retreat in moral and political philosophy.

36. About other views that are in some way 'relativist' I remain silent.

37. And, besides, the irrationalist cannot be answered nonquestionbeggingly *anyhow*, not if our answer presupposes the validity of reason. When thought through, the requirement of nonquestionbegging defensibility against all conceivable comers is ill-advised, and indeed incoherent. But once we see why

that is so, we should see also that reason cannot plausibly be held above perception or memory as a proper source of epistemic status.

38. 'But that bare assertion is so empty! Which are these virtues? What means this cogency? What else is involved?' To this reaction the response would have to be a very long story, if told in full, one that turns now longer, now shorter, with every advance in our understanding of ourselves and our thought and our environment and our origins, and the relations among all these. One's epistemic perspective is joined indispensably to one's broader worldview.

15

Scepticism, 'Externalism', and the Goal of Epistemology

Barry Stroud

Scepticism has been different things at different times in the history of philosophy, and has been put to different uses. In this century it has been understood primarily as a position—or threat—within the theory of knowledge. It says that nobody knows anything, or that nobody has good reason to believe anything. That view must be of central significance in epistemology, given that the goal of the enterprise is to explain how we know the things we think we do. It would seem that any satisfyingly positive theory of knowledge should imply the falsity of scepticism.

Scepticism need not always be taken as completely general. It has more typically been restricted to this or that particular kind of alleged knowledge or reasonable belief: we have no reason to believe anything about the future, for example, even if we know a great deal about the past and the present; we know nothing about the world around us, although we know what the course of our own experience is like; or I know what the physical world and my own thoughts and experiences are like, but I know nothing about the minds of other persons. Scepticism is most illuminating when restricted to particular areas of knowledge in this way because it then rests on distinctive and problematic features of the alleged knowledge in question, not simply on some completely general conundrum in the notion of knowledge itself, or in the very idea of reasonable belief. It is meant to be a theory about human beings as they actually are, and about the knowledge we think we actually have in the circumstances in which we find ourselves.

Reprinted by courtesy of the Editor of the Aristotelian Society from *Proceedings of the Aristotelian Society*, Supplementary Volume 68 (1994), pp. 291–307. © 1994.

Scepticism in the theory of knowledge involves much more than the bare assertion that no one knows anything or has any reason to believe anything of a particular kind. If all animate life were suddenly (or even gradually) wiped off the face of the earth no one would then know anything or have any reason to believe anything about the world, but that would not make scepticism about the external world true. A philosophical theorist wants to understand human knowledge as it is, as human beings and the world they live in actually are. But again not just any denial of human knowledge in a certain domain counts as philosophical scepticism. Human beings as they are right now do not know the causes of many kinds of cancer, or of AIDS, or the fundamental structure of matter. But universal ignorance in a particular domain does not make scepticism true of that domain. Scepticism holds that people as they actually are fail to know or have good reason to believe the sorts of things we all think we already know right now. Anti-scepticism, or a positive theory of knowledge, holds the opposite. It would explain how human beings, equipped as they are and living in the world they live in, do in fact know the sorts of things they think they do.

Theories of knowledge which conflict in this way nevertheless typically share many assumptions about human beings and their cognitive and perceptual resources. It is agreed on all sides, for example, that if human beings know things about the world around them, they know them somehow on the basis of what they perceive by means of the senses. The dispute then turns on whether and how what the senses provide can give us knowledge or good reason to believe things about the world. Knowledge of matters which go beyond perception to the independent world is seen, at least temporarily, as problematic. A successful positive theory of knowledge would explain how the problem is solved so that we know the things we think we know about the world after all.

It must be admitted, I think, that what many philosophers have said about perceptual knowledge is pretty clearly open to strong sceptical objections. That is, *if* the way we know things about the world is the way many philosophers have said it is, *then* a good case can be made for the negative sceptical conclusion that we do not really know such things after all. That is why scepticism remains such a constant threat. If you don't get your description of the human condition right, if you describe human perception and cognition and reasoning in certain natural but subtly distorted ways, you will leave human beings as you describe them incapable of the very knowledge you are trying to account for. A sceptical conclusion will be derivable from the very description which serves to pose the epistemological problem. Thus did the ancient sceptics argue, conditionally, against the Stoics: "if human knowledge is arrived at in the way you say it is, there could be no such thing as human knowledge at all." Even if true, that does not of course show that scepticism is correct. It shows at most that human knowledge or the human condition must be understood in some other way. The threat of scepticism is what keeps the theory of knowledge going.

The point is that scepticism and its competitors among more positive theories of knowledge are all part of the same enterprise. They offer conflicting answers to what is for all of them a common question or set of questions. The task is to understand all human knowledge of a particular kind, or all reasonable belief concerning a certain kind of matter of fact. Scepticism is one possible outcome of that task. In that sense, scepticism, like its rivals, is a general theory of human knowledge. But it is not a satisfactory theory or outcome. It is paradoxical. It represents us as having none of the knowledge or good reasons we ordinarily think we've got. No other theory or answer is satisfactory either if it does not meet and dispel the threat of scepticism. I think many philosophical theories of knowledge have failed to do that, despite what their defenders have claimed for them.

In fact, I find the force and resilience of scepticism in the theory of knowledge to be so great, once the epistemological project is accepted, and I find its consequences to be so paradoxical, that I think the best thing to do now is to look much more closely and critically at the very enterprise of which scepticism or one of its rivals is the outcome: the task of the philosophical theory of knowledge itself. Its goal is not just any understanding of human knowledge; it seeks to understand knowledge in a certain way. Both scepticism and its opposites claim to understand human knowledge in that special way, or from that special philosophical point of view. I would like to inquire what that way of understanding ourselves and our knowledge is, or is supposed to be. I wonder whether there is a coherent point of view from which we could get a satisfactory understanding of ourselves of the kind we apparently aspire to. Many would dismiss scepticism as absurd on the grounds that there is no such point of view, or that we could never get ourselves into the position of seeing that it is true if it were true. But to adopt a more positive theory of knowledge instead is still to offer a description of the human condition from that same special position or point of view. If we cannot get into that position and see that scepticism is true, can we be sure that we can get into it and see that scepticism is false?

The coherence and achievability of what we aspire to in the epistemological enterprise tends to be taken for granted, or left unexplored. But that question is prior to the question whether scepticism or one or another of its positive competitors is the true theory of human knowledge. What does a true theory of knowledge do? What does a philosophical theorist of knowledge seek?

These are large and complicated questions to which we obviously cannot hope to get a definitive answer today. Distinguishing them from the question of the relative merits of scepticism and its competitors might nonetheless help to locate the target of Ernest Sosa's opposition to something he calls 'scepticism'. He gives that label to the view that "there is no way to attain full philosophical understanding of our knowledge" or that "a fully general theory of knowledge is impossible."[1] That is obviously not what I have just called 'scepticism', which is itself a fully general theory of knowledge. Sosa considers a two-step argument for the view he has in mind which would show exhaustively that any general theory of knowledge possessing a certain feature would be what he calls "impossible," and that any general theory lacking that feature would be "impossible" too. So there couldn't be a fully general theory of knowledge. The conclusion certainly does follow from those two premisses, but Sosa doubts the second premiss. He thinks some theories which lack the feature in question have not been shown to be defective in the way the original argument was meant to show. The surviving theories are what he calls 'externalist'.

Theories of the first type hold that a belief acquires the status of knowledge only by "being based on some justification, argument, or reason."[2] That requirement is what makes them "impossible," according to Sosa, because in order to succeed they would have to show that our acceptance of the things we think we know is justified in each case by good inferences or arguments which are not circular or infinitely regressive. That is what it would take to 'legitimate' those beliefs, and that cannot be done. Every inference has to start from something, so without circular or regressive reasoning there must always be something whose acceptance by us is left unsupported by inference, and so cannot be accounted for as knowledge by theories of this type. But a fully general theory of knowledge must account for everything we know. Sosa concludes that there could be no fully "general, legitimating, philosophical understanding of all one's knowledge."[3] This is equivalent, I believe, to saying that no such theory avoids the conclusion that we know nothing. What he is saying of theories of this first type is that if, in order to know things, we had to satisfy what those theories say are conditions of knowledge, then we would not know anything, since we cannot satisfy those conditions. So theories of the first type depict us as knowing nothing. They cannot be distinguished, in their consequences, from the view that I (but not Sosa) have called 'scepticism'.

I take it to be the main point of Sosa's paper to show that certain "externalist, reliabilist" theories escape that fate. They can be fully general and still succeed where theories of other kinds fail. He thinks there is "a very wide and powerful current of thinking [which] would sweep away externalism root and branch,"[4] and he wants to resist that "torrent of thought".[5] He concentrates here on the reasons he thinks William Alston and I have given for thinking that, as he puts it, "externalism will leave us ultimately dissatisfied."[6] He appears to equate that charge with what he calls the "unacceptability in principle"[7] of 'externalism'.

What exactly are these objections? For my part, I do think there is a way in which 'externalism' would leave us "ultimately dissatisfied" as an answer to the completely general philosophical question of how any knowledge of the world is possible. I tried to indicate what I have in mind in the paper that Sosa refers to and discusses.[8] But I do not suggest that 'externalism' is unsatisfying because it cannot avoid depicting us as knowing nothing about the world and so is indistinguishable from the view that I call 'scepticism'. Nor would I argue that it is inconsistent or viciously circular or internally deficient in some other way which prevents it in principle from being true or acceptable. Sosa says the objections are "grounded in what seem to be demands inherent to the traditional epistemological project itself,"[9] and I think his efforts to meet the objections are intended to defend not only 'externalist' theories but also by implication that very epistemological project as well. My own doubts about 'externalism' could perhaps be said to be grounded in or at least connected with demands inherent to that project, but that is because they are doubts not only about 'externalism' but about the coherence or feasibility of the general epistemological project itself. That question is what I think should be our primary target, not just one or another of the answers offered to it. We need to examine more critically what we want or hope for from the traditional epistemological project of understanding human knowledge in general.

Alston's objections might well have a different source. I suspect that in opposing externalism as he does he is working towards what he sees as a more adequate theory of knowledge, perhaps one which would recognize some beliefs as evident or *prima facie* justified in a way that externalism cannot explain. But to support a theory that competes with pure externalism as the right answer to the philosophical question is not to bring that whole philosophical project itself into question. Although I think there are many points on which we would agree, I shall therefore leave Alston to one side. That leaves me with the question: does Sosa's defence of externalism show that it does not have that feature which I think means it must always leave us dissatisfied, and so by implication that the goal of epistemology must always leave us dissatisfied as well, or does he really accept the point and not regard it as a deficiency in his externalist theory?

The question is complicated because Sosa sees opposition to externalism as coming from some competing philosophical conception or theory of knowledge. His defence amounts to arguing that any theory from which the objections could come must be a theory of his first general type, and so can be discredited "for simple, demonstrable logical reasons."[10] If it is a conflict between competing theories of knowledge, 'externalism' must win, since it does not have the fatal defect those other theories have. In order to bring out my doubts about the kind of satisfaction offered by 'externalism' I can grant that point. I would like to reveal something that I think remains unsatisfying about 'externalism' even if it is the best philosophical theory of knowledge there is or could be. I do not want to put a better theory in its place; I want to ask what a philosophical theory of knowledge is supposed to be, even at its best. Revealing the unsatisfactoriness of even the best answer to the philosophical question can perhaps help draw attention to its unsatisfiable demands.

We aspire in philosophy to see ourselves as knowing all or most of the things we think we know and to understand how all that knowledge is possible. We want an explanation, not just of this or that item or piece of knowledge, but of knowledge, or knowledge of a certain kind, *in general*. Take all our knowledge of the world of physical objects around us, for example. A satisfactory theory or explanation of that knowledge must have several features. To be satisfyingly positive it must depict us as knowing all or most of the things of that sort that we think we know. It must explain, given what it takes to be the facts of human perception, how we nonetheless know the sorts of things we think we know about the world. To say simply that we see, hear, and touch the things around us and in that way know what they are like, would leave nothing even initially problematic about that knowledge. Rather than explaining how, it would simply state that we know. There is nothing wrong with that; it is true, but it does not explain how we know even in those cases in which (as we would say) we are in fact seeing or hearing or touching an object. That is what we want in a philosophical explanation of our knowledge. How, given what perception provides us with even in such cases, do we thereby know what the objects in question are like? What needs explanation is the connection between our perceiving what we do and our knowing the things we do about the physical objects around us. How does the one lead to, or amount to, the other?

Suppose there is an 'externalist, reliabilist' theory of the kind Sosa has in mind which accounts for this. I mean suppose there are truths about the world and the human condition which link human perceptual states and cognitive mechanisms with further states of knowledge and reasonable belief, and which imply that human beings acquire their beliefs about the physical world through the operation of belief-forming mechanisms which are on the whole reliable in the sense of giving them mostly true beliefs. Let us not pause over details of the formulation of such truths, although they are of course crucial and have not to this day been put right by anybody, as far as I know. If there are truths of this kind, although no one has discovered them yet, that fact alone obviously will do us no good as theorists who want to understand human knowledge in this philosophical way. At the very least we must believe some such truths; their merely being true would not be enough to give us any illumination or satisfaction. But our merely happening to believe them would not be enough either. We seek understanding of certain aspects of the human condition, so we seek more than just a set of beliefs about it; we want to know or have good reason for thinking that what we believe about it is true. This is why I say, as Sosa quotes me: "we need some reason to accept a theory of knowledge if we are going to rely on that theory to understand how our knowledge is possible."[11]

Sosa does not dispute that as a condition of success for understanding human knowledge. He disputes my going on to say that "no form of 'externalism' can give a satisfactory account"[12] of our having such a reason to accept it and so understanding our knowledge of the world in purely 'externalist' terms. He thinks my only support for that second claim comes from what he calls a *'metaepistemic requirement'*[13] which does not follow from the conditions of success admitted so far. It comes, he thinks, from "a deeply held intuition that underlies a certain way of thinking about epistemology."[14] He thinks I have an 'anti-externalist' conception of knowledge according to which 'what is important in epistemology is justification,' which in turn requires "appeal to *other* beliefs that constitute one's reasons for holding the given belief."[15] That is what can only lead in a circle or down an infinite regress, and so in Sosa's terms it is an "impossible" theory of knowledge. Without that requirement, he thinks, the objection vanishes.

Now I want to say that I do not accept any of that. As far as I know, I do not hold an 'anti-externalist' theory of knowledge with which I seek to oppose 'externalism'. I do not think that everything a person knows requires justification which involves appeal to other beliefs, and so on. I think that what I am drawing attention to about 'externalism' is something that can be recognized by anyone who has a good idea of what the general epistemological project is after. Of course, it could be that I am unwittingly imposing the 'anti-externalist' requirement that Sosa's diagnosis says I am. He thinks I must be; I don't think I am. But rather than searching my soul, which I am sure would be of limited general interest, let me again present for public assessment the way I think 'externalism' must leave us dissatisfied. I find in .any case that Sosa has not really considered the reasons I actually gave.

We agree that an "externalist" theorist of knowledge must know or have good reason to believe that his explanation of our knowledge of the physical world around us is correct in order to understand in that way how that knowledge is possible. How will he know or have good reason to believe that? Well, his theory is in part a theory of the conditions under which people in fact know or have good reasons to believe things about the world. If that theory is true in particular of the theorist's own acceptance of that theory, then the theorist has what his own theory says is knowledge of or reasonable belief in the truth of that theory. I believe this is the situation Sosa is describing when he says: "We can legitimately and with rational justification arrive at a belief that a certain set of faculties or doxastic practices are those that we employ *and* are reliable."[16] He thinks there is "no obstacle in principle"[17] to our achieving such a state. I do not disagree with that.

That Sosa thinks the resistance to 'externalism' must be based on some such obstacle in principle is suggested by his immediately going on to ask "why could we not conceivably attain thereby a general understanding of how we know whatever we do know?"[18] It is clear that his question at that point is rhetorical. His idea is that if we can have what an 'externalist' theory calls good reason to believe our 'externalist' theory, it could thereby give us a satisfactory general understanding of our knowledge. For me his question is not rhetorical. I think we can see why, even with what counts for an 'externalist' as good reason to believe his theory, there would remain something ineliminably unsatisfactory about the position a theorist would then be in for gaining a philosophical understanding of his knowledge of the physical world in general.

The difficulty I have in mind does not show up in understanding the knowledge which other people, not myself, have about the world. I understand others' knowledge by connecting their beliefs in the right way with what I know to be true in the world they live in. I can discover that others get their beliefs through the operation of belief-forming mechanisms which I can see to be reliable in the sense of producing beliefs which are largely true. But each of us as theorists of knowledge is also a human being to whom our theory of knowledge is meant to apply, so we must understand ourselves as knowers, just as we understand others. *All* human knowledge of the world is what we want to understand.

If I ask of my own knowledge of the world around me how it is possible, I can explain it along 'externalist' lines by showing that it is a set of beliefs I have acquired through perception by means of belief-forming mechanisms which are reliable. Suppose that is what I believe about the connection between my perceptions and the beliefs I acquire about the world. As we saw, my merely happening to believe such a story would not be enough for me to be said to understand in that way how that knowledge is possible. I must know or have good reason to believe that that story is true of me. As a good 'externalist', I do of course believe that I do. I think that I acquired my belief in my 'externalist' explanation of human knowledge by means of perception and of the operation of the same reliable belief-forming mechanisms which give me and others all our other knowledge of the world around us. So I think I do know or have good reason to believe my theory; I believe that I fulfil the conditions which that very theory says are sufficient for knowing or having good reason to believe it. Do I now have a satisfactory understanding of my knowledge of the world? Have I answered to my own satisfaction the philosophical question of how my knowledge of the world is possible? I want to say No.

It is admittedly not easy to describe the deficiency in a few words. It is not that there is some internal defect or circularity in the 'externalist' theory that I believe. Nor is there any obstacle to my believing that theory or even to my having good reasons in the 'externalist' sense to believe it. *If* the theory is true, and *if* I did acquire my belief in it in the way I think I did, *then* I do know or have a good reason to believe it to be true. To appreciate what I still see as a deficiency, or as less than what one aspires to as a philosophical theorist of knowledge, let us consider the merits of a different and conflicting, but still 'externalist', account of our knowledge of the world.

I have in mind a fictional 'externalist' whom I shall call "Descartes." The theory of our knowledge of the world which he accepts says that there is a beneficent, omnipotent, and omniscient God who guarantees that whatever human beings carefully and clearly and distinctly perceive to be true is true. The real René Descartes held a closely similar theory, but he tried to prove demonstratively that it is true. He was accused of arguing in a circle. My 'externalist' Descartes offers no proofs. He believes that when people carefully and clearly and distinctly perceive things to be true, they are true; God makes sure of that. That is how people come to know things. He also acknowledges that what he himself needs in order to know or have good reason to believe his own theory of knowledge is to fulfil the conditions it says are sufficient for knowing or having good reason to believe something: to acquire belief in it by carefully and clearly and distinctly perceiving it to be true while God guarantees that it is true. Suppose he examines the origins of his own theory and carefully and clearly and distinctly perceives that he did acquire his belief in it in just that way. Does he now have a satisfactory understanding of his knowledge of the world? Has he got what he can see to be a satisfactory answer to the philosophical question of how his knowledge of the world is possible? I want to say No.

Your seeing and sharing my reservations about the adequacy of 'externalism' and so about the feasibility of the epistemological project depend on your finding the position of this 'externalist' Descartes unsatisfactory in a certain way as an understanding of his knowledge. The question is what is wrong with it. I think most of us will say first that what is wrong is that his theory is simply not true; there is no divine guarantor of the truth of even our most carefully arrived-at beliefs, and he is therefore wrong to think that he acquired his belief in his theory in that way. Even if that is so, is it the only deficiency in his position? I think it is not.

We cannot deny that he does believe his explanation of human knowledge, and does believe that he came to believe that theory by a procedure which his theory says is reliable, so we have to admit that *if* his theory and his account of how he came to believe it were true, *then* he would know or have good reason to believe his explanation of knowledge. But if we say that the falsity of his theory is the only deficiency in his position we would have to admit that if his theory and his belief about how he came to believe it were true, then he would have a satisfactory understanding of all of his knowledge of the world. That implies that whether he understands how his knowledge is possible or not depends only on whether the theory which he holds about how he came to believe it is true or not. If it is true, he does understand his knowledge; if it is not, he does not. An 'externalist' theorist of this fictional kind who reflects on his position could still always ask: "I wonder whether I understand how my knowledge of the world is possible? I have a lot of beliefs about it. If what I believe about it is true, I do; if it is not, I don't. Of course, I believe all of it is true, so I believe that I do understand my knowledge. But I wonder whether I do." I think anyone who can get into only that position with respect to his alleged knowledge of the world has not achieved the kind of satisfaction which the traditional epistemological project aspires to. He has not got into a position from which he can see all of his knowledge of the world all at once in a way that accounts for it as reliable or true.

Sosa's 'externalist, reliabilist', I believe, can get himself into no better position for understanding himself. If what distinguishes his position from that of my 'externalist' Descartes is only that his theory is in fact true while that fictional character's theory is false, then he too will be in a position to say no more about himself than "If what I believe about my knowledge is true, I do understand it; if it is not, I do not. I think I do, but I wonder whether I understand my knowledge or not?" This is where the difficulty of describing the deficiency in his position comes in. It will not be true to say simply that although he believes his theory, he has no reason to believe it. If we imagine that his 'externalist' theory and his account of how he came to believe it are in fact true, as I have been conceding, then in that sense he does have good reason to believe his explanation of human knowledge. But still his own view of his position can look no better to him than the fictional 'externalist' Descartes's position looks to him.

It would be to no avail at this point for him to try to improve his position by asking himself whether he knows or has good reason to believe that he does know or have good reason to believe his theory. Answering that question would be a matter of coming by what he believes is a procedure that his theory says is reliable to the belief that he knows or has good reason to believe his theory. Again, if he did come to believe that in that way, and his theory is in fact true, he will in fact know or have good reason to believe a second-order claim about the goodness of his reasons for believing his theory. But still he could then make only the same sort of conditional assertion about his position one level up, as it were, as he made earlier. The 'externalist' Descartes could do the same. He could carefully and clearly and distinctly perceive that he came to believe his theory to be true of himself by what that very theory says is a way of coming to know or have good reason to believe. He could then come to a similarly true conditional verdict about his position. Both he and Sosa's 'externalist' could say at most: "If the theory I hold is true, I do know or have good reason to believe that I know or have good reason to believe it, and I do understand how I know the things I do.' I think that in each case we can see a way in which the satisfaction the theorist seeks in understanding his knowledge still eludes him. Given that all of his knowledge of the world is in question, he will still find himself able to say only "I might understand my knowledge, I might not. Whether I do or not all depends on how things in fact are in the world I think I've got knowledge of."

Those of us who are inclined to think that Sosa's 'externalist's' theory is in fact true and the fictional Descartes's theory false will say that he does know and perhaps that he does understand his knowledge and that the fictional Descartes does not. But that does not show that that theorist's position gives him a satisfactory understanding of his own knowledge. As I said, the difficulty does not show up in one's understanding all of someone else's knowledge of the world; it is only when each of us seeks to understand our own knowledge of the world in general that we reach this unsatisfactory position.

If we do recognize a certain ineliminable dissatisfaction in any such 'externalist' attempt at self-understanding, I do not think it is because of hidden attachment to an opposing 'internalist' theory which requires that everything we know must be justified by reasonable inference from something else we believe. We can be 'externalists' and still reach at best what I think is an unsatisfactory position, even if we do in fact have what 'externalism' regards as knowledge of or reasonable belief in that 'externalist' theory. I think the dissatisfaction, if we recognize it, is felt to come from the demands of the epistemological project itself, or perhaps we could say from the complete generality of the project. Whatever we seek, and what the theorists I have imagined appear to lack, is something that 'externalism' alone seems unable to explain or to account for.

Sosa grants that the epistemological goal can never be reached if the successful theory is expected to provide what he calls a "legitimating" account. He means by that an account which "specifies the reasons favouring one's beliefs,"[19] and he thinks no theory that is 'internalist' in his sense can do that without circularity or regress. But surely the goal of understanding how we know what we do does require that the successful account be legitimating at least in the sense of enabling us to understand that what we have got *is* knowledge of, or reasonable belief in, the world's being a certain way. We should be able to see that the view that I call 'scepticism' is not true of us, and we want to understand how we get the knowledge we can see that we've got. 'Externalism' implies that *if* such-and-such is true in the world, *then* human beings do know things about what the world is like. Applying that conditional proposition to ourselves, to our own knowledge of the world, to our own knowledge of how that knowledge is acquired, and so on, even when the antecedent and so the consequent are in fact both true, still leaves us always in the disappointingly second-best position I have tried to illustrate, however far up we go to higher and higher levels of reiterated knowledge or reasonable belief. We want to be in a position knowingly to detach that consequent about ourselves, and at the same time to know and so to understand how any or all of that knowledge of the world comes to be. And that would require appealing to or relying on part of our knowledge of the world in the course of explaining to ourselves how we come to have any knowledge of the world at all.

There are indications that Sosa acknowledges and accepts the situation I have tried to describe. Believing that our belief-forming mechanisms are reliable when they are in fact reliable, and coming by what are in fact those very mechanisms to believe that they are reliable, he says, is "the very best conceivable outcome"[20] of the epistemological project. "How could we possibly improve our epistemic situation?" he asks.[21] The thought that someone else could find his own 'epistemic situation' equally good on the basis of a competing theory of knowledge, he admits, might cause some dissatisfaction or discomfort, but he thinks that is "discomfort we must learn to tolerate."[22] He concedes that in explaining, even to ourselves, how we know our 'externalist' theory of knowledge to be true, we must appeal to that very theory, and so cannot avoid, as he puts it, "begging the question" or "arguing circularly"[23] in our attempts to account for our knowledge. But again, he asks, "once we understand this, what option is left to us except to go ahead and 'beg' that question?"[24] I think his thought is that without doing that, we would have no chance of answering the epistemological question at all. We have to "tolerate" the "discomfort" of relying on a "self-supporting argument"[25] for our theory simply because we could not arrive at a "successful and general theory of knowledge"[26] in any other way.

Here, perhaps, we approach something that Sosa and I can agree about. What I have tried to identify as a dissatisfaction that the epistemological project will always leave us with is for him something that simply has to be accepted if we are going to have a fully general theory of knowledge at all. He appears to think, as I do, that it is endemic to the epistemological project itself. We differ in what moral we draw from that thought.

I want to conclude that we should therefore re-examine the source of, and so perhaps find ourselves able to resist, the not-fully-satisfiable demand embodied in the epistemological question. I think its source lies somewhere within the familiar and powerful line of thinking by which all of our alleged knowledge of the world gets even temporarily split off all at once from what we get in perception, so we are presented with a completely general question of how perception so understood gives us knowledge of anything at all in the physical world. If that manoeuvre cannot really be carried off successfully, we have no completely general question about our knowledge of the world to answer. We could still ask how we know one sort of thing about the physical world, given that we know certain other things about it, but there would be no philosophical problem about all of our knowledge of the world in general. What then would 'externalism' or any other fully general theory of knowledge be trying to do?

Sosa wants his 'externalism', even with its admitted "discomfort," to serve as a bulwark against the 'relativism', 'contextualism', and 'scepticism' which he sees as rampant in our culture. I share his dark view of our times, but if those widely-invoked 'isms' are thought of as competing answers to a fully general question about our "epistemic situation" in the world, I think the resistance has to start farther back. It is what all such theories purport to be about, and what we expect or demand that any such theory should say about the human condition, that we should be examining, not just which one of them comes in first in the traditional epistemological sweepstake. In that tough competition, it still seems to me, scepticism will always win going away.

Notes

1. Sosa, 'Philosophical Scepticism and Epistemic Circularity', this volume.
2. Sosa, op. cit.

3. Sosa, op. cit.

4. Sosa, op. cit.

5. Sosa, op. cit.

6. Sosa, op. cit.

7. Sosa, op. cit.

8. 'Understanding Human Knowledge in General', in *Knowledge and Scepticism,* ed. M. Clay & K. Lehrer (Boulder, Colorado: Westview, 1989).

9. Sosa, op. cit.

10. Sosa, op. cit.

11. Sosa, op. cit. quoting from Clay & Lehrer (ed.), p.43.

12. Clay & Lehrer (ed.), p.43.

13. Sosa, 'Philosophical Scepticism and Epistemic Circularity', this volume.

14. Sosa, op. cit.

15. Sosa, op. cit.

16. Sosa, op. cit.

17. Sosa, op. cit.

18. Sosa, op. cit.

19. Sosa, op. cit.

20. Sosa, op. cit.

21. Sosa, op. cit.

22. Sosa, op. cit.

23. Sosa, op. cit.

24. Sosa, op. cit.

25. Sosa, op. cit.

26. Sosa, op. cit.

Part IV

Naturalized Epistemology

16

Epistemology Naturalized

W. V. Quine

Epistemology is concerned with the foundations of science. Conceived thus broadly, epistemology includes the study of the foundations of mathematics as one of its departments. Specialists at the turn of the century thought that their efforts in this particular department were achieving notable success: mathematics seemed to reduce altogether to logic. In a more recent perspective this reduction is seen to be better describable as a reduction to logic and set theory. This correction is a disappointment epistemologically, since the firmness and obviousness that we associate with logic cannot be claimed for set theory. But still the success achieved in the foundations of mathematics remains exemplary by comparative standards, and we can illuminate the rest of epistemology somewhat by drawing parallels to this department.

Studies in the foundations of mathematics divide symmetrically into two sorts, conceptual and doctrinal. The conceptual studies are concerned with meaning, the doctrinal with truth. The conceptual studies are concerned with clarifying concepts by defining them, some in terms of others. The doctrinal studies are concerned with establishing laws by proving them, some on the basis of others. Ideally the more obscure concepts would be defined in terms of the clearer ones so as to maximize clarity, and the less obvious laws would be proved from the more obvious ones so as to maximize certainty. Ideally the definitions would generate all the concepts from clear and distinct ideas, and the proofs would generate all the theorems from self-evident truths.

The two ideals are linked. For, if you define all the concepts by use

Reprinted from Quine, *Ontological Relativity and Other Essays* (New York: Columbia University Press, 1969), 68–90, by permission of the author and the publisher. Copyright 1969, Columbia University Press.

of some favored subset of them, you thereby show how to translate all theorems into these favored terms. The clearer these terms are, the likelier it is that the truths couched in them will be obviously true, or derivable from obvious truths. If in particular the concepts of mathematics were all reducible to the clear terms of logic; then all the truths of mathematics would go over into truths of logic; and surely the truths of logic are all obvious or at least potentially obvious, i.e., derivable from obvious truths by individually obvious steps.

This particular outcome is in fact denied us, however, since mathematics reduces only to set theory and not to logic proper. Such reduction still enhances clarity, but only because of the interrelations that emerge and not because the end terms of the analysis are clearer than others. As for the end truths, the axioms of set theory, these have less obviousness and certainty to recommend them than do most of the mathematical theorems that we would derive from them. Moreover, we know from Gödel's work that no consistent axiom system can cover mathematics even when we renounce self-evidence. Reduction in the foundations of mathematics remains mathematically and philosophically fascinating, but it does not do what the epistemologist would like of it: it does not reveal the ground of mathematical knowledge, it does not show how mathematical certainty is possible.

Still there remains a helpful thought, regarding epistemology generally, in that duality of structure which was especially conspicuous in the foundations of mathematics. I refer to the bifurcation into a theory of concepts, or meaning, and a theory of doctrine, or truth; for this applies to the epistemology of natural knowledge no less than to the foundations of mathematics. The parallel is as follows. Just as mathematics is to be reduced to logic, or logic and set theory, so natural knowledge is to be based somehow on sense experience. This means explaining the notion of body in sensory terms; here is the conceptual side. And it means justifying our knowledge of truths of nature in sensory terms; here is the doctrinal side of the bifurcation.

Hume pondered the epistemology of natural knowledge on both sides of the bifurcation, the conceptual and the doctrinal. His handling of the conceptual side of the problem, the explanation of body in sensory terms, was bold and simple: he identified bodies outright with the sense impressions. If common sense distinguishes between the material apple and our sense impressions of it on the ground that the apple is one and enduring while the impressions are many and fleeting, then, Hume held, so much the worse for common sense; the notion of its being the same apple on one occasion and another is a vulgar confusion.

Nearly a century after Hume's *Treatise,* the same view of bodies was espoused by the early American philosopher Alexander Bryan Johnson.[1] "The word iron names an associated sight and feel," Johnson wrote.

What then of the doctrinal side, the justification of our knowledge of truths about nature? Here, Hume despaired. By his identification of bodies with impressions he did succeed in construing some singular statements about bodies as indubitable truths, yes; as truths about impressions, directly known. But general statements, also singular statements about the future, gained no increment of certainty by being construed as about impressions.

On the doctrinal side, I do not see that we are further along today than where Hume left us. The Humean predicament is the human predicament. But on the conceptual side there has been progress. There the crucial step forward was made already before Alexander Bryan Johnson's day, although Johnson did not emulate it. It was made by Bentham in his theory of fictions. Bentham's step was the recognition of contextual definition, or what he called paraphrasis. He recognized that to explain a term we do not need to specify an object for it to refer to, nor even specify a synonymous word or phrase; we need only show, by whatever means, how to translate all the whole sentences in which the term is to be used. Hume's and Johnson's desperate measure of identifying bodies with impressions ceased to be the only conceivable way of making sense of talk of bodies, even granted that impressions were the only reality. One could undertake to explain talk of bodies in terms of talk of impressions by translating one's whole sentences about bodies into whole sentences about impressions, without equating the bodies themselves to anything at all.

This idea of contextual definition, or recognition of the sentence as the primary vehicle of meaning, was indispensable to the ensuing developments in the foundations of mathematics. It was explicit in Frege, and it attained its full flower in Russell's doctrine of singular descriptions as incomplete symbols.

Contextual definition was one of two resorts that could be expected to have a liberating effect upon the conceptual side of the epistemology of natural knowledge. The other resort is to the resources of set theory as auxiliary concepts. The epistemologist who is willing to eke out his austere ontology of sense impressions with these set-theoretic auxiliaries is suddenly rich: he has not just his impressions to play with, but sets of them, and sets of sets, and so on up. Constructions in the foundations of mathematics have shown that such set-theoretic aids are a powerful addition; after all, the entire glossary of concepts of classical mathematics is constructible from them. Thus equipped, our epistemologist may not need either to identify bodies with impressions or to settle for contextual definition; he may hope to find in some subtle construction of sets upon sets of sense impressions a category of objects enjoying just the formula properties that he wants for bodies.

The two resorts are very unequal in epistemological status. Contextual definition is unassailable. Sentences that have been given meaning as wholes are undeniably meaningful, and the use they make of their component terms is therefore meaningful, regardless of whether any translations are offered for those terms in isolation. Surely Hume and A. B. Johnson would have used contextual definition with pleasure if they had thought of it. Recourse to sets, on the other hand, is a drastic ontological move, a retreat from the austere ontology of impressions. There are philosophers who would rather settle for bodies outright than accept all these sets, which amount, after all, to the whole abstract ontology of mathematics.

This issue has not always been clear, however, owing to deceptive hints of continuity between elementary logic and set theory. This is why mathematics was once believed to reduce to logic, that is, to an innocent and unquestionable logic, and to inherit these qualities. And this is probably why Russell was content to resort to sets as well as to contextual definition when in *Our Knowledge of the External World* and elsewhere he addressed himself to the epistemology of natural knowledge, on its conceptual side.

To account for the external world as a logical construct of sense data—such, in Russell's terms, was the program. It was Carnap, in his *Der logische Aufbau der Welt* of 1928, who came nearest to executing it.

This was the conceptual side of epistemology; what of the doctrinal? There the Humean predicament remained unaltered. Carnap's constructions, if carried successfully to completion, would have enabled us to translate all sentences about the world into terms of sense data, or observation, plus logic and set theory. But the mere fact that a sentence is *couched* in terms of observation, logic, and set theory does not mean that it can be *proved* from observation sentences by logic and set theory. The most modest of generalizations about observable traits will cover more cases than its utterer can have had occasion actually to observe. The hopelessness of grounding natural science upon immediate experience in a firmly logical way was acknowledged. The Cartesian quest for certainty had been the remote motivation of epistemology, both on its conceptual and its doctrinal side; but that quest was seen as a lost cause. To endow the truths of nature with the full authority of immediate experience was as forlorn a hope as hoping to endow the truths of mathematics with the potential obviousness of elementary logic.

What then could have motivated Carnap's heroic efforts on the conceptual side of epistemology, when hope of certainty on the doctrinal side was abandoned? There were two good reasons still. One was that such constructions could be expected to elicit and clarify the sensory evidence for science, even if the inferential steps between sensory evidence and scientific doctrine must fall short of certainty. The other reason was that such constructions would deepen our understanding of our discourse about the world, even apart from questions of evidence; it would make all cognitive discourse as clear as observation terms and logic and, I must regretfully add, set theory.

It was sad for epistemologists, Hume and others, to have to acquiesce in the impossibility of strictly deriving the science of the external world from sensory evidence. Two cardinal tenets of empiricism remained unassailable, however, and so remain to this day. One is that whatever evidence there *is* for science *is* sensory evidence. The other, to which I shall return, is that all inculcation of meanings of words must rest ultimately on sensory evidence. Hence the continuing attractiveness of the idea of a *logischer Aufbau* in which the sensory content of discourse would stand forth explicitly.

If Carnap had successfully carried such a construction through, how could he have told whether it was the right one? The question would have had no point. He was seeking what he called a *rational reconstruction*. Any construction of physicalistic discourse in terms of sense experience, logic, and set theory would have been seen as satisfactory if it made the physicalistic discourse come out right. If there is one way there are many, but any would be a great achievement.

But why all this creative reconstruction, all this make-believe? The stimulation of his sensory receptors is all the evidence anybody has had to go on, ultimately, in arriving at his picture of the world. Why not just see how this construction really proceeds? Why not settle for psychology? Such a surrender of the epistemological burden to psychology is a move that was disallowed in earlier times as circular reasoning. If the epistemologist's goal is validation of the grounds of empirical science, he defeats his purpose by using psychology or other empirical science in the validation. However, such scruples against circularity have little point once we have stopped dreaming of deducing science from observations. If we are out simply to understand the link between observation and science, we are well advised to use any available information, including that provided by the very science whose link with observation we are seeking to understand.

But there remains a different reason, unconnected with fears of circularity, for still favoring creative reconstruction. We should like to be able to *translate* science into logic and observation terms and set theory. This would be a great epistemological achievement, for it would show all the rest of the concepts of science to be theoretically superfluous. It would legitimize them—to whatever degree the concepts of set theory, logic, and observation are themselves legitimate—by showing that everything done with the one apparatus could in principle be done with the other. If psychology itself could deliver a truly translational reduction of this kind, we should welcome it; but certainly it cannot, for certainly we did not grow up learning definitions of physicalistic language in terms of a prior language of set theory, logic, and observation. Here, then, would be good reason for persisting in a rational reconstruction: we want to establish the essential innocence of physical concepts, by showing them to be theoretically dispensable.

The fact is, though, that the construction which Carnap outlined in *Der logische Aufbau der Welt* does not give translational reduction either. It would not even if the outline were filled in. The crucial point comes where Carnap is explaining how to assign sense qualities to positions in physical space and time. These assignments are to be made in such a way as to fulfill, as well as possible, certain desiderata which he states, and with growth of experience the assignments are to be revised to suit. This plan, however illuminating, does not offer any key to *translating* the sentences of science into terms of observation, logic, and set theory.

We must despair of any such reduction. Carnap had despaired of it by 1936, when, in "Testability and Meaning,"[2] he introduced so-called *reduction forms* of a type weaker than definition. Definitions had shown always how to translate sentences into equivalent sentences. Contextual definition of a term showed how to translate sentences containing the term into equivalent sentences lacking the term. Reduction forms of Carnap's liberalized kind, on the other hand, do not in general give equivalences; they give implications. They explain a new term, if only partially, by specifying some sentences which are implied by sentences containing the term, and other sentences which imply sentences containing the term.

It is tempting to suppose that the countenancing of reduction forms in this liberal sense is just one further step of liberalization comparable to the earlier one, taken by Bentham, of countenancing contextual definition. The former and sterner kind of rational reconstruction might have been represented as a fictitious history in which we imagined our ancestors introducing the terms of physicalistic discourse on a phenomenalistic and set-theoretic basis by a succession of contextual definitions. The new and more liberal kind of rational reconstruction is a fictitious history in which we imagine our ancestors introducing those terms by a succession rather of reduction forms of the weaker sort.

This, however, is a wrong comparison. The fact is rather that the former and sterner kind of rational reconstruction, where definition reigned, embodied no fictitious history at all. It was nothing more nor less than a set of directions—or would have been, if successful—for accomplishing everything in terms of phenomena and set theory that we now accomplish in terms of bodies. It would have been a true reduction by translation, a legitimation by elimination. *Definire est eliminare.* Rational reconstruction by Carnap's later and looser reduction forms does none of this.

To relax the demand for definition, and settle for a kind of reduction that does not eliminate, is to renounce the last remaining advantage that we supposed rational reconstruction to have over straight psychology; namely, the advantage of translational reduction. If all we hope for is a reconstruction that links science to experience in explicit ways short of translation, the it would seem more sensible to settle for psychology. Better to discover how science is in fact developed and learned than to fabricate a fictitious structure to a similar effect.

The empiricist made one major concession when he despaired of deducing the truths of nature from sensory evidence. In despairing now even of translating those truths into terms of observation and logicomathematical auxiliaries, he makes another major concession. For suppose we hold, with the old empiricist Peirce, that the very meaning of a statement consists in the difference its truth would make to possible experience. Might we not formulate, in a chapter-length sentence in observational language, all the difference that the truth of a given statement might make to experience, and might we not then take all this as the translation? Even if the difference that the truth of the statement would make to experience ramifies indefinitely, we might still hope to embrace it all in the logical implications of our chapter-length formulation, just as we can axiomatize an infinity of theorems. In giving up hope of such translation, then, the empiricist is conceding that the empirical meanings of typical statements about the external world are inaccessible and ineffable.

How is this inaccessibility to be explained? Simply on the ground that the experiential implications of a typical statement about bodies are too complex for finite axiomatization, however lengthy? No; I have a different explanation. It is that the typical statement about bodies has no fund of experiential implications it can call its own. A substantial mass of theory, taken together, will commonly have experiential implications; this is how we make verifiable predictions. We may not be able to explain why we arrive at theories which make successful predictions, but we do arrive at such theories.

Sometimes also an experience implied by a theory fails to come off; and then, ideally, we declare the theory false. But the failure falsifies only a block of theory as a whole, a conjunction of many statements. The failure shows that one or more of those statements is false, but it does not show which. The predicted experiences, true and false, are not implied by any one of the component statements of the theory rather than another. The component statements simply do not have empirical meanings, by Peirce's standard, but a sufficiently inclusive portion of theory does. If we can aspire to a sort of *logischer Aufbau der Welt* at all, it must be to one in which the texts slated for translation into observational and logico-mathematical terms are mostly broad theories taken as wholes, rather than just terms or short sentences. The translation of a theory would be a ponderous axiomatization of all the experiential difference that the truth of the theory would make. It would be a queer translation, for it would translate the whole but one of the parts. We might better speak in such a case not of translation but simply of observational evidence for theories; and we may, following Peirce, still fairly call this the empirical meaning of the theories.

These considerations raise a philosophical question even about ordinary unphilosophical translation, such as from English into Arunta or Chinese. For, if the English sentences of a theory have their meaning only together as a body, then we can justify their translation into Arunta only together as a body. There will be no justification for pairing off the component English sentences with component Arunta sentences, except as these correlations make the translation of the theory as a whole come out right. Any translations of the English sentences into Arunta sentences will be as correct as any other, so long as the net empirical implications of the theory as a whole are preserved in translation. But it is to be expected that many different ways of translating the component sentences, essentially different individually, would deliver the same empirical implications for the theory as a whole; deviations in the translation of one component sentence could be compensated for in the translation of another component sentence. Insofar, there can be no ground for saying which of two glaringly unlike translations of individual sentences is right.[3]

For an uncritical mentalist, no such indeterminacy threatens. Every term and every sentence is a label attached to an idea, simple or complex, which is stored in the mind. When on the other hand we take a verification theory of meaning seriously, the indeterminacy would appear to be inescapable. The Vienna Circle espoused a verification theory of meaning but did not take it seriously enough. If we recognize with Peirce that the meaning of a sentence turns purely on what would count as evidence for its truth, and if we recognize with Duhem that theoretical sentences have their evidence not as single sentences but only as larger blocks of theory, then the indeterminacy of translation of theoretical sentences is the natural conclusion. And most sentences, apart from observation sentences, are theoretical. This conclusion, conversely, once it is embraced, seals the fate of any general notion of propositional meaning or, for that matter, state of affairs.

Should the unwelcomeness of the conclusion persuade us to abandon the verification theory of meaning? Certainly not. The sort of meaning that is basic to translation, and to the learning of one's own language, is necessarily empirical meaning and nothing more. A child learns his first words and sentences by hearing and using them in the presence of appropriate stimuli. These must be external stimuli, for they must act both on the child and on the speaker from whom he is learning.[4] Language is socially inculcated and controlled; the inculcation and control turn strictly on the keying of sentences to shared stimulation. Internal factors may vary *ad libitum* without prejudice to communication as long as the keying of language to external stimuli is undisturbed. Surely one has no choice but to be an empiricist so far as one's theory of linguistic meaning is concerned.

What I have said of infant learning applies equally to the linguist's learning of a new language in the field. If the linguist does not lean on related languages for which there are previously accepted translation practices, then obviously he had no data but the concomitances of native utterance and observable stimulus situation. No wonder there is indeterminacy of translation—for of course only a small fraction of our utterances report concurrent external stimulation. Granted, the linguist will end up with unequivocal translations of everything; but only by making many arbitrary choices—arbitrary even though unconscious—along the way. Arbitrary? By this I mean that different choices could still have made everything come out right that is susceptible in principle to any kind of check.

Let me link up, in a different order, some of the points I have made. The crucial consideration behind my argument for the indeterminacy of translation was that a statement about the world does not always or usually have a separable fund of empirical consequences that it can call its own. That consideration served also to account for the impossibility of an epistemological reduction of the sort where every sentence is equated to a sentence in observational and logico-mathematical terms. And the impossibility of that sort of epistemological reduction dissipated the last advantage that rational reconstruction seemed to have over psychology.

Philosophers have rightly despaired of translating everything into observational and logico-mathematical terms. They have despaired of this even when they have not recognized, as the reason for this irreducibility, that the statements largely do not have their private bundles of empirical consequences. And some philosophers have seen in this irreducibility the bankruptcy of epistemology. Carnap and the other logical positivists of the Vienna Circle had already pressed the term "metaphysics" into pejorative use, as connoting meaninglessness; and the term "epistemology" was next. Wittgenstein and his followers, mainly at Oxford, found a residual philosophical vocation in therapy: in curing philosophers of the delusion that there were epistemological problems.

But I think that at this point it may be more useful to say rather that epistemology still goes on, though in a new setting and a clarified status. Epistemology, or something like it, simply falls into place as a chapter of psychology and hence of natural science. It studies a natural phenomemon, viz., a physical human subject. This human subject is accorded a certain experimentally controlled input—certain patterns of irradiation in assorted frequencies, for instance—and in the fullness of time the subject delivers as output a description of the three-dimensional external world and its history. The relation between the meager input and the torrential output is a relation that we are prompted to study for somewhat the same reasons that always prompted epistemology; namely, in order to see how evidence relates to theory, and in what ways one's theory of nature transcends any available evidence.

Such a study could still include, even, something like the old rational reconstruction, to whatever degree such reconstruction is practicable; for imaginative constructions can afford hints of actual psychological processes, in much the way that mechanical simulations can. But a conspicuous difference between old epistemology and the epistemological enterprise in this new psychological setting is that we can now make free use of empirical psychology.

The old epistemology aspired to contain, in a sense, natural science; it would construct it somehow from sense data. Epistemology in its new setting, conversely, is contained in natural science, as a chapter of psychology. But the old containment remains valid too, in its way. We are studying how the human subject of our study posits bodies and projects his physics from his data, and we appreciate that our position in the world is just like his. Our very epistemological enterprise, therefore, and the psychology wherein it is a component chapter, and the whole of natural science wherein psychology is a component book—all this is our own construction or projection from stimulations like those we were meting out to our epistemological subject. There is thus reciprocal containment, though containment in different senses: epistemology in natural science and natural science in epistemology.

This interplay is reminiscent again of the old threat of circularity, but it is all right now that we have stopped dreaming of deducing science from sense data. We are after an understanding of science as an institution or process in the world, and we do not intend that understanding to be any better than the science which is its object. This attitude is indeed one that Neurath was already urging in Vienna Circle days, with his parable of the mariner who has to rebuild his boat while staying afloat in it.

One effect of seeing epistemology in a psychological setting is that it resolves a stubborn old enigma of epistemological priority. Our retinas are irradiated in two dimensions, yet we see things as three-dimensional without conscious inference. Which is to count as observation— the unconscious two-dimensional reception or the conscious three-dimensional apprehension? In the old epistemological context the conscious form had priority, for we were out to justify our knowledge of the external world by rational reconstruction, and that demands awareness. Awareness ceased to be demanded when we gave up trying to justify our knowledge of the external world by rational reconstruction. What to count as observation now can be settled in terms of the stimulation of sensory receptors, let consciousness fall where it may.

The Gestalt psychologists' challenge to sensory atomism, which seemed so relevant to epistemology forty years ago, is likewise deactivated. Regardless of whether sensory atoms or Gestalten are what favor the forefront of our consciousness, it is simply the stimulations of our sensory receptors that are best looked upon as the input to our cognitive mechanism. Old paradoxes about unconscious data and inference, old problems about chains of inference that would have to be completed too quickly—these no longer matter.

In the old anti-psychologistic days the question of epistemological priority was moot. What is epistemologically prior to what? Are Gestalten prior to sensory atoms because they are noticed, or should we favor sensory atoms on some more subtle ground? Now that we are permitted to appeal to physical stimulation, the problem dissolves; A is epistemologically prior to B if A is causally nearer than B to the sensory receptors. Or, what is in some ways better, just talk explicitly in terms of causal proximity to sensory receptors and drop the talk of epistemological priority.

Around 1932 there was debate in the Vienna Circle over what to count as observation sentences, or *Protolkollsätze*.[5] One position was that they had the form of reports of sense impressions. Another was that they were statements of an elementary sort about the external world, e.g., "A red cube is standing on the table." Another, Neurath's, was that they had the form of reports of relations between percipients and external things: "Otto now sees a red cube on the table." The worst of it was that there seemed to be no objective way of settling the matter: no way of making real sense of the question.

Let us now try to view the matter unreservedly in the context of the external world. Vaguely speaking, what we want of observation sentences is that they be the ones in closest causal proximity to the sensory receptors. But how is such proximity to be gauged? The idea may be rephrased this way: observation sentences are sentences which, as we learn language, are most strongly conditioned to concurrent sensory stimulation rather than to stored collateral information. Thus let us imagine a sentence queried for our verdict as to whether it is true or false, queried for our assent or dissent. Then the sentence is an observation sentence if our verdict depends only on the sensory stimulation present at the time.

But a verdict cannot depend on present stimulation to the exclusion of stored information. The very fact of our having learned the language evinces much storing of information, and of information without which we should be in no position to give verdicts on sentences however observational. Evidently then we must relax our definition of observation sentence to read thus: a sentence is an observation sentence if all verdicts on it depend on present sensory stimulation and on no stored information beyond what goes into understanding the sentence.

This formulation raises another problem: how are we to distinguish between information that goes into understanding a sentence and information that goes beyond? This is the problem of distinguishing between analytic truth, which issues from the mere meanings of words, and synthetic truth, which depends on more than meanings. Now I have long maintained that this distinction is illusory. There is one step toward such a distinction, however, which does make sense: a sentence that is true by mere meanings of words should be expected, at least if it is simple, to be subscribed to by all fluent speakers in the community. Perhaps the controversial notion of analyticity can be dispensed with, in our definition of observation sentence, in favor of this straightforward attribute of community-wide acceptance.

This attribute is of course no explication of analyticity. The community would agree that there have been black dogs, yet none who talk of analyticity would call this analytic. My rejection of the analyticity notion just means drawing no line between what goes into the mere understanding of the sentences of a language and what else the community sees eye-to-eye on. I doubt that an objective distinction can be made between meaning and such collateral information as is community-wide.

Turning back then to our task of defining observation sentences, we get this: an observation sentence is one on which all speakers of the language give the same verdict when given the same concurrent stimulation. To put the point negatively, an observation sentence is one that is not sensitive to differences in past experience within the speech community.

This formulation accords perfectly with the traditional role of the observation sentence as the court of appeal of scientific theories. For by our definition the observation sentences are the sentences on which all members of the community will agree under uniform stimulation. And what is the criterion of membership in the same community? Simply, general fluency of dialogue. This criterion admits of degrees, and indeed we may usefully take the community more narrowly for some studies than for others. What count as observation sentences for a community of specialists would not always so count for a larger community.

There is generally no subjectivity in the phrasing of observation sentences, as we are now conceiving them; they will usually be about bodies. Since the distinguishing trait of an observation sentence is intersubjective agreement under agreeing stimulation, a corporeal subject matter is likelier than not.

The old tendency to associate observation sentences with a subjective sensory subject matter is rather an irony when we reflect that observation sentences are also meant to be the intersubjective tribunal of scientific hypotheses. The old tendency was due to the drive to base science on something firmer and prior in the subject's experience; but we dropped that project.

The dislodging of epistemology from its old status of first philosophy loosed a wave, we saw, of epistemological nihilism. This mood is reflected somewhat in the tendency of Polányi, Kuhn, and the late Russell Hanson to belittle the role of evidence and to accentuate cultural relativism. Hanson ventured even to discredit the idea of observation, arguing that so-called observations vary from observer to observer with the amount of knowledge that the observers bring with them. The veteran physicist looks at some apparatus and sees an x-ray tube. The neophyte, looking at the same place, observes rather "a glass and metal instrument replete with wires, reflectors, screws, lamps, and pushbuttons."[6] One man's observation is another man's closed book of flight of fancy. The notion of observation as the impartial and objective source of evidence for science is bankrupt. Now my answer to the x-ray example was already hinted a little while back: what counts as an observation sentence varies with the width of community considered. But we can also always get an absolute standard by taking in all speakers of the language, or most.[7] It is ironical that philosophers, finding the old epistemology untenable as a whole, should react by repudiating a part which has only now moved into clear focus.

Clarification of the notion of observation sentence is a good thing, for the notion is fundamental in two connections. These two correspond to the duality that I remarked upon early in this essay: the duality between concept and doctrine, between knowing what a sentence means and knowing whether it is true. The observation sentence is basic to both enterprises. Its relation to doctrine, to our knowledge of what is true, is very much the traditional one: observation sentences are the repository of evidence for scientific hypotheses. Its relation to meaning is fundamental too, since observation sentences are the ones we are in a position to learn to understand first, both as children and as field linguists. For observation sentences are precisely the ones that we can correlate with observable circumstances of the occasion of utterance or assent, independently of variations in the past histories of individual informants. They afford the only entry to a language.

The observation sentence is the cornerstone of semantics. For it is, as we just saw, fundamental to the learning of meaning. Also, it is where meaning is firmest. Sentences higher up in theories have no empirical consequences they can call their own; they confront the tribunal of sensory evidence only in more or less inclusive aggregates. The observation sentence, situated at the sensory periphery of the body scientific, is the minimal verifiable aggregate; it has an empirical content all its own and wears it on its sleeve.

The predicament of the indeterminacy of translation has little bearing on observation sentences. The equating of an observation sentence of our language to an observation sentence of another language is mostly a matter of empirical generalization; it is a matter of identity between the range of stimulations that would prompt assent to the one sentence and the range of stimulations that would prompt assent to the other.[8]

It is no shock to the preconceptions of old Vienna to say that epistemology now becomes semantics. For epistemology remains centered as always on evidence, and meaning remains centered as always on verification; and evidence is verification. What is likelier to shock preconceptions is that meaning, once we get beyond observation sentences, ceases in general to have any clear applicability to single sentences; also that epistemology merges with psychology, as well as with linguistics.

This rubbing out of boundaries could contribute to progress, it seems to me, in philosophically interesting inquiries of a scientific nature. One possible area is perceptual norms. Consider, to begin with, the linguistic phenomenon of phonemes. We form the habit, in hearing the myriad variations of spoken sounds, of treating each as an approximation to one or another of a limited number of norms—around thirty altogether—constituting so to speak a spoken alphabet. All speech in our language can be treated in practice as sequences of just those thirty elements, thus rectifying small deviations. Now outside the realm of language also there is probably only a rather limited alphabet of perceptual norms altogether, toward which we tend unconsciously to rectify all perceptions. These, if experimentally identified, could be taken as epistemological building blocks, the working elements of experience. They might prove in part to be culturally variable, as phonemes are, and in part universal.

Again there is the area that the psychologist Donald T. Campbell calls evolutionary epistemology.[9] In this area there is work by Hüseyin Yilmaz, who shows how some structural traits of color perception could have been predicted from survival value.[10] And a more emphatically epistemological topic that evolution helps to clarify is induction, now that we are allowing epistemology the resources of natural science.[11]

Notes

1. A. B. Johnson, *A Treatise on Language* (New York, 1836; Berkeley, 1947).

2. Carnap, *Philosophy of Science* 3 (1936): 419–71; 4 (1937): 1–40.

3. See Quine, *Ontological Relativity* (New York, 1969), pp. 2ff.

4. See ibid., p. 28.

5. Carnap and Neurath in *Erkenntnis* 3 (1932): 204–28.

6. N. R. Hanson, "Observation and Interpretation," in S. Morgenbesser, ed., *Philosophy of Science Today* (New York: Basic Books, 1966).

7. This qualification allows for occasional deviants such as the insane or the blind. Alternatively, such cases might be excluded by adjusting the level of fluency of dialogue whereby we define sameness of language. (For prompting this note and influencing the development of this essay also in more substantial ways I am indebted to Burton Dreben.)

8. Cf. Quine, *Word and Object* (Cambridge, 1960), pp. 31–46, 68.

9. D. T. Campbell, "Methodological Suggestions from a Comparative Psychology of Knowledge Processes," *Inquiry* 2 (1959): 152–82.

10. Hüseyin Yilmaz, "On Color Vision and a New Approach to General Perception," in E. E. Bernard and M. R. Kare, eds., *Biological Prototypes and Synthetic Systems* (New York: Plenum, 1962); "Perceptual Invariance and the Psychophysical Law," *Perception and Psychophysics* 2 (1967): 533–538.

11. See Quine, "Natural Kinds," in *Ontological Relativity,* chap. 5.

17

Quine as Feminist: The Radical Import of Naturalized Epistemology

Louise M. Antony

The truth is always revolutionary.
—Antonio Gramsci

I. Introduction

Do we need a feminist epistemology? This is a very complicated question. Nonetheless it has a very simple answer: yes and no.

Of course, what I should say (honoring a decades-old philosophical tradition) is that a great deal depends on what we *mean* by "feminist epistemology." One easy—and therefore tempting—way to interpret the demand for a feminist epistemology is to construe it as nothing more than a call for more theorists *doing* epistemology. On this way of viewing things, calls for "feminist political science," "feminist organic chemistry," and "feminist finite mathematics" would all be on a par, and the need for any one of them would be justified in exactly the same way, viz., by arguing for the general need for an infusion of feminist consciousness into the academy.

Construed in this way, an endorsement of "feminist epistemology" is perfectly neutral with respect to the eventual content of the epistemological theories that feminists might devise. Would it turn out, for example, that feminists as a group reject individualism or foundationalism? Would they favor empiricism over rationalism? Would they endorse views that privileged intuition over reason or the subjective over the objective? We'd just have to wait and see. It must even be left open, at least at the outset, whether a feminist epistemology would

Reprinted by permission of Westview Press from L. M. Antony and C. Witt, eds., *A Mind of One's Own: Feminist Essays on Reason and Objectivity*, pp. 185–225. Copyright 1992 by Westview Press.

be discernibly and systematically different from epistemology as it currently exists, or whether there would instead end up being exactly the same variety among feminists as there is now among epistemologists in general.

Now it might appear that the project of developing a feminist epistemology in this sense is one that we can all happily sign on to, for who could object to trying to infuse the disciplines with feminist consciousness? But now I must honor a somewhat newer philosophical tradition than the one I honored earlier, and ask, "We, who?" For though the determined neutrality of this way of conceiving feminist epistemology—let me call it "bare proceduralism"—may give it the superficial appearance of a consensus position, it is in fact quite a partisan position. Even setting aside the fact that there are many people—yes, even some philosophers—who would rather be infused with bubonic plague than with feminist consciousness, it's clear that not everyone is going to like bare proceduralism. And ironically, it is its very neutrality that makes this an unacceptable reading of many, if not most, of the theorists who are currently calling for a feminist epistemology.[1]

To see the sticking point, consider the question of whether we should, as feminists, have an obligation to support *any* project whose participants represent themselves as feminists. Should we, for example, support the development of a "feminist sociobiology" or a "feminist military science" on the grounds that it's always a good idea to infuse a discipline, or a theory, with feminist consciousness, or on the grounds that there are people who are engaged in such projects who regard themselves as feminists and therefore have a claim on our sympathies? The answer to these questions, arguably, is no. Some projects, like the rationalization of war, may simply be *incompatible* with feminist goals; and some theories, like those with biological determinist presuppositions, may be *inconsistent* with the results of feminist inquiry to date.

Bare proceduralism, with its liberal, all-purpose, surely-there's-something-we-can-all-agree-on ethos, both obscures and begs the important question against those who believe that not all epistemological frameworks cohere—or cohere equally well—with the insights and aims of feminism. Specifically, it presupposes something that many feminist philosophers are at great pains to deny, namely the *prima facie* adequacy, from a feminist point of view, of those epistemological theories currently available within mainstream Anglo-American philosophy. At the very least, one who adopts the bare proceduralist standpoint with respect to feminist epistemology is making a substantive presupposition about where we currently stand in the process of feminist theorizing. To allow even that a feminist epistemology *might* utilize certain existing epistemological frameworks is to assert that feminist theorizing has not yet issued in substantive results regarding such frameworks.[2] Such a view, if not forthrightly expressed and explicitly defended, is disrespectful to the work of those feminists who claim to have already shown that those very epistemological theories are incompatible with feminism.

So we can't simply interpret the question, "Do we need a feminist epistemology?" in the bare proceduralist way and nod an enthusiastic assent. If we do, we'll be obscuring or denying the existence of substantive disagreements among feminists about the relation between feminism and theories of knowledge. One natural alternative to the bare proceduralist interpretation would be to try to give feminist epistemology a *substantive* sense—that is, take it to refer to a particular kind of epistemology or to a particular theory within epistemology, one that is specifically feminist.

But this won't work either, for two good reasons. First, there simply is no substantive consensus position among feminists working in epistemology, so that it would be hubris for anyone to claim that his or her epistemology was *the* feminist one.[3] Second, many feminists would find the idea that there *should* be such a single "feminist" position repellent. Some would dislike the idea simply for its somewhat totalitarian, "PC" ring. (Me, I'm not bothered by that—it seems to me that one should strive to be correct in all things, including politics.) Some theorists would argue that variety in feminist philosophical positions is to be expected at this point in the development of feminist consciousness, and that various intra- and inter-theoretic tensions in philosophical inquiry reflect unprocessed conflicts among deeply internalized conceptions of reality, of ourselves as human beings, and of ourselves as women.[4] Still others would see the expectation or hope that there will *ever* be a single, comprehensive, "true" feminist position as nothing but a remnant of outmoded, patriarchal ways of thinking.[5]

Thus, while individual feminist theorists may be advertising particular epistemological theories as feminist theories, general calls for the development of a feminist epistemology cannot be construed as advocacy for any particular one of these. But recognition of this fact does not throw us all the way back to the bare proceduralist notion. It simply means that in order to decide on the need for a feminist epistemology, we need to look at details—both with respect to the issues that feminism is supposed to have raised for the theory of knowledge and with respect to the specific epistemological theories that have been proffered as answering to feminist needs.

This is where the yes-and-no comes in. If we focus on the existence of what might be called a "feminist agenda" in epistemology—that is, if the question, "Do we need a feminist epistemology?" is taken to mean, "Are there specific questions or problems that arise as a result of feminist analysis, awareness, or experience that any adequate epistemology must accommodate?"—then I think the answer is clearly yes. But if, taking for granted the existence of such an agenda, the question is taken to be, "Do we need, in order to accommodate these questions, insights, and projects, a specifically feminist alternative to currently available epistemological frameworks?" then the answer, to my mind, is no.

Now it is on this point that I find myself in disagreement with many feminist philosophers. For despite the diversity of views within contemporary feminist thought, and despite the disagreements about even the desiderata for a genuinely feminist epistemology, one theoretical conclusion shared by almost all those feminists who explicitly advocate the development of a feminist epistemology is that existing epistemological paradigms—particularly those available within the framework of contemporary analytic philosophy—are fundamentally unsuited to the needs of feminist theorizing.

It is this virtual unanimity about the inadequacy of contemporary analytic epistemology that I want to challenge. There is an approach to the study of knowledge that promises enormous aid and comfort to feminists attempting to expose and dismantle the oppressive intellectual ideology of a patriarchal, racist, class-stratified society, and it is an approach that lies squarely within the analytic tradition. The theory I have in mind is Quine's "naturalized epistemology"—the view that the study of knowledge should be treated as the empirical investigation of knowers.

It's both unfortunate and ironic that Quine's work has been so uniformly neglected by feminists interested in the theory of knowledge, because although naturalized epistemology is nowadays as mainstream a theory as there is, Quine's challenges to logical positivism were radical in their time, and still retain an untapped radical potential today. His devastating critique of epistemological foundationalism bears many similarities to contemporary feminist attacks on "modernist" conceptions of objectivity and scientific rationality, and his positive views on the holistic nature of justification provide a theoretical basis for pressing the kinds of critical questions feminist critics are now raising.

Thus my primary aim in this essay is to highlight the virtues, from a feminist point of view, of naturalized epistemology. But—as is no doubt quite clear—I have a secondary, polemical aim as well. I want to confront head-on the charges that mainstream epistemology is irremediably phallocentric, and to counter the impression, widespread among progressives both within and outside of the academy, that there is some kind of natural antipathy between radicalism on the one hand and the methods and aims of analytic philosophy on the other. I believe that this impression is quite false, and its promulgation is damaging not only to individual feminists—especially women— working within the analytic tradition, but also to the prospects for an adequate feminist philosophy.

The "Bias" Paradox

I think the best way to achieve both these aims—defending the analytic framework in general and showcasing naturalized epistemology in particular—is to put the latter to work on a problem that is becoming increasingly important within feminist theory. The issue I have in mind is the problem of how properly to conceptualize *bias*. There are several things about this issue that make it particularly apt for my purposes.

In the first place, the issue provides an example of the way in which feminist analysis can generate or uncover serious epistemological questions, for the problem about bias that I want to discuss will only be recognized as a problem by individuals who are critical, for one reason or another, of one standard conception of objectivity. In the second place, because of the centrality of this problem to feminist theory, the ability of an epistemological theory to provide a solution offers one plausible desideratum of a theory's adequacy as a feminist epistemology. Last of all, because the notions of bias and partiality figure so prominently in feminist critiques of mainstream analytic epistemology, discussion of this issue will enable me to address directly some of the charges that have led some feminist theorists to reject the analytic tradition.

But what is the problem? Within certain theoretical frameworks, the analysis of the notion of "bias" is quite straightforward. In particular, strict empiricist epistemology concurs with liberal political theory in analyzing bias as the mere possession of belief or interest prior to investigation. But for anyone who wishes to criticize the liberal/ empiricist ideal of an "open mind," the notion of bias is enormously problematic and threatens to become downright paradoxical.

Consider feminist theory: On the one hand, it is one of the central aims of feminist scholarship to expose the male-centered assumptions and interests—the male *biases,* in other words—underlying so much of received "wisdom." But on the other hand, there's an equally important strain of feminist theory that seeks to challenge the ideal of pure objectivity by emphasizing both the ubiquity and the value of certain kinds of partiality and interestedness. Clearly, there's a tension between those feminist critiques that accuse science or philosophy of displaying male bias and those that reject the ideal of impartiality.

The tension blossoms into paradox when critiques of the first sort are applied to the concepts of objectivity and impartiality themselves. According to many feminist philosophers, the flaw in the ideal of impartiality is supposed to be that the ideal itself is biased: Critics charge either that the concept of "objectivity" serves to articulate a masculine or patriarchal viewpoint (and possibly a pathological one),[6] or that it has the ideological function of protecting the rights of those in power, especially men.[7] But how is it possible to criticize the partiality of the concept of objectivity without presupposing the very value under attack? Put baldly: If we don't think it's good to be *im*partial, then how can we object to men's being *partial?*

The critiques of "objectivity" and "impartiality" that give rise to this paradox represent the main source of feminist dissatisfaction with existing epistemological theories. It's charged that mainstream epistemology will be forever unable to either acknowledge or account for the partiality and locatedness of knowledge, because it is wedded to precisely those ideals of objective or value-neutral inquiry that ultimately and inevitably subserve the interests of the powerful. The valorization of impartiality within mainstream epistemology is held to perform for the ruling elite the critical ideological function of *denying the existence of partiality itself.*[8]

Thus Lorraine Code, writing in the *APA Newsletter on Feminism and Philosophy,*[9] charges that mainstream epistemology (or what she has elsewhere dubbed "malestream" epistemology[10]) has "defined 'the epistemological project' so as to make it illegitimate to ask questions about the identities and specific circumstances of these knowers." It has accomplished this, she contends, by promulgating a view of knowers as essentially featureless and interchangeable, and by donning a "mask of objectivity and value-neutrality." The transformative potential of a feminist—as opposed to a malestream—epistemology lies in its ability to tear off this mask, exposing the "complex power structure of vested interest, dominance, and subjugation" that lurks behind it.

But not only is it not the case that contemporary analytic epistemology is committed to such a conception of objectivity, it was analytic epistemology that was largely responsible for initiating the critique of the empiricistic notions Code is attacking. Quine, Goodman, Hempel, Putnam, Boyd, and others within the analytic tradition have all argued that a certain received conception of objectivity is untenable as an ideal of epistemic practice. The detailed critique of orthodox empiricism that has developed within the analytic tradition is in many ways more pointed and radical than the charges that have been leveled from without.

Furthermore, these philosophers, like many feminist theorists, have emphasized not only the *ineliminability* of bias but also the *positive value* of certain forms of it. As a result, the problems that arise for a naturalized epistemology are strikingly similar to those that beset the feminist theories mentioned above: Once we've acknowledged the necessity and legitimacy of partiality, *how do we tell the good bias from the bad bias?*

What kind of epistemology is going to be able to solve a problem like this? Code asserts that the specific impact of feminism on epistemology has been "to move the question '*Whose* knowledge are we talking about?' to a central place in epistemological discussion,"[11] suggesting that the hope lies in finding an epistemological theory that assigns central importance to consideration of the nature of the subjects who actually do the knowing. I totally agree: No theory that abjures empirical study of the cognizer, or of the actual processes by which knowledge develops, is ever going to yield insight on this question.

But more is required than this. If we as feminist critics are to have any basis for distinguishing the salutary from the pernicious forms of bias, we can't rest content with a *description* of the various ways in which the identity and social location of a subject make a difference to her beliefs. We need, in addition, to be able to make *normative* distinctions among various processes of belief-fixation as well. Otherwise, we'll never escape the dilemma posed by the bias paradox: either endorse pure impartiality or give up criticizing bias.[12]

It is here that I think feminist philosophy stands to lose the most by rejecting the analytic tradition. The dilemma will be impossible to escape, I contend, for any theory that eschews the notion of *truth*—for any theory, that is, that tries to steer some kind of middle course between absolutism and relativism. Such theories inevitably leave themselves without resources for making the needed normative distinctions, because they deprive themselves of any conceptual tools for distinguishing the grounds of a statement's truth from the explanation of a statement's acceptance.

Naturalized epistemology has the great advantage over epistemological frameworks outside the analytic tradition (I have in mind specifically standpoint and postmodern epistemologies) in that it permits an appropriately realist conception of truth, viz., one that allows a conceptual gap between epistemology and metaphysics, between the world as we see it and the world as it is.[13] Without appealing to at least this minimally realist notion of truth, I see no way to even state the distinction we ultimately must articulate and defend. Quite simply, an adequate solution to the paradox must enable us to say the following: What makes the *good* bias good is that it facilitates the search for truth, and what makes the *bad* bias bad is that it impedes it.

Now that my absolutist leanings are out in the open, let me say one more thing about truth that I hope will forestall a possible misunderstanding of my project here. I do believe in truth, and I have *never* understood why people concerned with justice have given it such a bad rap. Surely one of the goals of feminism is to *tell the truth* about women's lives and women's experience. Is institutionally supported discrimination not a *fact?* Is misogynist violence not a *fact?* And isn't the existence of ideological denial of the first two facts *itself* a fact? What in the world else could we be doing when we talk about these things, *other* than asserting that the world actually *is* a certain way?

Getting at the truth is complicated, and one of the things that complicates it considerably is that powerful people frequently have strong motives for keeping less powerful people from getting at the truth. It's one job of a critical epistemology, in my view, to expose this fact, to make the mechanisms of such distortions transparent. But if we, as critical epistemologists, lose sight of what we're after, if we concede that there's nothing at stake other than the matter of whose "version" is going to prevail, then our projects become as morally bankrupt and baldly self-interested as Theirs.

This brings me to the nature of the current discussion. I would like to be clear that in endorsing the project of finding a "feminist epistemology," I do not mean to be advocating the construction of a serviceable epistemological ideology "for our side." And when I say that I think naturalized epistemology makes a good feminist epistemology, I don't mean to be suggesting that the justification for the theory is instrumental. A good *feminist* epistemology must be, in the first place, a good epistemology, and that means being a theory that is likely to be *true*. But of course I would not think that naturalized epistemology was likely to be true unless I also thought it explained the facts. And among the facts I take to be central are the long-ignored experiences and wisdom of women.

In the next section, I will explain in more detail the nature of the charges that have been raised by feminist critics against contemporary analytic epistemology. I'll argue that the most serious of these charges are basically misguided—that they depend on a misreading of the canonical figures of the Enlightenment as well as of contemporary epistemology. In the last section, I'll return to the bias paradox and try to show why a naturalized approach to the study of knowledge offers some chance of a solution.

II. What Is Mainstream Epistemology and Why Is It Bad?

One difficulty that confronts anyone who wishes to assess the need for a "feminist alternative" in epistemology is the problem of finding out exactly what such an epistemology would be an alternative to. What is "mainstream" epistemology anyway? Lorraine Code is more forthright than many in her willingness to name the enemy. According to her, "mainstream epistemology," the proper object of feminist critique, is "post-positivist empiricist epistemology: the epistemology that still dominates in Anglo-American philosophy, despite the best efforts of socialist, structuralist, hermeneuticist, and other theorists of knowledge to deconstruct or discredit it."[14]

By the "epistemology that still dominates in Anglo-American philosophy," Code would have to be referring to the set of epistemological theories that have developed within the analytic paradigm, for analytic philosophy has been, in fact, the dominant philosophical paradigm in the English-speaking academic world since the early twentieth century.[15] This means, at the very least, that the agents of sexism within academic philosophy—the individuals who have in fact been the ones to discriminate against women as students, job applicants, and colleagues—have been, for the most part, analytic philosophers, a fact that on its own makes the analytic paradigm an appropriate object for feminist scrutiny.

But this is not the main reason that Code and others seek to "deconstruct or discredit" analytic epistemology. The fact that the analytic paradigm has enjoyed such an untroubled hegemony within this country during the twentieth century—the period of the most rapid growth of American imperial power—suggests to many radical social critics that analytic philosophy fills an ideological niche. Many feminist critics see mainstream analytic philosophy as the natural metaphysical and epistemological complement to liberal political theory, which, by obscuring real power relations within the society, makes citizens acquiescent or even complicit in the growth of oppression, here and abroad.

What is it about analytic philosophy that would enable it to play this role? Some have argued that analytic or "linguistic" philosophy, together with its cognate fields (such as formal linguistics and computationalist psychology), is inherently male, "phallogocentric."[16] Others have argued that the analytic paradigm, because of its emphasis on abstraction and formalization and its valorization of elite skills, may be an instrument of cognitive control, serving to discredit the perspectives of members of nonprivileged groups.[17]

But most of the radical feminist critiques of "mainstream" epistemology (which, as I said, must denote the whole of analytic epistemology) are motivated by its presumed allegiance to the conceptual structures and theoretical commitments of the Enlightenment, which provided the general philosophical background to the development of modern industrialized "democracies."[18] By this means, "mainstream" epistemology becomes identified with "traditional" epistemology, and this traditional epistemology becomes associated with political liberalism. Feminist theorists like Alison Jaggar and Sandra Harding, who have both written extensively about the connection between feminist political analysis and theories of knowledge, have encouraged the idea that acceptance of mainstream epistemological paradigms is tantamount to endorsing liberal feminism. Jaggar contends that the connection lies in the radically individualistic conception of human nature common to both liberal political theory and Enlightenment epistemology. In a chapter entitled "Feminist Politics and Epistemology: Justifying Feminist Theory," she writes:

> Just as the individualistic conception of human nature sets the basic problems for the liberal political tradition, so it also generates the problems for the tradition in epistemology that is associated historically and conceptually with liberalism. This tradition begins in the 17th century

with Descartes, and it emerges in the 20th century as the analytic tradition. Because it conceives humans as essentially separate individuals, this epistemological tradition views the attainment of knowledge as a project for each individual on her or his own. The task of epistemology, then, is to formulate rules to enable individuals to undertake this project with success.[19]

Harding, in a section of her book called "A Guide to Feminist Epistemologies," surveys what she sees as the full range of epistemological options open to feminists. She imports the essentially conservative political agenda of liberal feminism, which is focused on the elimination of formal barriers to gender equality, into mainstream epistemology, which she labels "feminist empiricism": "*Feminist empiricism* argues that sexism and androcentrism are social biases correctable by stricter adherence to the existing methodological norms of scientific inquiry."[20] Harding takes the hallmark of feminist empiricism (which on her taxonomy is the only alternative to the feminist standpoint and postmodernist epistemologies) to be commitment to a particular conception of objectivity, which, again, is held to be part of the legacy of the Enlightenment. In her view, acceptance of this ideal brings with it faith in the efficacy of "existing methodological norms of science" in correcting biases and irrationalities within science, in the same way that acceptance of the liberal ideal of impartiality brings with it faith in the system to eliminate political and social injustice.

In Harding's mind, as in Jaggar's, this politically limiting conception of objectivity is one that can be traced to traditional conceptions of the knowing subject, specifically to Enlightenment conceptions of "rational man." The message, then, is that mainstream epistemology, because it still operates with this traditional conception of the self, functions to limit our understanding of the real operations of power, and of our place as women within oppressive structures. A genuine feminist transformation in our thinking therefore requires massive overhaul, if not outright repudiation, of central aspects of the tradition.

This is clearly the message that political scientist Jane Flax gleans from her reading of feminist philosophy; she argues that feminist theory ought properly to be viewed as a version of postmodern thought, since postmodern theorists and feminist theorists are so obviously engaged in a common project:

Postmodern philosophers seek to throw into radical doubt beliefs still prevalent in (especially American) culture but derived from the Enlighten-

ment . . . ;[21] feminist notions of the self, knowledge and truth are too
contradictory to those of the Enlightenment to be contained within
its categories. The way to feminist future(s) cannot lie in reviving or
appropriating Enlightenment concepts of the person or knowledge.[22]

But there are at least two serious problems with this argument. The
first is that the "tradition" that emerges from these critiques is a gross
distortion and oversimplification of the early modern period. The
critics' conglomeration of all classical and Enlightenment views into a
uniform "traditional" epistemology obscures the enormous amount of
controversy surrounding such notions as knowledge and the self during
the seventeenth and eighteenth centuries, and encourages crude misun-
derstandings of some of the central theoretical claims. Specifically,
this amalgamation makes all but invisible a debate that has enormous
relevance to discussions of bias and objectivity, viz., the controversy
between rationalists and empiricists about the extent to which the
structure of the mind might constrain the development of knowledge.[23]

The second problem is that the picture of analytic epistemology
that we get once it's allied with this oversimplified "traditional"
epistemology is downright cartoonish. When we look at the actual
content of the particular conceptions of objectivity and scientific
method that the feminist critics have culled from the modern period,
and which they subsequently attach to contemporary epistemology, it
turns out that these conceptions are precisely the ones that have been
the focus of *criticism* among American analytic philosophers from
the 1950s onward. The feminist critics' depiction of "mainstream"
epistemology utterly obscures this development in analytic epistemology, and in glossing over the details of the analytic critique of positivism, misses points that are of crucial relevance to any truly radical
assault on the liberal ideology of objectivity.[24]

The second problem is partly a consequence of the first. The
feminist critics, almost without exception, characterize mainstream
epistemology as "empiricist." But one of the chief accomplishments
of the analytic challenge to positivism was the demonstration that a
strictly empiricistic conception of knowledge is untenable. As a result,
much of analytic epistemology has taken a decidedly rationalistic turn.
Neglect of the rationalist/empiricist debate and misunderstanding of
rationalist tenets make the critics insensitive to these developments
and blind to their implications.

But the misreading of contemporary epistemology is also partly just
a matter of the critics' failure to realize the extent to which analytic

philosophy represents a *break* with tradition. I do not mean to deny that there were *any* important theoretical commitments common to philosophers of the early modern period. One such commitment, shared at least by classical rationalists and empiricists, and arguably by Kant, was an epistemological meta-hypothesis called "externalism." This is the view that the proper goal of epistemological theory is the rational *vindication* of human epistemic practice. But if externalism is regarded as the hallmark of "traditional epistemology," then the identification of analytic epistemology with traditional epistemology becomes all the more spurious.

It was the main burden of Quine's critique of positivism to demonstrate the impossibility of an externalist epistemology, and his suggested replacement, "naturalized epistemology," was meant to be what epistemology could be once externalist illusions were shattered. As a result of the analytic critique of externalism, the notions of objectivity and rationality available to contemporary analytic epistemologists are necessarily more complicated than the traditional conceptions they replace. This is so even for epistemologists who would not identify themselves as partisans of naturalized epistemology.

In what follows, I'll discuss in turn these two problems: first, the mischaracterization of the tradition, and then the caricature of contemporary analytic epistemology.

Rationalism v. Empiricism: The Importance of Being Partial

What I want to show first is that the "traditional epistemology" offered us by Jaggar and Flax grafts what is essentially a rationalist (and in some respects, specifically Cartesian) theory of *mind* onto what is essentially an empiricist conception of *knowledge*. This is a serious error. Although Jaggar and Flax claim that there are deep connections between the one and the other, the fact of the matter is that they are solidly opposed. The conception of objectivity that is ultimately the object of radical critique—perfect impartiality—is only supportable as an epistemic ideal on an empiricist conception of *mind*. Thus, I'll argue, the rationalistic conception of the self attacked by Jaggar and Flax as unsuitable or hostile to a feminist point of view actually provides the basis for a critique of the view of knowledge they want ultimately to discredit.

Much of what is held to be objectionable in "traditional epistemology" is supposed to derive from the tradition's emphasis on *reason*. But different traditional figures emphasized reason in different ways.

Only the rationalists and Kant were committed to what I'll call "cognitive essentialism," a feature of the "traditional" conception of mind that comes in for some of the heaviest criticism. I take cognitive essentialism to be the view (1) that there are certain specific properties the possession of which is both distinctive of and universal among human beings, (2) that these properties are cognitive in nature, (3) that our possession of these properties amounts to a kind of innate knowledge, and (4) that our status as moral agents is connected to the possession of these properties. Empiricists denied all these claims—in particular, they denied that reason had anything but a purely instrumental role to play in either normative or nonnormative activity, and tended to be opposed to any form of essentialism, cognitive or otherwise.

Although the purely instrumental conception of reason is also criticized by feminist scholars, cognitive essentialism is the focus of one specific set of feminist concerns. It is held to be suspect on the grounds that such a doctrine could easily serve to legitimate the arrogant impulses of privileged Western white men: first to canonize their own culture- and time-bound speculations as revelatory of the very norms of human existence, and then simultaneously to deny the very properties deemed "universal" to the majority of human beings on the planet.

Here's how it is supposed to work: Cognitive essentialism is supposed to engender a kind of fantasy concerning actual human existence and the actual prerequisites of knowledge. Because of its emphasis on *cognitive* characteristics, it's argued, the view permits privileged individuals to ignore the fact of their embodiment, and with that, the considerable material advantages they enjoy in virtue of their class, gender, and race.[25] To the extent that the characteristics they find in themselves are the result of their particular privileges instead of a transcendent humanity, the fantasy provides a basis for viewing less-privileged people—who well may lack such characteristics—as inherently less human. But since these characteristics have been lionized as forming the essence of moral personhood, the fantasy offers a rationale for viewing any differences between themselves and others as negative deviations from a moral norm.

Recall, for example, that the particular elements of Enlightenment thought that Flax finds inimical to feminist theory and praxis are the alleged universality, transcendence, and abstractness assigned to the faculty of reason:

> The notion that reason is divorced from "merely contingent" existence still predominates in contemporary Western thought and now appears to

mask the embeddedness and dependence of the self upon social relations, as well as the partiality and historical specificity of this self's existence. . . .

In fact, feminists, like other postmodernists, have begun to suspect that all such transcendental claims reflect and reify the experience of a few persons—mostly White, Western males.[26]

But moreover, cognitive essentialism is supposed to lead to what Jaggar calls "individualism,"[27] the view that individual human beings are epistemically self-sufficient, that human society is unnecessary or unimportant for the development of knowledge. If the ideal "man of reason" is utterly without material, differentiating features, then the ideal knower would appear to be *pure* rationality, a mere calculating mechanism, a person who has been stripped of all those particular aspects of self that are of overwhelming human significance. Correlatively, as it is precisely the features "stripped off" the self by the Cartesian method that "traditional" epistemology denigrates as distorting influences, the ideally objective cognizer is also the man of reason. Knowledge is then achieved, it appears, not by active engagement with one's world and with the people in it, but by a pristine transcendence of the messy contingencies of the human condition.[28]

Lending support to Lorraine Code's grievance against "traditional" epistemology, Jaggar thus insists that it is this abstract and detached individualism that underwrites a solipsistic view of the construction of knowledge and precludes assigning any epistemological significance to the situation of the knower.

Because it conceives humans as essentially separate individuals, this epistemological tradition views the attainment of knowledge as a project for each individual on his or her own. The task of epistemology, then, is to formulate rules to enable individuals to undertake this project with success.[29]

It is here that the link is supposed to be forged between the Cartesian/Kantian conception of the self and the particular conception of objectivity—objectivity as pure neutrality—that is thought to be pernicious.

But the individualism Jaggar takes to unite rationalists and empiricists is not in fact a view that *anyone* held. She derives it from a fairly common—indeed, almost canonical—misreading of the innate ideas debate. Significantly, Jaggar acknowledges the existence of disagreements within the early modern period, but avers that such issues as

divided rationalists from empiricists are differences that make no
difference. Both were foundationalists, she points out, and though the
foundation for rationalists was self-evident truths of reason and the
foundation for empiricists was reports of sensory experience, "in
either case, . . . the attainment of knowledge is conceived as essentially
a solitary occupation that has no necessary social preconditions."[30]

The reading, in other words, is that whereas the empiricists thought
all knowledge came from experience, the rationalists thought *all knowledge came from reason*. But the second element of this interpretation
is simply wrong. It was no part of Descartes's project (much less
Kant's) to assert the self-sufficiency of reason. Note that a large part
of the goal of the exercise of hyperbolic doubt in the *Meditations* was
to establish the reliability of sensory experience, which Descartes took
to be essential to the development of adequate knowledge of the world.
And although he maintained the innateness of many ideas, including
sensory ideas, he carefully and repeatedly explained that he meant by
this only that human beings were built in such a way that certain
experiences would trigger these ideas and no others.[31]

Furthermore, Descartes himself explicitly endorses two of the very
epistemic values his position is supposed to preclude. Not only does
he clearly reject the sort of epistemic individualism Jaggar deplores,
but he strongly upholds the necessity of acquainting oneself with the
variety of human experience in order to form a just conception of the
world. Expressing his contempt for the contradictions and sophistries
of his learned and cloistered teachers, he recounts how, as soon as he
was old enough to "emerge from the control of [his] tutors," he
"entirely quitted the study of letters."

> And resolving to seek no other science than that which could be found in
> myself, *or at least in the great book of the world* [my emphasis], I
> employed the rest of my youth in travel, in seeing courts and armies, in
> intercourse with men of diverse temperaments and conditions, in collect-
> ing varied experiences, in proving myself in the various predicament in
> which I was placed by fortune, and under all circumstances bringing my
> mind to bear on the things which came before it, so that I might derive
> some profit from my experience.[32]

And far from recommending the divestiture of one's particular concerns as sound epistemic practice, Descartes affirms the importance
of concrete engagement in finding the truth, pointing to the degradation
of knowledge that can result from disinterestedness.

For it seemed to me that I might meet with much more truth in the reasonings that each man makes on the matters that specifically concern him, and the issue of which would very soon punish him if he made a wrong judgment, than in the case of those made by a man of letters in his study touching speculations which lead to no result, and which bring about no other consequences to himself excepting that he will be all the more vain the more they are removed from common sense, since in this case it proves him to have employed so much the more ingenuity and skill in trying to make them seem probable.[33]

The bottom line is that rationalists, Descartes especially, did not hold the view that experience was inessential or even that it was unimportant; nor did they hold the view that the best epistemic practice is to discount one's own interests. The misreading that saddles Descartes with such views stems from a popular misconception about the innate ideas debate.

The disagreement between rationalists and empiricists was not simply about the existence of innate ideas. Both schools were agreed that the mind was natively structured and that that structure partially determined the shape of human knowledge. What they disagreed about was the *specificity* of the constraints imposed by innate mental structure. The rationalists believed that native structure placed quite specific limitations on the kinds of concepts and hypotheses the mind could form in response to experience, so that human beings were, in effect, natively *biased* toward certain ways of conceiving the world. Empiricists, on the other hand, held that there were relatively few native constraints on how the mind could organize sensory experience, and that such constraints as did exist were *domain-general* and *content-neutral*.

According to the empiricists, the human mind was essentially a mechanism for the manipulation of sensory data. The architecture of the mechanism was supposed to ensure that the concepts and judgments constructed out of raw sense experience accorded with the rules of logic. This did amount to a minimal constraint on the possible contents of human thought—they had to be logical transforms of sensory primitives—but it was a highly general one, applying to every subject domain in precisely the same way. Thus, on this model, any one hypothesis should be as good as any other as far as the mind is concerned, as long as both hypotheses are logically consistent with the sensory evidence.[34] This strict empiricist model of mind, as it turns out, supports many of the elements of epistemology criticized by

Code, Jaggar, and others (e.g., a sharp observation/theory distinction, unmediated access to a sensory "given," and an algorithmic view of justification). I'll spell this out in detail in the next section. For present purposes, however, the thing to note is that the model provides clear warrant for the particular conception of the ideal of objectivity—perfect neutrality—that is the main concern of Jaggar and the others and that is supposed to follow from cognitive essentialism. Here's how.

Because the mind itself, on the empiricist model, makes no substantive contribution to the contents of thought, knowledge on this model is *entirely* experience-driven: All concepts and judgments are held to reflect regularities in an individual's sensory experience. But one individual cannot see everything there is to see—one's experience is necessarily limited, and there's always the danger that the regularities that form the basis of one's own judgments are not general regularities, but only artifacts of one's limited sample. (There is, in other words, a massive restriction-of-range problem for empiricists.) The question then arises how one can tell whether the patterns one perceives are present in nature generally, or are just artifacts of one's idiosyncratic perspective.

The empiricists' answer to this question is that one can gauge the general validity of one's judgments by the degree to which they engender reliable expectations about sensory experience. But although this answer addresses the problem of how to tell whether one's judgments are good or bad, it doesn't address the problem of how to get good judgments in the first place. Getting good judgments means getting good data—that is, exposing oneself to patterns of sensations that are representative of the objective distribution of sensory qualities throughout nature.

This idea immediately gives rise to a certain ideal (some would say fantasy) of epistemic location—the best spot from which to make judgments would be that spot which is *least particular*. Sound epistemic practice then becomes a matter of constantly trying to maneuver oneself into such a location—trying to find a place (or at least come as close as one can) where the regularities in one's own personal experience match the regularities in the world at large. A knower who could be somehow stripped of all particularities and idiosyncrasies would be the best possible knower there is.

This is not, however, a fantasy that would hold any particular appeal for a rationalist, despite the image of detachment evoked by a cursory reading of the *Meditations*. The rationalists had contended all along that sensory experience *by itself* was insufficient to account for the richly detailed body of knowledge that human beings manifestly possessed, and thus that certain elements of human knowledge—what classical rationalists called *innate ideas*—must be natively present, a part of the human essence.

Because the rationalists denied that human knowledge was a pure function of the contingencies of experience, they didn't need to worry nearly as much as the empiricists did about epistemic location. If it is the structure of mind, rather than the accidents of experience, that largely determines the contours of human concepts, then we can relax about at least the broad parameters of our knowledge. We don't have to worry that idiosyncratic features of our epistemic positions will seriously distort our worldviews, because the development of our knowledge is not dependent upon the patterns that happen to be displayed in our particular experiential histories. The regularities we "perceive" are, in large measure, regularities that we're *built* to perceive.

"Pure" objectivity—if that means giving equal weight to every hypothesis consistent with the data, or if it means drawing no conclusions beyond what can be supported by the data—is thus a nonstarter as an epistemic form from a rationalist's point of view. The rationalists were in effect calling attention to the *value* of a certain kind of partiality: if the mind were not natively biased—i.e., disposed to take seriously certain kinds of hypotheses and to disregard or fail to even consider others—then knowledge of the sort that human beings possess would itself be impossible. There are simply too many ways of combining ideas, too many different abstractions that could be performed, too many distinct extrapolations from the same set of facts, for a pure induction machine to make much progress in figuring out the world.

The realization that perfect neutrality was not necessarily a good thing, and that bias and partiality are potentially salutary, is thus a point that was strongly present in the early modern period, *pace* Jaggar and Flax. There was no single "traditional" model of mind; the model that can properly be said to underwrite the conceptions of rationality and objectivity that Jaggar brings under feminist attack is precisely a model to which Descartes and the other rationalists were *opposed,* and, ironically, the one that, on the face of it, assigns the most significance to experience. And although it is the cognitive essentialists who are charged with deflecting attention away from epistemically significant characteristics of the knower, it was in fact these same essentialists, in explicit opposition to the empiricists, who championed the idea that human knowledge was necessarily "partial."

Hume, Quine, and the Break with Tradition

Let me turn now to the second serious problem with the feminist criticisms of "mainstream" epistemology: To the extent that there really is a "tradition" in epistemology, it is a tradition that has been explicitly rejected by contemporary analytic philosophy.

If the rationalists solved one problem by positing innate ideas, it was at the cost of raising another. Suppose that there are, as the rationalists maintained, innate ideas that perform the salutary function of narrowing down to a manageable set the hypotheses that human minds have to consider when confronted with sensory data. That eliminates the problem faced by the empiricists of filtering out idiosyncratic "distortions." But now the question is, How can we be sure that these biases—so helpful in getting us to *a* theory of the world—are getting us to the *right* theory of the world? What guarantees that our minds are inclining us in the right direction? Innate ideas lead us somewhere, but do they take us where we want to go?

The rationalists took this problem very seriously. A large part of their project was aimed at validating the innate constraints, at showing that these mental biases did not lead us astray. Descartes's quest for "certainty" needs to be understood in this context: The method of hyperbolic doubt should be viewed not as the efforts of a paranoid to free himself forever from the insecurity of doubt, but as a theoretical exercise designed to show that the contours imposed on our theories by our own minds were proper reflections of the topography of reality itself.

It is at this point that we're in a position to see what rationalists and empiricists actually had in common—not a conception of mind, not a theory of how knowledge is constructed, but a theory of *theories* of knowledge. If there is a common thread running through Enlightenment epistemologies, it is this: a belief in the possibility of providing a *rational* justification of the processes by which human beings arrive at theories of the world. For the empiricists, the trick was to show how the content of all knowledge could be reduced to pure reports of sensory experience; for the rationalists, it was showing the indubitability of the innate notions that guided and facilitated the development of knowledge. Philosophers in neither group were really on a quest for certainty—all the wanted was a reliable map of its boundaries.

But if one of the defining themes of the modern period was the search for an externalist justification of epistemic practice, then *Hume* must be acknowledged to be the first postmodernist. Hume, an empiricist's empiricist discovered a fatal flaw in his particular proposal for justifying human epistemic practice. He realized that belief in the principle of induction—the principle that says that the future will resemble the past or that similar things will behave similarly—could not be rationally justified. It was clearly not a truth of reason, since its denial was not self-contradictory. But neither could it be justified by experience: Any attempt to do so would be circular, because the practice of using past experience as evidence about the future is itself only warranted if one accepts the principle of induction.

Hume's "skeptical solution" to his own problem amounted to an abandonment of the externalist hopes of his time. Belief in induction, he concluded, was a *custom,* a tendency of mind ingrained by nature, one of "a species of natural instincts, which no reasoning or process of the thought and understanding is able, either to produce or to prevent."[35] For better or worse, Hume contended, we're stuck with belief in induction—we are constitutionally incapable of doubting it and conceptually barred from justifying it. The best we can do is to *explain* it.

Hume's idea was thus to offer as a replacement for the failed externalist project of rational justification of epistemic practice, the *empirical* project of characterizing the cognitive nature of creatures like ourselves, and then figuring out how such creatures, built to seek knowledge in the ways we do, could manage to survive and flourish. In this way, he anticipated to a significant degree the "postmodernist" turn taken by analytic philosophy in the twentieth century as the result of Quine's and others' critiques of externalism's last gasp—logical positivism.

Before fast-forwarding into the twentieth century, let me summarize what I take to be the real lessons of the modern period—lessons that, I've argued, have been missed by many feminist critiques of "traditional" epistemology. First, there is the essentially rationalist insight that perfect objectivity is not only impossible but undesirable, that certain kinds of "bias" or "partiality" are necessary to make our epistemic tasks tractable. Second, there is Hume's realization that externalism won't work, that we can never manage to offer a justification of epistemic norms without somehow presupposing the very norms we wish to justify. See this, if you will, as the beginning of the postmodern recognition that theory always proceeds from an "embedded" location, that there is no transcendent spot from which we can inspect our own theorizing.

The rationalist lesson was pretty much lost and the import of Hume's insight submerged by the subsequent emergence and development of neo-empiricist philosophy. This tradition, which involved primarily the British empiricists Mill and Russell, but also Wittgenstein and the Vienna Circle on the Continent, culminated in the school of thought known as logical positivism.[36] The positivists' project was, in some ways, an externalist one. They hoped to develop criteria that would enforce a principled distinction between empirically significant and empirically meaningless sentences. In the minds of some positivists (Schlick, arguably, and Ayer), this criterion would help to vindicate scientific practice by helping to distinguish science from "metaphysics," which was for positivists, a term of abuse.

The positivists were perfectly well aware of Hume's dilemma about the status of the principle of induction—similar problems about even more fundamental principles of logic and mathematics had come to light since his time. But the positivists in effect attempted to rehabilitate epistemological externalism by means of a bold move. They took all the material that was needed to legitimize scientific practice but that could not be traced directly to sensory experience, and relegated it to the *conventions* of human language. This tack had, at least *prima facie,* some advantages over Hume's nativist move: If our epistemic norms are a matter of convention, then (1) there's no longer any question of explaining how we got them—they're there because we *put* them there; and (2) there's no need to justify them because the parameter of evaluation for conventions is not truth but *utility*.

The positivists thus embarked on a program they called "rational reconstruction"—they wanted to show, in detail, how any empirically meaningful claim could be reduced, by the successive application of semantic and logical rules, to statements purely about sensory experience. If such reconstructions could be shown to be possible at least in principle, then all theoretical disagreements could be shown to be susceptible to resolution by appeal to the neutral court of empirical experience. And in all of this, the positivists were committed to basically the same series of assumptions that warranted the view of objectivity that I earlier associated with classical empiricism.

But there were two things absolutely essential to the success of this project. First, there had to be a viable distinction that could be drawn between statements whose truth depended on empirical contingencies (the contentful claims of a theory that formed the substance of the theory) and statements that were true "by convention" and thus part of the logical/semantic structure of the theory. Second, it would have to be shown that the reduction of empirically contentful statements to specific sets of claims about sensory experience could be carried out. But in the early 1950s, Quine (together with Hempel, Goodman, Putnam, and others) began producing decisive arguments against precisely these assumptions.[37] The ensuing changes in analytic epistemology were nothing short of radical.

Quine's main insight was that individual statements do not have any specific consequences for experience if taken individually—that it is only in conjunction with a variety of other claims that experiential consequences can even be derived. It follows from this that no single experience or observation can decisively refute any theoretical claim or resolve any theoretical dispute, and that all experimental tests of hypotheses are actually tests of *conjunctions* of hypotheses. The second insight—actually a corollary of the first point—was that no principled distinction can be drawn among statements on the basis of the grounds of their truth—there can be no distinction between statements made true or false by experience and those whose truth value depends entirely on semantic or logical conventions.

The implications of these two insights were far-reaching. Quine's arguments against the "two dogmas of empiricism" entailed, in the first place, that the confirmation relation could not be hierarchical, as the foundationalist picture required, but must rather be holistic. Because theories have to face "the tribunal of sensory experience as a corporate body" (to use Quine's military-industrial metaphor), there can be no evidentially foundational set of statements that asymmetrically confirm all the others—every statement in the theory is linked by some justificatory connections to every other.

It also meant that responses at the theoretical level to the acquisition of empirical data were not fully dictated by logic. If experimental tests were always tests of *groups* of statements, then if the prediction fails, logic will tell us only that *something* in the group must go, but not *what*. If logic plus data don't suffice to determine how belief is modified in the face of empirical evidence, then there must be, in addition to logic and sensory evidence, *extra-empirical* principles that partially govern theory selection. The "justification" of these principles can only be pragmatic—we are warranted in using them just to the extent that they work.[38]

But to say this is to say that epistemic norms—a category that must include any principle that in fact guides theory selection—are themselves subject to empirical disconfirmation. And indeed, Quine embraces this consequence, explicitly extending the lesson to cover not only pragmatic "rules of thumb," but to rules of logic and language as well. In short, any principle that facilitates the development of knowledge by narrowing down our theoretical options becomes itself a part of the theory, and a part that must be defended on the same basis as any other part. So much for the fact/value distinction.

The reasoning above represents another of the many routes by which Quine's attack on foundationalism can be connected with his critique of the analytic/synthetic distinction, so central to positivist projects. With the demonstration that any belief, no matter how apparently self-evident, could in principle be rejected on the basis of experience, Quine effectively destroyed the prospects for any "first philosophy"—any Archimedean fixed point from which we could inspect our own epistemic practice and pronounce it sound.

But his critique also pointed the way (as Hume's "skeptical solution" did to the problem of induction) to a different approach to the theory of knowledge. Epistemology, according to Quine, had to be "naturalized," transformed into the empirical study of the actual processes—not "rational reconstructions" of those processes—by which human cognizers achieve knowledge.[39] If we accept this approach, several consequences follow for our understanding of knowledge and of the norms and properly govern its pursuit.

The first lesson is one that I believe may be part of what the feminist critics are themselves pointing to in their emphasis on the essential locatedness of all knowledge claims. The lesson is that all theorizing *takes some knowledge for granted.* Theorizing about theorizing is no exception. The decision to treat epistemology as the empirical study of the knower requires us to presume that we can, at least for a class of clear cases, distinguish epistemic success from epistemic failure. The impossibility of the externalist project shows us that we cannot expect to learn *from our philosophy* what counts as knowledge and how much of it we have; rather, we must begin with the assumption that we know certain things and figure out how that happened.

This immediately entails a second lesson. A naturalized approach to knowledge requires us to give up the idea that our own epistemic practice is transparent to us—that we can come to understand how knowledge is obtained either by *a priori* philosophizing or by casual introspection. It requires us to be open to the possibility that the processes that we actually rely on to obtain and process information about the world are significantly different from the ones our philosophy told us had to be the right ones.

Let me digress to point out a tremendous irony here, much remarked upon in the literature on Quine's epistemology and philosophy of mind. Despite his being the chief evangelist of the gospel that everything is empirical, Quine's own philosophy is distorted by his a prioristic commitment to a radically empiricistic, instrumentalist theory of psychology, namely psychological behaviorism. Quine's commitment to this theory—which holds that human behavior can be adequately explained without any reference to mental states or processes intervening between environmental stimuli and the organism's response—is largely the result of his philosophical antipathy to intentional objects, together with a residual sympathy for the foundationalist empiricism that he himself was largely responsible for dismantling.

Chomsky, of course, was the person most responsible for pointing out the in-principle limitations of behaviorism, by showing in compelling detail the empirical inadequacies of behaviorist accounts of the acquisition of language.[40] Chomsky also emphasized the indefensibility of the a prioristic methodological constraints that defined empiricistic accounts of the mind, appealing to considerations that Quine himself marshaled in his own attacks on instrumentalism in nonpsychological domains.[41]

Chomsky's own theory of language acquisition did not differ from the behaviorist account only, or even primarily, in its mentalism. It was also rationalistic: Chomsky quite self-consciously appealed to classical rationalistic forms of argument about the necessity of mental partiality in establishing the empirical case for his strong nativism. Looking at the actual circumstances of language acquisition, and then at the character of the knowledge obtained in those circumstances, Chomsky argued that the best explanation of the whole process is one that attributes to human beings a set of innate biases limiting the kinds of linguistic hypotheses available for their consideration as they respond to the welter of data confronting them.[42]

Chomsky can thus be viewed, and is viewed by many, as a naturalized epistemologist *par excellence*. What his work shows is that a naturalized approach to epistemology—in this case, the epistemology of language—yields an *empirical* vindication of rationalism. Since Chomsky's pathbreaking critique of psychological behaviorism, and the empiricist conception of mind that underlies it, nativism in psychology has flourished, and a significant degree of rationalism has been imported into contemporary epistemology.

A casual student of the analytic scene who has read only Quine could, of course, be forgiven for failing to notice this, given Quine's adamant commitment to an empiricist conception of mind; this may explain why so many of the feminist critics of contemporary epistemology seem to identify analytic epistemology with empiricism and to ignore the more rationalistic alternatives that have developed out of the naturalized approach. But I think, too, that the original insensitivity to the details of the original rationalist/empiricist controversy plays a role. Anyone who properly appreciates the import of the rationalist defense of the value of partiality will, I think, see where Quine's rejection of externalism is bound to lead.

So let's do it. I turn now to the feminist critique of objectivity and the bias paradox.

III. Quine as Feminist: What Naturalized Epistemology Can Tell Us About Bias

I've argued that much of the feminist criticism of "mainstream" epistemology depends on a misreading of both contemporary analytic philosophy, and of the tradition from which it derives. But it's one thing to show that contemporary analytic philosophy is not what the feminist critics think it is, and quite another to show that the contemporary analytic scene contains an epistemology that can serve as an adequate *feminist* epistemology. To do this, we must return to the epistemological issues presented to us by feminist theory and see how naturalized epistemology fares with respect to them. I want eventually to show how a commitment to a naturalized epistemology provides some purchase on the problem of conceptualizing bias, but in order to do that, we must look in some detail at those feminist arguments directed against the notion of objectivity.

Capitalist Science and the Ideal of Objectivity

As we've seen, one of the most prominent themes in feminist epistemology and feminist philosophy of science concerns the alleged ideological function of a certain conception of objectivity. Many feminist critics see a connection between radical (i.e., nonliberal) critiques of science and feminist critiques of "received" epistemology. Such critics take as their starting point the observation that science, as it has developed within industrialized capitalist societies like the United

States, is very much an instrument of oppression: Rather than fulfilling its Enlightenment promise as a liberatory and progressive force, institutionalized science serves in fact to sustain and even to enhance existing structures of inequality and domination.[43]

Although all feminists agree that part of the explanation of this fact must be that modern science has been distorted by the sexist, racist, and classist biases it inherits from the society in which it exists, feminist theorists divide on the issue of whether some "deeper" explanation is required. Alison Jaggar's "liberal feminists" and Sandra Harding's "feminist empiricists" hold that society and science are both potentially self-correcting—that more equitable arrangements of power and more scrupulous enforcement of the rules of fairness would turn science back to its natural progressive course.

But Harding and Jaggar, together with Lorraine Code and Evelyn Fox Keller, disagree with this liberal analysis. They contend that the modern scientific establishment has not simply inherited its oppressive features from the inequitable society that conditions it. Rather, they claim, a large part of the responsibility for societal injustices lies deep within science itself, in the conception of knowledge and knowers that underlies "scientific method." These critics charge that the very ideals to which Western science has traditionally aspired—particularly rationality and objectivity—serve to sanction and promote a form of institutionalized inquiry uniquely suited to the needs of patriarchy. Thus, it's argued, feminist critique must not stop at exposing cases in which science has broken its own rules; it must press on to expose the androcentric bias inherent in the rules themselves.

Thus Evelyn Fox Keller claims that any critique that does not extend to the rules of scientific method allies itself with political liberalism in virtue of its epistemology. Any such critique, she argues, "can still be accommodated within the traditional framework by the simple argument that the critiques, if justified, merely reflect the fact that [science] is not sufficiently scientific." In contrast, there is "the truly radical critique that attempts to locate androcentric bias . . . in scientific ideology itself. The range of criticism takes us out of the liberal domain and requires us to question the very assumptions of rationality that underlie the scientific enterprise."[44]

All this seems to set a clear agenda for feminist philosophers who wish to be part of the struggle for a genuinely radical social transformation: If one's going to go deeper politically and criticize the presuppositions of liberal political theory, then one must coordinately

go deeper *conceptually* and criticize the presuppositions of the epistemology and metaphysics that underwrite the politics.

But does this argument work? I think that it doesn't. To see why,
we need to look more closely at the epistemological position that the
feminist critics take to be allied with liberalism and look in more detail
at the argument that is supposed to show that such a view of knowledge
is oppressive.

The "traditional" epistemology pictured in the work of Flax, Code,
and Jaggar, I've argued, is an unvigorous hybrid of rationalist and
empiricist elements, but the features that are supposed to limit it from
the point of view of feminist critique of science all derive from the
empiricist strain. Specifically, the view of knowledge in question
contains roughly the following elements:

(1) it is strongly foundationalist: It is committed to the view that there
is a set of epistemically privileged beliefs, from which all knowledge is in
principle derivable.

(2) it takes the foundational level to be constituted by reports of
sensory experience, and views the mind as a mere calculating device,
containing no substantive contents other than what results from experience.

(3) as a result of its foundationalism and its empiricism, it is committed
to a variety of sharp distinctions: observation/theory, fact/value, context
of discovery/context of justification.

This epistemological theory comes very close to what Hempel has
termed "narrow inductivism,"[45] but I'm just going to call it the
"Dragnet" theory of knowledge. To assess the "ideological potential"
of the Dragnet theory, let's look first at some of the epistemic values
and attitudes the theory supports.

To begin with, because of its empiricistic foundationalism, the view
stigmatizes both inference and theory. On this view, beliefs whose
confirmation depends upon logical relations to other beliefs bear a
less direct, less "objective" connection to the world than reports of
observations, which are supposed to provide us transparent access to
the world. To "actually see" or "directly observe" is better, on this
conception, than to infer, and an invidious distinction is drawn between the "data" or "facts" (which are incontrovertible) on the one
hand and "theories" and "hypotheses" (unproven conjectures) on
the other.

Second, the view supports the idea that any sound system of beliefs
can, in principle, be rationally reconstructed. That is, a belief worth

having is either itself a fact or can be assigned a position within a clearly articulated confirmational hierarchy erected on fact. With this view comes a denigration of the epistemic role of hunches and intuitions. Such acts of cognitive impulse can be difficult to defend "rationally" if the standards of defense are set by a foundationalist ideal. When a hunch can't be defended, but the individual persists in believing it anyway, that's *ipso facto* evidence of irresponsibility or incompetence. Hunches that happen to pay off are relegated to the context of discovery and are viewed as inessential to the justification of the ensuing belief. The distinction between context of discovery and context of justification itself follows from foundationalism: As long as it's possible to provide a rational defense of a belief *ex post facto* by demonstrating that it bears the proper inferential relation to established facts, we needn't give any thought to the circumstances that actually gave rise to that belief. Epistemic location becomes, to that extent, evidentially irrelevant.

Finally, the Dragnet theory is going to lead to a certain conception of how systematic inquiry ought to work. It suggests that good scientific practice is relatively mechanical: that data gathering is more or less passive and random, that theory construction emerges from the data in a relatively automatic way, and that theory testing is a matter of mechanically deriving predictions and then subjecting them to decisive experimental tests. Science (and knowledge-seeking generally) will be good *to the extent that* its practitioners can conform to the ideal of objectivity.

This ideal of objective method requires a good researcher, therefore, to put aside all prior beliefs about the outcome of the investigation, and to develop a willingness to be carried wherever the facts may lead. But other kinds of discipline are necessary, too. Values are different in kind from facts, on this view, and so are not part of the confirmational hierarchy. Values (together with the emotions and desires connected with them) become, at best, epistemically irrelevant and, at worst, disturbances or distortions. Best to put them aside, and try to go about one's epistemic business in as calm and disinterested a way as possible.

In sum, the conception of ideal epistemic practice yielded by the Dragnet theory is precisely the conception that the feminist critics disdain. Objectivity, on this view (I'll refer to it from now on as "Dragnet objectivity"), is the result of complete divestiture—divestiture of theoretical commitments, of personal goals, of moral values, of hunches and intuitions. We'll get to the truth, sure as taxes, provided everyone's willing to be rational and to play by the (epistemically relevant) rules. Got an especially knotty problem to solve? Just the facts, ma'am.

Now let's see how the Dragnet theory of knowledge, together with the ideal of objectivity it supports, might play a role in the preservation of oppressive structures.

Suppose for the sake of argument that the empirical claims of the radical critics are largely correct. Suppose, that is, that in contemporary U.S. society institutionalized inquiry does function to serve the specialized needs of a powerful ruling elite (with trickle-down social goods permitted insofar as they generate profits or at least don't impede the fulfillment of ruling-class objectives). Imagine also that such inquiry is very costly, and that the ruling elite strives to socialize those costs as much as possible.

In such a society, there will be a great need to obscure this arrangement. The successful pursuit of the agendas of the ruling elite will require a quiescent—or, as it's usually termed, "stable"—society, which would surely be threatened if the facts were known. Also required is the acquiescence of the scientists and scholars, who would like to view themselves as autonomous investigators serving no masters but the truth and who would deeply resent the suggestion (as anyone with any self-respect would) that their honest intellectual efforts subserve any baser purpose.

How can the obfuscation be accomplished? One possibility would be to promote the idea that science is organized for the sake of *public* rather than *private* interests. But the noble lie that science is meant to make the world a better place is a risky one. It makes the public's support for science contingent upon science's producing tangible and visible public benefits (which may not be forthcoming) and generates expectations of publicity and accountability that might lead to embarrassing questions down the road.

An altogether more satisfactory strategy is to promote the idea that science is *value-neutral*—that it's organized for the sake of *no* particular interests at all! Telling people that science serves only the truth is safer than telling people that science serves *them,* because it not only hides the truth about who benefits, but deflects public attention away from the whole question. Belief in the value-neutrality of science can thus serve the conservative function of securing *unconditional* public support for what are in fact ruling-class initiatives. Any research agenda whatsoever—no matter how pernicious—can be readily legitimated on the grounds that it is the natural result of the self-justifying pursuit of truth, the more or less inevitable upshot of a careful look at the facts.

It will enhance the lie that science is objective, to augment it with the lie that scientists as individuals are especially "objective," either by nature or by dint of their scientific training. If laypersons can be brought to believe this, then the lie that scientific practice can transcend its compromised setting becomes somewhat easier to swallow. And if *scientists* can be brought to embrace this gratifying self-image, then the probability of *their* acquiescence in the existing system will be increased. Scientists will find little cause for critical reflection on their own potential biases (since they will believe that they are more able than others to put aside their own interests and background beliefs in the pursuit of knowledge), and no particular incentive to ponder the larger question of who actually is benefiting from their research.[46]

Now in such a society, the widespread acceptance of a theory of knowledge like the Dragnet theory would clearly be a good thing from the point of view of the ruling elite. By fostering the epistemic attitudes it fosters, the Dragnet theory helps confer special authority and status on science and its practitioners and deflects critical attention away from the material conditions in which science is conducted. Furthermore, by supporting Dragnet objectivity as an epistemic ideal, the theory prepares the ground for reception of the ideology of the objectivity of science.

In a society in which people have a reason to believe that science is successful in yielding knowledge, the Dragnet theory and the ideology of objectivity will in fact be mutually reinforcing. If one believes that science must be objective to be good, then if one independently believes that science is good, one must also believe that science *is objective!* The Dragnet theory, taken together with propagandistic claims that science is value-neutral, etc., offers an *explanation* of the fact that science leads to knowledge. Against the background belief that knowledge is actually structured the way the Dragnet theory says it is, the *success* of science seems to confirm the ideology.

We can conclude from all this that the Dragnet theory, along with the ideal of objectivity it sanctions, has clear ideological value, in the sense that their acceptance may play a causal role in people's acceptance of the ideology of scientific objectivity.

But we cannot infer from this fact either that the Dragnet theory is false or that its ideals are flawed. Such an inference depends on conflating what are essentially *prescriptive* claims (claims about how science ought to be conducted) with *descriptive* claims (claims about how science is in fact conducted). It's one thing to embrace some particular ideal of scientific method and quite another to accept ideologically useful assumptions about the satisfaction of that ideal within existing institutions.[47]

Note that in a society such as the one I've described, the ideological value of the Dragnet theory depends crucially on how successfully it can be promulgated *as a factual characterization* of the workings of the intellectual establishment. It's no use to get everyone to believe simply that it would be a good thing if scientists *could* put aside their prior beliefs and their personal interests; people must be brought to believe that scientists largely *succeed* in such divestitures. The ideological cloud of Dragnet objectivity thus comes not so much from the belief that science *ought* to be value-free, as from the belief that it *is* value-free. And of course it's precisely the fact that science is *not* value-free in the way it's proclaimed to be that makes the ideological ploy necessary in the first place.

If science as an institution fails to live up to its own ideal of objectivity, then the character of existing science entails nothing about the value of the ideal, nor about the character of some imagined science which *did* live up to it. In fact, notice that the more we can show that compromised science is *bad* science (in the sense of leading to false results), the less necessary we make it to challenge the Dragnet theory itself. A good part of the radical case, after all, is made by demonstrating the ways in which scientific research has been *distorted* by some of the very factors a Dragnet epistemologist would cite as inhibitors of epistemic progress: prejudiced beliefs, undefended hunches, material desires, ideological commitments.

There's no reason, in short, why a Dragnet theorist couldn't come to be convinced of the radical analysis of the material basis of science. Such a person might even be expected to experience a special kind of outrage at discovering the way in which the idea of objectivity is ideologically exploited in the service of special interests, much the way many peace activists felt when they first learned of some of the realities masked by U.S. officials' pious avowals of their commitment to "human rights" and "democracy."

A materialist analysis of institutionalized science leads to awareness of such phenomena as the commoditization of knowledge, the "rationalization" of scientific research, and the proletarianization of scientists. Such phenomena make the limits of liberal reformism perfectly clear: Not even the most scrupulous adherence to prescribed method on the part of individual scientists could by itself effect the necessary transformations. But it's possible for even a Dragnet theorist to ac-knowledge these limits, and to do so without giving up the ideal of neutral objectivity.

I began by considering the claim, defended by several feminist theorists, that "traditional" epistemology limits the possibilities for exposing the machinations of the elite because it endorses the rules of the elite's game. On the contrary, I've argued; since a big part of the lie that needs exposing is the fact that capitalist science *doesn't follow* its own rules, the task of exposing the ideology of scientific objectivity needn't change the rules. A radical critique of science and society, *even if* it implicates certain ideals, *does not require repudiation of those ideals.*

Naturalized Epistemology and the Bias Paradox

What I think I've shown so far is that if our only desideratum on an adequate critical epistemology is that it permits us to expose the real workings of capitalist patriarchy, then the Dragnet theory will do just fine, *pace* its feminist critics. But I certainly do not want to defend that theory; nor do I want to defend as an epistemic ideal the conception of objectivity as neutrality. In fact, I want to join feminist critics in rejecting this ideal. But I want to be clear about the proper basis for criticizing it.

There are, in general, two strategies that one can find in the episte-mological literature for challenging the ideal of objectivity as impartial-ity. (I leave aside for the moment the question of why one might want to challenge an epistemic ideal, though this question will figure importantly in what follows.) The first strategy is to prove the *impossi-bility* of satisfying the ideal—this involves pointing to the *ubiquity* of bias. The second strategy is to try to demonstrate the *undesirability* of satisfying the ideal—this involves showing the *utility* of bias. The second strategy is employed by some feminist critics, but often the first strategy is thought to be sufficient, particularly when it's pursued together with the kind of radical critique of institutionalized science discussed above. Thus Jaggar, Code, and others emphasize the essen-tial locatedness of every individual knower, arguing that if all knowl-edge proceeds from some particular perspective, then the transcendent standpoint suggested by the ideology of objectivity is unattainable. All knowledge is conditioned by the knower's location, it is claimed; if we acknowledge that, then we cannot possibly believe that anyone is "objective" in the requisite sense.

But the appeal to the *de facto* partiality of all knowledge is simply not going to justify rejecting the ideal of objectivity, for three reasons. In the first place, the wanted intermediate conclusion—that Dragnet objectivity is impossible—does not follow from the truism that all knowers are located. The Dragnet conception of impartiality is perfectly compatible with the fact that all knowers start from some particular place. The Dragnet theory, like all empiricist theories, holds that knowledge is a strict function of the contingencies of experience. It therefore entails that differences in empirical situation will lead to differences in belief, and to that extent validates the intuition that all knowledge is partial.[48] Thus the neutrality recommended by the Dragnet theory does not enjoin cognizers to abjure the particularities of their own experience, only to honor certain strictures in drawing conclusions from that experience. Impartiality is not a matter of where you are, but rather how well you do from where you sit.

In the second place, even if it could be shown to be impossible for human beings to achieve perfect impartiality, that fact in itself would not speak against Dragnet objectivity *as an ideal*. Many ideals—particularly moral ones—are unattainable but that does not make them useless, or reveal them to be inadequate as ideals.[49] The fact—and I have no doubt that it is a fact—that no one can fully rid oneself of prejudices, neurotic impulses, selfish desires, and other psychological detritus does not impugn the oral or the cognitive value of attempting to do so. Similarly, the fact that no one can fully abide by the cognitive strictures imposed by the standards of strict impartiality doesn't entail that one oughtn't to try. The real test of the adequacy of a norm is not whether it can be realized, but (arguably) whether we get closer to what we want if we try to realize it.

But the third and the most serious problem with this tack is that it is precisely the one that is going to engender the bias paradox. Notice that the feminist goal of exposing the structures of interestedness that constitute patriarchy and other forms of oppression requires doing more than just demonstrating that particular interests are being served. It requires criticizing that fact, showing that there's something wrong with a society in which science selectively serves the interests of one dominant group. And it's awfully hard to see how such a critical stand can be sustained without some appeal to the value of impartiality.

A similar problem afflicts the variation on this strategy that attempts to base a critique of the norm of objectivity on the androcentric features of its *source*. Even if it could be established that received epistemic norms originated in the androcentric fantasies of European white males (and I meant to give some reason to question this in section II), how is that fact supposed to be elaborated into a *critique* of those norms? All knowledge is partial—let it be so. How then does the particular partiality of received conceptions of objectivity diminish their worth?

The question that must be confronted by anyone pursuing this strategy is basically this: If bias is ubiquitous and ineliminable, then what's the good of exposing it? It seems to me that the whole thrust of feminist scholarship in this area has been to demonstrate that androcentric biases have distorted science and, indeed, distorted the search for knowledge generally. But if biases are distorting, and if we're all biased in one way or another, then it seems there could be no such thing as an *undistorted* search for knowledge. So what are we complaining about? Is it just that we want it to be distorted in *our* favor, rather than in theirs? We must say something about the badness of the biases we expose or our critique will carry no normative import at all.

We still have to look at the second of the two strategies for criticizing the ideal of objectivity, but this is a good place to pick up the question I bracketed earlier on: *Why* might one want to challenge an epistemic ideal? If my arguments have been correct up to this point, then I have shown that many of the arguments made against objectivity are not only unsound but ultimately self-defeating. But by now the reader must surely be wondering why we need *any* critique of the notion of objectivity as neutrality. If radical critiques of the ideology of scientific objectivity are consistent with respect for this ideal, and if we need some notion of objectivity anyway, why not this one?

The short answer is this: because the best empirical theories of knowledge and mind do not sanction pure neutrality as sound epistemic policy.

The fact is that the Dragnet theory is *wrong*. We know this for two reasons: First, the failure of externalism tells us that its foundationalist underpinnings are rotten, and second, current work in empirical psychology tells us that its empiricist conception of the mind is radically incorrect. But if the Dragnet theory is wrong about the structure of knowledge and the nature of the mind, then the main source of warrant for the ideal of epistemic neutrality is removed. It becomes an open question whether divestiture of emotions, prior beliefs, and moral commitments hinders, or aids, the development of knowledge.

The fact that we find ourselves wondering about the value of a proposed epistemic ideal is itself a consequence of the turn to a naturalized epistemology. As I explained in section II, Quine's critique of externalism entailed that epistemic norms themselves were among the presuppositions being subjected to empirical test in the ongoing process of theory confirmation. This in itself authorizes the project of *criticizing* norms—it makes coherent and gives point to a project which could be nothing but an exercise in skepticism, to an externalist's way of thinking.

Naturalized epistemology tells us that there is no presuppositionless position from which to assess epistemic practice, that we must take some knowledge for granted. The only thing to do, then, is to begin with whatever it is we think we know, and try to figure out how we came to know it: Study knowledge by studying the knower. Now if, in the course of such study, we discover that much of human knowledge is possible only because our knowledge seeking does not conform to the Dragnet model, then we will have good empirical grounds for rejecting perfect objectivity as an epistemic ideal. And so we come back to the second of the two strategies I outlined for challenging the ideal of objectivity. Is there a case to be made against the desirability of epistemic neutrality? Indeed there is, on the grounds that a genuinely open mind, far from leading us closer to the truth, would lead to epistemic chaos.

As I said in section II, empirical work in linguistics and cognitive science is making it increasingly clear how seriously mistaken the empiricist view of the mind actually is. From Chomsky's groundbreaking research on the acquisition of language, through David Marr's theory of the computational basis of vision, to the work of Susan Carey, Elizabeth Spelke, Barbara Landau, Lila Gleitman, and others in developmental psychology, the evidence is mounting that inborn conceptual structure is a crucial factor in the development of human knowledge.[50]

Far from being the streamlined, uncluttered logic machine of classical empiricism, the mind now appears to be much more like a bundle of highly specialized modules, each natively fitted for the analysis and manipulation of a particular body of sensory data. General learning strategies of the sort imagined by classical empiricists, if they are employed by the mind at all, can apply to but a small portion of the cognitive tasks that confront us. Rationalism vindicated.

But if the rationalists have turned out to be right about the structure of the mind, it is because they appreciated something that the empiricists missed—the value of partiality for human knowers. Whatever might work for an ideal mind, operating without constraints of time or space, it's clear by now that complete neutrality of the sort empiricists envisioned would not suit human minds in human environments. A completely "open mind," confronting the sensory evidence we confront, could never manage to construct the right systems of knowledge we construct in the short time we take to construct them. From the point of view of an *unbiased* mind, the human sensory flow contains both too much information and too little: too much for the mind to generate *all* the logical possibilities, and too little for it to decide among even the relatively few that *are* generated.

The problem of paring down the alternatives is the defining feature of the human epistemic condition. The problem is partly solved, I've been arguing, by one form of "bias"—native conceptual structure. But it's important to realize that this problem is absolutely endemic to human knowledge seeking, whether we're talking about the subconscious processes by which we acquire language and compute sensory information, or the more consciously accessible processes by which we explicitly decide what to believe. The everyday process of forming an opinion would be grossly hampered if we were really to consider matters with anything even close to an "open mind."

This point is one that Quine has emphasized over and over in his discussions of the underdetermination of theory by data. If we had to rely on nothing but logic and the contingencies of sensory experience, we could never get anywhere in the process of forming an opinion, because we would have *too many choices*. There are an infinite number of distinct and incompatible hypotheses consistent with any body of data, never mind that there are always more data just around the corner, and never mind that we're logically free to reinterpret the "data" to save our hypotheses. If we really had to approach data gathering and theory building with a perfectly open mind, we wouldn't get anywhere.

This insight is also borne out by the history of science. As Thomas Kuhn has pointed out, science is at its least successful during the periods in its history when it most closely resembles the popular models of scientific objectivity. During a discipline's "pre-paradigm" phase, when there is no consensus about fundamental principles, nor even about what to count as the central phenomena, research is anarchic and unproductive. But progress accelerates dramatically when a discipline enters its mature period, marked by the emergence of a theory—a paradigm—capable of organizing the phenomena in a compelling enough way that it commands near-universal acceptance.

Kuhn emphasizes that one of the chief benefits a paradigm brings with it is a degree of closure about foundational issues, instilling in members of the community a principled and highly functional unwillingness to reconsider basic assumptions. The paradigm not only settles important empirical controversies, but also decides more methodological matters—what are the acceptable forms of evidence, what is the right vocabulary for discussing things, what are the proper standards for judging research. The fact is that all of these matters are disputable in principle—but a paradigm relieves its adherents of the considerable burden of having constantly to dispute them.

But what this means is that the practice and attitudes of scientists working within a paradigm will systematically deviate from the popular ideal of scientific objectivity: They will approach their research with definite preconceptions, and they will be reluctant to entertain hypotheses that conflict with their own convictions. Kuhn's point, however, is that the existence of such closed-mindedness among working scientists—what he calls "the dogmatism of mature science"—is not to be regretted; that it is actually beneficial to the course of scientific development: "Though preconception and resistance to innovation could very easily choke off scientific progress, their omnipresence is nonetheless symptomatic of characteristics upon which the continuing vitality of research depends."[51]

Once we appreciate these aspects of mature science, we can explain a great deal about how a fantasy of the pure objectivity of science can take hold independently of any ideological purposes such a fantasy might serve. (This is important if we want a serious, nuanced story about how ideologies work.) The fact that certain tenets of theory are, for all practical purposes, closed to debate can render invisible their actual status as hypotheses. Deeply entrenched theoretical principles, like the laws of thermodynamics or the principle of natural selection, become established "facts."[52] Similarly, the high degree of theoretical background required to translate various numbers and images into observations or data is forgotten by people accustomed to performing the requisite inferences on a daily basis.

Consensus and uniformity thus translate into objectivity. The more homogeneous an epistemic community, the more objective it is likely to regard itself, and, if its inquiries are relatively self-contained, the more likely it is to be viewed as objective by those outside the community. This suggests one fairly obvious explanation for the general perception that the physical sciences are more objective than the social sciences: Sociology, political science, economics, and psychology are disciplines that still lack paradigms in Kuhn's technical sense.

Because there is still public debate in these fields about basic theoretical and methodological issues, there can be no credible pretense by any partisan of having hold of the unvarnished truth.

The kind of bias that Kuhn is here identifying is, of course, different in several important respects from the kinds of biases that classical rationalists and contemporary cognitive psychologists are concerned with. For one thing, the biases that come with belief in a paradigm are acquired rather than innate; for another, there is an important social component in one case but not in the other. The lesson, however, is still the same: Human beings would know less, not more, if they were to actualize the Dragnet ideal.

What all this means is that a naturalized approach to knowledge provides us with *empirical* grounds for rejecting pure neutrality as an epistemic ideal, and for valuing those kinds of "biases" that serve to trim our epistemic jobs to manageable proportions. But it also seems to mean that we have a new route to the bias paradox—if biases are now not simply ineliminable, but downright *good,* how is it that *some* biases are *bad?*

I'm going to answer this question, honest, but first let me show how bad things really are. It's possible to see significant analogies between the function of a paradigm within a scientific community, and what is sometimes called a "worldview" within other sorts of human communities. Worldviews confer some of the same cognitive benefits as paradigms, simplifying routine epistemic tasks, establishing an informal methodology of inquiry, etc., and they also offer significant social benefits, providing a common sense of reality and fostering a functional sense of normalcy among members of the community.

But what about those outside the community? A shared language, a set of traditions and mores, a common sense of what's valuable and why—the very things that bind some human beings together in morally valuable ways—function simultaneously to exclude those who do not share them. Moreover, human communities are not homogeneous. In a stratified community, where one group of people dominates others, the worldview of the dominant group can become a powerful tool for keeping those in the subordinate groups in their places.

The real problem with the liberal conceptions of objectivity and neutrality begins with the fact that while they are unreliable, it's possible for those resting comfortably in the center of a consensus to find that fact invisible. Members of the dominant group are given no reason to question their own assumptions: Their world-view acquires, in their minds, the status of established fact. Their opinions are transformed into what "everybody" knows.[53] Furthermore, these privileged individuals have the power to promote and elaborate their own worldview in public forums while excluding all others, tacitly setting limits to the range of "reasonable" opinion.[54]

Because of the familiarity of its content, the "objectivity" of such reportage is never challenged. If it were, it would be found woefully lacking *by liberal standards*. That's because the liberal ideal of objectivity is an *unreasonable* one; it is not just unattainable, but unattainable by a long measure. But because the challenge is *only* mounted against views that are aberrant, it is *only* such views that will ever be demonstrated to be "non-objective," and thus *only* marginal figures that will ever be charged with bias.[55]

Lorraine Code makes a similar point about the unrealistic stringency of announced standards for knowledge.[56] She rightly points out that most of what we ordinarily count as knowledge wouldn't qualify as such by many proposed criteria. I would go further and say that as with all unrealistically high standards, they tend to support the status quo—in this case, received opinion—by virtue of the fact that they will only be invoked in "controversial" cases, i.e., in case of challenge to familiar or received or "expert" opinion. Since the standards are unreasonably high, the views tested against them will invariably be found wanting; since the only views so tested will be unpopular ones, their failure to pass muster serves to add additional warrant to prevailing prejudices, as well as a patina of moral vindication to the holders of those prejudices, who can self-righteously claim to have given "due consideration" to the "other side."

But what are we anti-externalist, naturalized epistemologists to say about this? We can't simply condemn the members of the dominant class for their "bias," for their lack of "open-mindedness" about our point of view. To object to the hegemony of ruling-class opinion on this basis would be to tacitly endorse the discredited norm of neutral objectivity. "Biased" they are, but then, in a very deep sense, so are we. The problem with ruling-class "prejudices" cannot be the fact that they are deeply held beliefs, or beliefs acquired "in advance" of the facts—for the necessity of such *kinds* of belief is part of the human epistemic condition.

The real problem with the ruling-class worldview is not that it is biased; it's that it is false. The epistemic problem with ruling-class people is not that they are closed-minded; it's that they hold too much power. The recipe for radical epistemological action then becomes simple: Tell the truth and get enough power so that people have to listen. Part of telling the truth, remember, is telling the truth about how knowledge is actually constructed—advocates of feminist epistemology are absolutely correct about that. We do need to dislodge those attitudes about knowledge that give unearned credibility to elements of the ruling-class worldview, and this means dislodging the hold of the Dragnet theory of knowledge. But we must be clear: The Dragnet theory is not false because it's pernicious; it's pernicious because it is false.

Whether we are talking in general about the ideology of scientific objectivity, or about particular sexist and racist theories, we must be willing to talk about truth and falsity. If we criticize such theories primarily on the basis of their ideological function, we risk falling prey to the very illusions about objectivity that we are trying to expose. I think this has happened to some extent within feminist epistemology. Because so much of feminist criticism has been oblivious to the rationalistic case that can be made against the empiricistic conception of mind at work in the Dragnet theory, empiricistic assumptions continue to linger in the work of even the most radical feminist epistemologists. This accounts, I believe, for much of the ambivalence about Dragnet objectivity expressed even by those feminist critics who argue most adamantly for its rejection.

This ambivalence surfaces, not surprisingly, in discussions about what to do about bad biases, where positive recommendations tend to fall perfectly in line with the program of liberal reformism. Lorraine Code's discussion of stereotypical thinking provides a case in point.[57] Code emphasizes, quite correctly, the degree to which stereotypical assumptions shape the interpretation of experience, both in science and in everyday life. But despite her recognition of the "unlikelihood of pure objectivity,"[58] the "unattainability of pure theory-neutrality,"[59] and her acknowledgment of the necessary role of background theory *in science,* her recommendations for reforming everyday epistemic practice are very much in the spirit of liberal exhortations to open-mindedness. She sees a difference between a scientist's reliance on his or her paradigm, and ordinary dependence on stereotypes:

> It is not possible for practitioners to engage in normal science without paradigms to guide their recognition of problems, and their problem-solving endeavours. Stereotype-governed thinking is different in this respect, for it is both possible and indeed desirable to think and to know in a manner *not* governed by stereotypes.[60]

But it's by no means clear that it *is* possible. I sense that Code has not appreciated the depth of human reliance on theories that cannot be shown to be "derived from the facts alone." In characterizing certain kinds of background belief and certain forms of "hasty generalization" *as stereotypes,* she is presupposing a solution to the very problem that must be solved: viz., telling which of the background theories that we routinely bring to bear on experience are *reliable* and which ones are not.

The liberal epistemological fantasy, still somewhat at work here, is that there will be formal marks that distinguish good theories from bad. The empiricist version of this fantasy is that the formal mark consists in a proper relation between theory and "fact." In this case, the good theories are supposed to be the ones that derive in the proper way from the data, whereas the bad ones—the biases, the prejudices, the stereotypes—are the ones that antedate the data. But once we realize that theory infects observation and that confirmation is a multidirectional relation, we must also give up on the idea that the good theories are going to look different from the bad theories. They can't be distinguished on the basis of their formal relation to the "facts," because (1) there are no "facts" in the requisite sense, and (2) there are too many good biases whose relation to the data will appear as tenuous as those of the bad ones.

But what's the alternative?

A naturalized approach to knowledge, because it requires us to give up *neutrality* as an epistemic ideal, also requires us to take a different attitude toward bias. We know that human knowledge requires biases; we also know that we have no possibility of getting *a priori* guarantees that our biases incline us in the right direction. What all this means is that the "biasedness" of biases drops out as a parameter of epistemic evaluation. There's only one thing to do, and it's the course always counseled by a naturalized approach: *We must treat the goodness or badness of particular biases as an empirical question.*

A naturalistic study of knowledge tells us biases are good when and to the extent that they facilitate the gathering of *knowledge*—that is, when they lead us to the truth. Biases are bad when they lead us *away* from the truth. One important strategy for telling the difference between good and bad biases is thus to evaluate the overall theories in which the biases figure. This one point has important implications for feminist theory in general and for feminist attitudes about universalist or essentialist theories of human nature in particular.

As we saw in section II, much of the feminist criticism raised against cognitive essentialism focused on the fact that rationalist and Kantian theories of the human essence were all devised by men, and based, allegedly, on exclusively male experience. Be that so—it would still follow from a naturalized approach to the theory of knowledge that it is an *empirical* question whether or not 'androcentrism' of that sort leads to bad theories. Partiality does not in general compromise theories; as we feminists ourselves have been insisting, all theorizing proceeds from *some* location or other. We must therefore learn to be cautious of claims to the effect that particular forms of partiality will inevitably and systematically influence the outcome of an investigation. Such claims must be treated as empirical hypotheses, subject to investigation and challenge, rather than as enshrined first principles.

So what about universalist or essentialist claims concerning human nature? I have argued that there really are no grounds for regarding such claims as antipathetic to feminist aspirations or even to feminist insights regarding the importance of embodiment or the value of human difference. Suggestions that essentialist theories reify aspects of specifically male experience, I argued, involve a serious misunderstanding of the rationalist strategy. But notice that even if such charges were true, the real problem with such theories should be their *falseness,* rather than their androcentrism. A theory that purports to say what human beings are like essentially must apply to *all human beings;* if it does not, it is wrong, whatever its origins.

In fact, I think there is excellent evidence for the existence of a substantial human nature and virtually no evidence for the alternative, the view that there is no human essence. But what's really important is to recognize that the latter view is as much a substantive empirical thesis as the Cartesian claim that we are essentially rational language-users. We need to ask ourselves *why* we ought to believe that human selves are, at the deepest level, "socially constructed"—the output of a confluence of contingent factors.[61]

Another thing that a naturalized approach to knowledge offers us is the possibility of *an empirical theory of biases.* As we've already seen, there are different kinds of biases—some are natively present, some are acquired. An empirical study of biases can refine the taxonomy and possibly tell us something about the reliability and the corrigibility of biases of various sorts. It may turn out that we can on this basis get something like a principled sorting of biases into good ones and bad ones, although it will be more likely that we'll learn that even a "good" bias can lead us astray in certain circumstances.[62]

One likely upshot of an empirical investigation of bias is a better understanding of the processes by which human beings design research programs. What we decide to study and how we decide to study it are matters in which unconscious biases—tendencies to see certain patterns rather than others, to attend to certain factors rather than others, to act in accordance with certain interests rather than others—play a crucial role. We can't eliminate the biases—we shouldn't want to, for we'd have no research programs left if we did—but we can identify the particular empirical presuppositions that lie behind a particular program of research so that we can subject them, if necessary, to empirical critique.

One important issue is the *saliency* of certain properties. Every time a study is designed, a decision is made, tacitly or explicitly, to pay attention to some factors and to ignore others. These "decisions" represent tacit or explicit hypotheses about the likely connection between various aspects of the phenomena under study, hypotheses that can be subjected to empirical scrutiny.

Imagine a study purporting to investigate the development of human language by examining a sample of two hundred preschoolers. Must the sample, to be a valid basis for extrapolation, contain boys and girls? Must it be racially mixed? How one answers this question will depend on the empirical assumptions one makes about the likely connection between parameters like gender and race, on the one hand, and the language faculty on the other. To think that gender or race must be controlled for in such studies is to make a substantive empirical conjecture—in this case, it is to deny the rationalistic hypothesis that human beings' biological endowment includes a brain structured in a characteristic way, and to make instead the assumption that cognitive development is sensitive to the kinds of differences that we *socially* encode as gender and race.

Such an assumption, laid out this baldly, seems pretty dubious. Indeed, it's hard to see what such an assumption is doing other than reflecting sexist, racist, and classist beliefs to the effect that social groupings are determined by biological groupings. Realizing this is a necessary first step to countering the genuinely pernicious "essentialist" theories of Jensen, Herrnstein, and the human sociobiologists and to exposing the racism and sexism inherent in their programs of "research." Such "research" is precisely at odds with rationalist methodology, which only invokes human essences as a way of explaining human *commonalities*—and then, only when such commonalties cannot plausibly be explained by regularities in the environment.

Consider, for example, the claims that blacks are "innately" less intelligent than whites.[63] In the first place, we must point out, as we do, that race is not a biological kind, but rather a *social* kind. That is to say that while there may be a biological explanation for the presence of each of the characteristics that constitute racial criteria—skin color, hair texture, and the like—the *selection of those characteristics as criteria* of membership in some category is *conventionally* determined. Here is where the empiricist notion of "nominal essence" has some work to do: race, in contrast to some other categories, *is* socially constructed.

The second step is to point out that if such classifications as race fail to reflect deep regularities in human biology, and reflect instead only historically and culturally specific interests, then there is no reason, *apart from racist ones,* to investigate the relation between race and some presumably biological feature of human beings. Again, it takes an extreme form of empiricism to believe that brute correlations between one arbitrarily selected characteristic and another constitutes *science*—but even from such a perspective it must be an arbitrary choice to investigate one set of such correlations rather than another. Why intelligence and *race?* Why not intelligence and number of hair follicles?

It is this point that really gives the lie to Herrnstein's repugnant invocation of "scientific objectivity" in defense of his racist undertakings.[64] The fact that there is no empirical grounding for the selection of race as a theoretical parameter in the study of intelligence utterly defeats the disingenuous defense that such "science" as Herrnstein is engaged in is simply detached fact gathering—callin' 'em like he sees 'em. The decision to use race as an analytical category betrays a host of substantive assumptions that would be exceedingly hard to defend once made explicit. How could one defend the proposition that race and intelligence are connected without confronting the embarrassing fact that there's no biologically defensible definition of "race"? And how could one defend the proposition that human "mating strategies" will receive their explanation at the biological level, without having to explicitly argue against the wealth of competing explanations available at the social and personal/intentional levels?[65]

In sum, a naturalized approach to knowledge requires us, as feminists and progressives, to be critical of the saliency such categories as gender and race have *for us.* The fact that such parameters have been egregiously overlooked in cases where they are demonstrably relevant shouldn't make us think automatically that they are always theoretically significant. The recognition that selection of analytical categories is an empirical matter, governed by both background theory and consideration of the facts, is in itself part of the solution to the paradox of partiality.

The naturalized approach proceeds by showing the empirical inadequacy of the theory of mind and knowledge that makes perfect neutrality seem like a good thing. But at the same time that it removes the warrant for one epistemic ideal, it gives support for new norms, ones that will enable us to criticize some biases without presupposing the badness of bias in general. The naturalized approach can therefore vindicate all of the insights feminist theory has produced regarding the ideological functions of the concept of objectivity without undercutting the critical purpose of exposing androcentric and other objectionable forms of bias, when they produce oppressive falsehoods.

The End

I began this essay by asking whether we need a "feminist" epistemology, and I answered that we did, as long as we understood that need to be the need for an epistemology informed by feminist insight, and responsive to the moral imperatives entailed by feminist commitments. But I've argued that we do not necessarily need a conceptual transformation of epistemological theory in order to get a feminist epistemology in this sense. We need, in the first instance, a *political* transformation of the society in which theorizing about knowledge takes place. We've got to stop the oppression of women, eliminate racism, redistribute wealth, and *then* see what happens to our collective understanding of knowledge.

My bet? That some of the very same questions that are stimulating inquiry among privileged white men, right now in these sexist, racist, capitalist-imperialist times, are *still* going to be exercising the intellects and challenging the imaginations of women of color, gay men, physically handicapped high school students, etc.

I'm not saying that we should stop doing epistemology until after the revolution. That would of course be stupid, life being short. What I am saying is that those of us who think we know what feminism is, must guard constantly against the presumptuousness we condemn in others, of claiming as Feminist the particular bit of ground upon which we happen to be standing. We need to remember that part of what unites philosophers who choose to characterize their own work as "feminist" is the conviction that philosophy ought to matter—that it should make a positive contribution to the construction of a more just, humane, and nurturing world than the one we currently inhabit.

I have argued that contemporary analytic philosophy is capable of making such a contribution and that it is thus undeserving of the stigma "malestream" philosophy. But there's more at stake here than the abstract issue of mischaracterization. Attacks on the analytic tradition as "androcentric," "phallogocentric," or "male-identified" are simultaneously attacks on the feminist credentials of those who work within the analytic tradition. And the stereotyping of contemporary analytic philosophy—the tendency to link it with views (like the Dragnet theory) to which it is in fact antipathetic—has turned feminists away from fruitful philosophical work, limiting our collective capacity to imagine genuinely novel and transformative philosophical strategies.

I acknowledge both the difficulty and the necessity of clarifying the implications of feminist theory for other kinds of endeavors. It's important, therefore, for feminist theorists to continue to raise critical challenges to particular theories and concepts. But surely this can be done without the caricature, without the throwaway refutations, in a way that is more respectful of philosophical differences.

Let's continue to argue with each other by all means. But let's stop arguing about which view is more feminist, and argue instead about which view is more likely to be true. Surely we can trust the dialectical process of feminists discussing these things with other feminists to yield whatever "feminist epistemology" we need.[66]

Notes

1. A possible exception may be Jean Grimshaw, who comes closer than any other thinker I've encountered to endorsing what I'm calling a "bare proceduralist" conception of feminist philosophy: "There is no particular view, for example, of autonomy, of morality, of self, no one characterisation of women's activities which can be appealed to in any clear way as the woman's (or feminist) view. But I think nevertheless that feminism makes a difference to philosophy. The difference it makes is that women, in doing philosophy, have often raised new problems, problematised issues in new ways and moved to the centre questions which have been marginalised or seen as unimportant or at the periphery." From Grimshaw, *Philosophy and Feminist Thinking* (Minneapolis: University of Minnesota Press, 1986), p. 260.

2. Naomi Scheman made this point in a letter to members of the Committee on the Status of Women of the American Philosophical Association in 1988, when she and I were serving on the committee. Her letter was partly a

response to a letter of mine raising questions about whether our charge as a committee should include the promotion of "feminist philosophy."

3. For discussions of epistemological frameworks available to feminists, see Sandra Harding, *The Science Question in Feminism,* (Ithaca, N.Y.: Cornell University Press, 1986), especially pp. 24–29; Mary Hawkesworth, "Feminist Epistemology: A Survey of the Field," *Women and Politics* 7 (1987): 112–124; and Hilary Rose, "Hand, Brain, and Heart: A Feminist Epistemology for the Natural Sciences," *Signs* 9, 11 (1983): 73–90.

4. See Mary E. Hawkesworth, "Knowers, Knowing, Known: Feminist Theory and Claims of Truth," *Signs* 14, 3 (1989): 533–557.

5. See, for example, Sandra Harding: "I have been arguing for open acknowledgement, even enthusiastic appreciation, of certain tensions that appear in the feminist critiques. I have been suggesting that these reflect valuable alternative social projects which are in opposition to the coerciveness and regressiveness of modern science. . . . [S]table and coherent theories are not always the ones to be most highly desired; there are important understandings to be gained in seeking the social origins of instabilities and incoherences in our thoughts and practices—understandings that we cannot arrive at if we repress recognition of instabilities and tensions in our thought" (*Science Question in Feminism,* pp. 243–244).

6. See Naomi Scheman, "Othello's Doubt/Desdemona's Death: The Engendering of Skepticism," in *Power, Gender, Values,* ed. Judith Genova (Edmonton, Alberta: Academic Printing and Publishing, 1987). See also Evelyn Fox Keller, "Cognitive Repression in Physics," *American Journal of Physics* 47 (1979): 718–721; and "Feminism and Science," in *Sex and Scientific Inquiry,* ed. S. Harding and J. O'Barr (Chicago: University of Chicago Press, 1987), pp. 233–246, reprinted in *The Philosophy of Science,* ed. by Richard Boyd, Philip Gaspar, and John Trout (Cambridge, Mass.: MIT Press, 1991).

7. For example, see Catharine A. MacKinnon, *Towards a Feminist Theory of the State* (Cambridge, Mass.: Harvard University Press, 1989).

8. This is not quite right—the ideology of 'objectivity' is perfectly capable of charging those *outside* the inner circle with partiality, and indeed, such charges are also crucial to the preservation of the status quo. More on this below.

9. Lorraine Code, "The Impact of Feminism on Epistemology," *APA Newsletter on Feminism and Philosophy,* ed. by Morwenna Griffiths and Margaret Whitford (Bloomington: Indiana University Press, 1988), pp. 189ff.

10. Lorraine Code, "Experience, Knowledge, and Responsibility," in *Feminist Perspectives in Philosophy,* ed. by Morwenna Griffiths and Margaret Whitford (Bloomington: Indiana University Press, 1988), pp. 189ff.

11. Code, "Impact of Feminism on Epistemology," p. 25.

12. It might be objected that there is a third option—that we could criticize those biases that are biases against our interests and valorize those that

promote our interests. But if we are in fact left with only this option, then we are giving up on the possibility of any medium of social change other than power politics. This is bad for two reasons: (1) as moral and political theory, egoism should be repugnant to any person ostensibly concerned with justice and human well-being; and (2) as tactics, given current distributions of power, it's really stupid.

13. I have defended a kind of non-realist conception of truth, but one which maintains this gap. See my "Can Verificationists Make Mistakes?" *American Philosophical Quarterly* 24, 3 (July 1987): 225–236. For a defense of a more robustly realist conception of truth, see Michael Devitt, *Realism and Truth* (Princeton, N.J.: Princeton University Press, 1984). (A new edition is in press.)

14. Code, "Impact of Feminism on Epistemology," p. 25.

15. Significantly, these theories are not all empiricist, and the theories that are most "postpositivist" are the least empiricist of all. I'll have much more to say about this in what follows.

16. See, e.g., Helen Cixous, "The Laugh of the Medusa," tr. by Keith Cohen and Paula Cohen, *Signs* 1, 4 (1976): 875–893; Luce Irigaray, "Is the Subject of Science Sexed?" tr. by Carol Mastrangelo Bove, *Hypatia* 2, 3 (Fall 1987): 65–87; and Andrea Nye, "The Inequalities of Semantic Structure: Linguistics and Feminist Philosophy," *Metaphilosophy* 18, 3–4 (July/October 1987): 222–240. I must say that for the sweepingness of Nye's claims regarding "linguistics" and "semantic theory," her survey of work in these fields is, to say the least, narrow and out-of-date.

17. See, e.g., Ruth Ginzberg, "Feminism, Rationality, and Logic" and "Teaching Feminist Logic," *APA Newsletter on Feminism and Philosophy* 88, 2 (March 1989): 34–42 and 58–65.

18. Note that the term "Enlightenment" itself does not have any single, precise meaning, referring in some contexts to only the philosophers (and *philosophes*) of eighteenth-century France, in other contexts to any philosopher lying on the trajectory of natural-rights theory in politics, from Hobbes and Locke through Rousseau, and in still other contexts to all the canonical philosophical works of the seventeenth and eighteen centuries, up to and including Kant. I shall try to use the term "early modern philosophy" to denote seventeenth-century rationalism and empiricism, but I may slip up.

19. In Alison Jaggar, *Feminist Politics and Human Nature* (Totowa, N.J.: Rowman and Allenheld, 1983), p. 355.

20. In Harding, *Science Question in Feminism,* p. 24.

21. Jane Flax, "Postmodernism and Gender Relations in Feminist Theory," *Signs* 12, 4 (Summer 1987): 624.

22. Ibid., p. 627.

23. Never mind Kant, who, apart from this note, I'm going to pretty much ignore. Virtually nothing that Flax cites as constitutive of the Enlightenment legacy can be easily found in Kant. He was not a dualist, at least not a Cartesian dualist; his opinions regarding the possible existence of a mind-

independent reality were complicated (to say the least), but he clearly thought that it would be impossible for human beings to gain knowledge of such a world if it *did* exist; and the reading of the Categorical Imperative—how does it go? "Treat others as ends-in-themselves, never merely as means"?—that has Kant coming out as ignorant or neglectful of human difference seems to me to be positively Orwellian.

24. Harding is an exception, since she acknowledges Quine, though nothing after Quine. Code does allude to there being some changes in mainstream epistemology since the heyday of positivism, but she says that the changes are not of the right nature to license the questions she thinks are central to feminist epistemology. The only contemporary analytic epistemologist Code ever cites in either of her two books is Alvin Goldman, whom she does not discuss.

This is ironic, because Goldman has been one of the chief advocates of a version of epistemology called reliabilism, that makes the actual circumstances of belief production an essential part of their justification. See his *Epistemology and Cognition* (Cambridge, Mass.: Harvard University Press, 1986). It is also terribly unfair. Goldman takes it to be a truism that knowledge has a social component and that the study of knowledge requires consideration of the social situation of the knower: "Most knowledge is a cultural product, channeled through language and social communication. So how could epistemology *fail* to be intertwined with studies of culture and social systems?" I do not believe Goldman deserves the opprobrium Code heaps upon him.

Jaggar, too, acknowledges that positivism has lost favor, but says nothing about the shape of the theories that have succeeded it. See Jaggar, *Feminist Politics*.

25. Cognitive essentialism generally gets associated with another thesis singled out for criticism—namely, dualism, the view that the mind is separate from the body and that the self is to be identified with the mind. Although dualism is not exclusively a rationalist view (Locke is standardly classified as a dualist), it is most closely associated with Descartes, and it is Descartes's *a priori* argument for dualism in the *Meditations* that seems to draw the most fire. Cartesian dualism is seen as providing a metaphysical rationale for dismissing the relevance of material contingencies to the assessment of knowledge claims, because it separates the knowing subject from the physical body, and because it seems to assert the sufficiency of disembodied reason for the attainment of knowledge.

In fact, dualism is a red herring. It's an uncommon view in the history of philosophy. Many people classically characterized as dualists, like Plato, were surely not Cartesian dualists. And on top of that, the dualism does not work. Being a dualist is neither necessary nor sufficient for believing that the human essence is composed of cognitive properties.

26. Flax, "Postmodernism," p. 262.

27. "Individualism" as Jaggar uses it is rather a term of art. It has a variety of meanings within philosophical discourse, but I don't know of any standard

use within epistemology that matches Jaggar's. In the philosophy of mind, the term denotes the view that psychological states can be individuated for purposes of scientific psychology, without reference to objects or states outside the individual. This use of the term has *nothing* to do with debates in political theory about such issues as individual rights or individual autonomy. A liberal view of the moral/political individual can work just as well (or as poorly) on an anti-individualist psychology (such as Hilary Putnam's or Tyler Burge's) as on an individualist view like Jerry Fodor's.

28. See also Naomi Scheman's essay in *A Mind of One's Own* (Westview, 1993).

29. Jaggar, "Postmodernism," p. 355.

30. Ibid.

31. See, for example, the excerpts from *Notes Directed against a Certain Program,* in Margaret Wilson, ed., *The Essential Descartes* (New York: Mentor Press, 1969).

32. Ibid., p. 112.

33. Ibid. One passage from one work should, of course, not be enough to convince anyone, and Descartes is clearly fictionalizing his own history to some extent (like who doesn't?). I do not have the space here to provide a full defense of my interpretation, but I invite you to read the *Discourse* on your own.

34. A little qualification is necessary here: The empiricist's requirement that all concepts be reducible to sensory simples does count as a substantive restriction on the possible contents of thought, but it's one which is vitiated by the reductionist semantic theory favored by empiricists, which denies the meaningfulness of any term which cannot be defined in terms of sensory primitives. See the discussion of this point in Jerry Fodor, *Modularity of Mind: An Essay on Faculty Psychology* (Cambridge, Mass.: MIT Press, 1983).

Also, the empiricists did allow a kind of "bias" in the form of innate standards of similarity, which would permit the mind to see certain ideas as inherently resembling certain others. This innate similarity metric was needed to facilitate the operation of *association,* which as the mechanism for generating more complex and more abstract ideas out of the sensory simples. But the effects of a bias such as this were vitiated by the fact that associations could also be forged by the contiguity of ideas in experience, with the result once more that no effective, substantive limits were placed on the ways in which human beings could analyze the data presented them by sensory experience.

35. David Hume, *An Enquiry Concerning Human Understanding* (Indianapolis: Hackett, 1977), p. 30. For a different assessment of Hume's potential contributions to a feminist epistemology, see Annette Baier's essay in *A Mind of One's Own.*

36. I have been much chastised by serious scholars of early-twentieth-century analytic philosophy (specifically Warren Goldfarb, Neil Tennant, and Philip Kitcher) for here reinforcing the myth that logical positivism was a uniform "school of thought." I guess I should thank them. The view that I am

labeling "positivism" is the usual received view of the movement, but it may have belonged to only some of the more flatfooted and marginal members of the group (like A. J. Ayer) and certainly was not the view of the most important philosopher in the movement, Rudolf Carnap.

Still, the version of positivism I am outlining is the version that Quine attributed to his predecessors, and the version that he was reacting against. Moreover, even if Carnap was not an externalist in the sense of seeking a metaphysical vindication of scientific practice (as Michael Friedman argues in "The Re-evaluation of Logical Positivism," *Journal of Philosophy* 88, 10 [October 1991]: 505–519), he still was committted to a sharp separation between contentful and merely analytic statements, which is enough to generate the kinds of difficulties that I'm claiming beset positivism generally. My thanks to Marcia Homiak for calling my attention to the Friedman article.

37. Here are some of the most important works: W.V.O. Quine, "Two Dogmas of Empiricism," reprinted in Quine, *From a Logical Point of View* (Cambridge, Mass.: Harvard University Press 1953); Carl G. Hempel, "Problems and Changes in the Empiricist Criterion of Meaning," *Revue Internationale de Philosophie* 11 (1950): 41–63, and "Empiricist Criteria of Cognitive Significance: Problems and Changes," in Hempel, *Aspects of Scientific Explanation and Other Essays in the Philosophy of Science* (New York: Free Press, 1965); Nelson Goodman, *Fact, Fiction, and Forecast* (Cambridge, Mass.: Harvard University Press, 1955); and Hilary Putnam, "What Theories Are Not," reprinted in Putnam, *Mathematics, Matter, and Method: Philosophical Papers, Vol. I* (Cambridge: Cambridge University Press, 1975).

38. Quine and J. S. Ullian catalog these principles—which they refer to as the "virtues" of hypotheses—in an epistemological primer called "The Web of Belief (New York: Random House, 1970). Quine and Ullian employ a strikingly Humean strategy in trying to explain the epistemological value of the virtues.

39. W.V.O. Quine, "Epistemology Naturalized," in Quine, *Ontological Relativity and Other Essays* (New York: Columbia University Press, 1969), pp. 69–90.

40. See Noam Chomsky, "Review of B. F. Skinner's *Verbal Behavior,*" *Language* 35, 1 (1959): 53–68.

41. See Noam Chomsky, "Quine's Empirical Assumptions," in *Words and Objections: Essays on the Work of W. V. Quine,* ed. by D. Davidson and J. Hintikka (Dordrecht: D. Reidel, 1969). See also Quine's response to Chomsky in the same volume.

I discuss the inconsistency between Quine's commitment to naturalism and his *a prioristic* rejection of mentalism and nativism in linguistics in "Naturalized Epistemology and the Study of Language," in *Naturalistic Epistemology: A Symposium of Two Decades,* ed. by Abner Shimony and Debra Nails (Dordrecht: D. Reidel, 1987), pp. 235–257.

42. For an extremely helpful account of the Chomskian approach to the

study of language, see David Lightfoot's *The Language Lottery: Toward a Biology of Grammars* (Cambridge, Mass.: MIT Press, 1984).

43. I take this to be an established fact. There's a mountainous body of scholarship on this issue, much of it the result of feminist concerns about specific ways in which women have been excluded from and damaged by institutionalized science. The whole area of biological determinist theorists provides an excellent case study of the ways in which science both supports and is distorted by social stratification. *Genes and Gender II*, ed. by Ruth Hubbard and Marian Lowe (New York: Gordion Press, 1979), is a collection of now classic articles critically examining alleged biological and ethological evidence for the genetic basis of gender differences. For a more current analysis of similar research in neurophysiology and endocrinology, see Helen Longino, *Science as Social Knowledge* (Princeton, N.J.: Princeton University Press, 1990), ch. 6. Two excellent general discussions of the interactions among politics, economics, ideology, and science as exemplified by the growth of biological determinist theories are Stephen Jay Gould, *The Mismeasure of Man* (New York: W. W. Norton, 1981); and R. C. Lewontin, Steven Rose, and Leon J. Kamin, *Not in Our Genes* (New York: Pantheon Books, 1984).

44. Evelyn Fox Keller, "Feminism and Science," in Boyd, Gaspar, and Trout, eds., *Philosophy of Science*, p. 281. In this passage, Keller is also remarking on the tendency of (what she views as) the liberal critiques to focus on the "softer" biological and social sciences, and to leave alone the "harder" sciences of math and physics.

45. Carl R. Hempel, *Philosophy of Natural Science* (Englewood Cliffs, N.J.: Prentice-Hall, 1966). See especially pp. 10–18.

46. There's a good case to be made that scientists actually have *disincentives* to ponder such questions. The structure of incentives in academia necessitates rapid generation and publication of research, and research requires securing long-term funding, usually from a government agency or a private corporate foundation. Scientific research is thus heavily compromised at the outset, whatever the ideals and values of the individual scientists. For a detailed discussion of the ways in which academic and economic pressures systematically erode "objectivity" in science, see William Broad and Nicholas Wade, *Betrayers of the Truth: Fraud and Deceit in the Halls of Science* (New York: Simon and Schuster, 1982).

47. This follows from a general point emphasized by Georges Rey in personal conversation: It's important in general to distinguish people's theories of human institutions from the actual character of those institutions.

48. This despite the fact that the Dragnet theory supports a strong context of discovery/context of justification distinction. On empiricist theories, the justification of an individual's belief is ultimately a relation between the belief and the sensory experience of that individual. Location matters, then, because the same belief could be justified for one individual and unjustified for another, precisely because of the differences in their experiences.

49. This is not to say that there are no puzzling issues about moral ideals that are in some sense humanly unattainable. One such issue arises with respect to the ideals of altruism and supererogation, ideals which it would be, arguably, *unhealthy* for human beings to fully realize. See Larry Blum, Marcia Homiak, Judy Housman, and Naomi Scheman, "Altruism and Women's Oppression," in *Women and Philosophy,* ed. by Carol C. Gould and Marx W. Wartofsky (New York: G. P. Putnam, 1980), pp. 222–247. On the question of whether it would be good for human beings to fully realize *any* moral ideal, see Susan Wolf, "Moral Saints," *The Journal of Philosophy* 79, 8 (August 1982): 419–439.

50. Jerry Fodor, *Modularity of Mind* (Cambridge, Mass.: MIT Press, 1983); Noam Chomsky, *Reflections on Language* (New York: Random House, 1975); David Marr, *Vision: A Computational Investigation Into the Human Representation and Processing of Visual Information* (San Francisco: W. H. Freeman, 1982); Susan Carey, *Conceptual Change in Childhood* (Cambridge, Mass.: MIT Press, 1985); Elizabeth Spelke, "Perceptual Knowledge of Objects in Infancy," in J. Mehler, E. C. T. Walker, and M. Garrett, eds., *Perspectives on Mental Representations* (Hillsdale, N.Y.: Erlbaum, 1982); Barbara Landau and Lila Gleitman, *Language and Experience: Evidence from the Blind Child* (Cambridge, Mass.: Harvard University Press, 1985); Steven Pinker, *Learnability and Cognition: The Acquisition of Argument Structure* (Cambridge, Mass.: MIT Press, 1989).

51. Thomas S. Kuhn, "The Function of Dogma in Scientific Research," (1963), reprinted in Janet A. Kourany, ed., *Scientific Knowledge* (Belmont, Calif.: Wadsworth, 1987), pp. 253–265. Quotation is from p. 254.

52. This phenomenon affects even as sensitive and sophisticated a critic of science as Stephen Jay Gould. Responding to creationist charges that evolution is "just a theory," Gould insists: "Well, evolution *is* a theory. It is also a fact. And facts and theories are different things, not rungs in a hierarchy of increasing certainty. Facts are the world's data. Theories are structures of ideas that explain and interpret facts. Facts do not go away while scientists debate rival theories for explaining them. . . . [H]uman beings evolved from apelike ancestors whether they did so by Darwin's proposed mechanism or by some other, yet to be discovered." Stephen Jay Gould, "Evolution as Fact and Theory," *Hen's Teeth and Horse's Toes* (New York: W. W. Norton, 1980), pp. 253–262. Quotation from p. 254.

Gould's point, I believe, is that the world is as it is independently of our ability to understand it—a position I share. But if facts are part of the mind-independent world, they cannot also be "the world's data." *"Data"* is the name we give to that *part* of our theory about which we can achieve a high degree of interpersonal and intertheoretic agreement; however, there can be as much contention about "the data" as about "the theory." Gould concedes as much in the next paragraph when he writes: "Moreover, 'fact' does not mean 'absolute certainty.'. . . In science, 'fact' can only mean 'confirmed to such a degree that it would be perverse to withhold provisional assent.' " If *that's* what "facts" are, then they can and do sometimes "go away while scientists debate rival theories for explaining them." Ibid., p. 255.

53. Notice that we don't have to assume here that anyone is knowingly telling lies. Clearly, in the real world, members of the ruling elite *do* consciously lie, and they do it a lot. But here I'm trying to point out that some of the mechanisms that can perpetuate oppressive structures are epistemically legitimate.

54. See Edward Herman and Noam Chomsky, *Manufacturing Consent* (New York: Pantheon, 1988); Noam Chomsky, *Necessary Illusions: Thought Control in Democratic Society* (Boston: South End Press, 1989), esp. ch. 3 ("The Bounds of the Expressible"); and Martin A. Lee and Norman Solomon, *Unreliable Sources: A Guide to Detecting Bias in News Media* (New York: Carol Publishing Group, 1990).

55. This explains some of what's going on in the so-called "debate" about so-called "political correctness." Most of what's going on involves pure dishonesty and malice, but to the extent that there are some intelligent and relatively fair-minded people who find themselves worrying about such issues as the "politicization" of the classroom, or about "ideological biases" among college professors, these people are reacting to the *unfamiliarity* of progressive perspectives. Those foundational beliefs that are very common within the academy—belief in a (Christian) god, in the benignity of American institutions, in the viability of capitalism—generally go without saying and are thus invisible. *Our* worldviews are unfamiliar, and so must be articulated and acknowledged. Precisely because we are willing and able to do that, while our National Academy of Scholars colleagues are not, we become open to the charge of being "ideological."

It's the very fact that there are so *few* leftist, African-Americans, Hispanic, openly gay, feminist, female persons in positions of academic authority that accounts for all this slavish nonsense about our "taking over."

56. Lorraine Code, "Credibility: A Double Standard," in *Feminist Perspectives,* ed. Code, Mullett, and Overall, pp. 65–66.

57. Ibid.

58. Ibid., p. 71.

59. Ibid., p. 73.

60. Ibid., p. 72.

61. Ironically, the preference among many feminist theorists for "thin" theories of the self, like postmodernist constructivist theories, is itself a vestige of an incompletely exorcised empiricism in contemporary feminist thought. It is a specifically empiricist position that the groupings of objects into kinds effected by human cognition are not keyed to "real essences," but are rather reflections of superficial regularities in experience that persist only because of their pragmatic utility.

62. We know, for example, that some of the built-in rules that make it

possible for the human visual system to pick out objects from their backgrounds—so-called structure from motion rules—also make us subject to certain specific kinds of visual illusions. See A. L. Yuille and S. Ullman, "Computational Theories of Low-Level Vision," in *Visual Cognition and Action,* ed. by Daniel N. Osherson, Stephen M. Kosslyn, and John M. Hollerbach, vol. 2 of *An Invitation to Cognitive Science,* ed. by Daniel N. Osherson (Cambridge, Mass.: MIT Press, 1990), pp. 5–39.

63. I am here reiterating the arguments Chomsky mounted against Hernstein's apologia for Jensen's theory of race and intelligence. See Noam Chomsky, "Psychology and Ideology," reprinted in Chomsky, *For Reasons of State* (New York: Random House, 1973), pp. 318–369; excerpted and reprinted as "The Fallacy of Richard Herrnstein's IQ," in *The IQ Controversy,* ed. by Ned Block and Gerald Dworkin (New York: Random House, 1976), pp. 285–298.

64. See Herrnstein's reply to Chomsky, "Whatever Happened to Vaudeville?" in Block and Dworkin, eds., *IQ Controversy,* esp. pp. 307–309.

65. These considerations also help defeat the charge, hurled against critics of biological determinist theories, that we progressives are the ones guilty of "politicizing" the debate about nature and nurture. The Herrnsteins and E. O. Wilsons of this world like to finesse the meticulously arrayed empirical criticisms of their work by accusing their critics of the most pathetic kind of wishful thinking—"Sorry if you don't *like* what my utterly objective and bias-free research has proven beyond a shadow of a doubt. You must try to be big boys and girls and learn to cope with the unpleasant truth." For examples, see Herrnstein, "Whatever Happened to Vaudeville?" in Block and Dworkin, eds., *IQ Controversy;* and E. O. Wilson, "Academic Vigilantism and the Political Significance of Sociobiology," reprinted in *The Sociobiology Debate,* ed. by Arthur L. Caplan (New York: Harper and Row, 1978), pp. 291–303.

66. Much of the preliminary work for this essay was done during a fellowship year at the National Humanities Center, and I wish to thank both the center and the Andrew J. Mellon Foundation for this support. The essay is based on a presentation I gave at the Scripps College Humanities Institute Conference, "Thinking Women: Feminist Scholarship in the Humanities," in March 1990. I want to thank the institute, especially Norton Batkin, for the invitation to think about these issues. I also want to thank my co-participants at the conference, especially Naomi Scheman, to whom I owe a special debt. I have enjoyed an enormous amount of stimulating and challenging conversation and correspondence with Naomi about all the issues in this essay. It's a tribute to her sense of intellectual fairness and her commitment to feminist praxis that she and I have managed to conduct such an extended dialogue about these issues, given the intensity of our disagreements. I also want to make it clear that while I had the benefit of reading Naomi's essay before completing my own, I did not finish mine in time for her to react to any of the points I raise here.

Many other people have helped me with this essay. I want to thank Judith Ferster, Suzanne Graver, Charlotte Gross, Sally Haslanger, Barbara Metcalf, and Andy Reath for hours of valuable conversation. Marcia Homiak, Alice Kaplan, and Georges Rey supplied extremely useful comments on earlier drafts; David Auerbach did all that *and* extricated me from an eleventh-hour computer crisis, and I thank them heartily. Very special thanks to my co-editor, Charlotte Witt, for her excellent philosophical and editorial advice and for her abundant patience and good sense. I cannot express my thanks to Joe Levine for all he's done, intellectually and personally, to help me complete this project. Thanks as well to my children, Paul and Rachel, for their patience during all the times I was out consorting with my muse.

18

Epistemic Folkways and Scientific Epistemology

Alvin Goldman

I

What is the mission of epistemology, and what is its proper methodology? Such meta-epistemological questions have been prominent in recent years, especially with the emergence of various brands of "naturalistic" epistemology. In this paper, I shall reformulate and expand upon my own meta-epistemological conception most fully articulated in Goldman (1986), retaining many of its former ingredients while reconfiguring others. The discussion is by no means confined, though, to the meta-epistemological level. New substantive proposals will also be advanced and defended.

Let us begin, however, at the meta-epistemological level, by asking what role should be played in epistemology by our ordinary epistemic concepts and principles. By some philosophers' lights, the sole mission of epistemology is to elucidate commonsense epistemic concepts and principles: concepts like knowledge, justification, and rationality, and principles associated with these concepts. By other philosophers' lights, this is not even part of epistemology's aim. Ordinary concepts and principles, the latter would argue, are fundamentally naive, unsystematic, and uninformed by important bodies of logic and/or mathematics. Ordinary principles and practices, for example, ignore or violate the probability calculus, which ought to be the cornerstone of epistemic rationality. Thus, on the second view, proper epistemology must neither *end* with naive principles of justification or rationality, nor even *begin* there.

Reprinted by permission from Alvin Goldman, *Liaisons*, pp. 155–75. MIT Press, 1991.

My own stance on this issue lies somewhere between these ex-
tremes. To facilitate discussion, let us give a label to our commonsense
epistemic concepts and norms; let us call them our *epistemic folkways*.
In partial agreement with the first view sketched above, I would hold
that *one* proper task of epistemology is to elucidate our epistemic
folkways. Whatever else epistemology might proceed to do, it should
at least have its roots in the concepts and practices of the folk. If these
roots are utterly rejected and abandoned, by what rights would the
new discipline call itself 'epistemology' at all? It may well be desirable
to reform or transcend our epistemic folkways, as the second of the
views sketched above recommends. But it is essential to preserve
continuity; and continuity can only be recognized if we have a satisfactory characterization of our epistemic folkways. Actually, even if one
rejects the plea for continuity, a description of our epistemic folkways
is in order. How would one know what to criticize, or what needs to
be transcended, in the absence of such a description? So a first mission
of epistemology is to describe or characterize our folkways.

Now a suitable description of these folk concepts, I believe, is likely
to depend on insights from cognitive science. Indeed, identification of
the semantic contours of many (if not all) concepts can profit from
theoretical and empirical work in psychology and linguistics. For this
reason, the task of describing or elucidating folk epistemology is a
scientific task, at least a task that should be informed by relevant
scientific research.

The second mission of epistemology, as suggested by the second
view above, is the formulation of a more adequate, sound, or systematic set of epistemic norms, in some way(s) transcending our naive
epistemic repertoire. How and why these folkways might be transcended, or improved upon, remains to be specified. This will partly
depend on the contours of the commonsense standards that emerge
from the first mission. On my view, epistemic concepts like knowledge
and justification crucially invoke psychological faculties or processes.
Our folk understanding, however, has a limited and tenuous grasp of
the processes available to the cognitive agent. Thus, one important
respect in which epistemic folkways should be transcended is by
incorporating a more detailed and empirically based depiction of
psychological mechanisms. Here too epistemology would seek assistance from cognitive science.

Since both missions of epistemology just delineated lean in important respects on the deliverances of science, specifically cognitive
science, let us call our conception of epistemology *scientific epistemol-*

ogy. Scientific epistemology, we have seen, has two branches: *descriptive* and *normative*. While descriptive scientific epistemology aims to describe our ordinary epistemic assessments, normative scientific epistemology continues the practice of making epistemic judgments, or formulating systematic principles for such judgments.[1] It is prepared to depart from our ordinary epistemic judgments, however, if and when that proves advisable. (This overall conception of epistemology closely parallels the conception of metaphysics articulated in *Liaisons*, chapters 2 and 3. The descriptive and normative branches of scientific epistemology are precise analogues of the descriptive and prescriptive branches of metaphysics, as conceptualized there.) In the remainder of this paper, I shall sketch and defend the particular forms of descriptive and normative scientific epistemology that I favor.

II

Mainstream epistemology has concentrated much of its attention on two concepts (or terms): knowledge and justified belief. Chapter 8 of *Liaisons* primarily illustrates the contributions that cognitive science can make to an understanding of the former; this essay focuses on the latter. We need not mark this concept exclusively by the phrase 'justified belief'. A family of phrases pick out roughly the same concept: 'well-founded belief', 'reasonable belief', 'belief based on good grounds', and so forth. I shall propose an account of this concept that is in the reliabilist tradition, but departs at a crucial juncture from other versions of reliabilism. My account has the same core idea as Ernest Sosa's *intellectual virtues* approach, but incorporates some distinctive features that improve its prospects.[2]

The basic approach is, roughly, to identify the concept of justified belief with the concept of belief obtained through the exercise of intellectual virtues (excellences). Beliefs acquired (or retained) through a chain of "virtuous" psychological processes qualify as justified; those acquired partly by cognitive "vices" are derogated as unjustified. This, as I say, is a *rough* account. To explain it more fully, I need to say things about the psychology of the epistemic evaluator, the possessor and deployer of the concept in question. At this stage in the development of semantical theory (which, in the future, may well be viewed as part of the "dark ages" of the subject), it is difficult to say just what the relationship is between the meaning or "content" of concepts and the form or structure of their mental representation. In the present case, however, I believe that an account of the form of representation can contribute to our understanding of the content, although I am unable to formulate these matters in a theoretically satisfying fashion.

The hypothesis I wish to advance is that the epistemic evaluator has a mentally stored set, or list, of cognitive virtues and vices. When asked to evaluate an actual or hypothetical case of belief, the evaluator considers the processes by which the belief was produced, and matches these against his list of virtues and vices. If the processes match virtues only, the belief is classified as justified. If the processes are matched partly with vices, the belief is categorized as unjustified. If a belief-forming scenario is described that features a process not on the evaluator's list of either virtues or vices, the belief may be categorized as neither justified nor unjustified, but simply *non*justified. Alternatively (and this alternative plays an important role in my story), the evaluator's judgment may depend on the (judged) *similarity* of the novel process to the stored virtues and vices. In other words, the "matches" in question need not be perfect.

This proposal makes two important points of contact with going theories in the psychology of concepts. First, it has some affinity to the *exemplar* approach to concept representation (cf. Medin and Schaffer 1978; Smith and Medin 1981; Hintzman 1986). According to that approach, a concept is mentally represented by means of representations of its positive instances, or perhaps types of instances. For example, the representation of the concept *pants* might include a representation of a particular pair of faded blue jeans and/or a representation of the type *blue jeans*. Our approach to the concept of justification shares the spirit of this approach insofar as it posits a set of examples of virtues and vices, as opposed to a mere abstract characterization—e.g., a definition—of (intellectual) virtue or vice. A second affinity to the exemplar approach is in the appeal to a similarity, or matching, operation in the classification of new target cases. According to the exemplar approach, targets are categorized as a function of their similarity to the positive exemplars (and dissimilarity to the foils). Of course, similarity is invoked in many other approaches to concept deployment as well (see E. E. Smith 1990). This makes our account of justification consonant with the psychological literature generally, whether or not it meshes specifically with the exemplar approach.

Let us now see what this hypothesis predicts for a variety of cases. To apply it, we need to make some assumptions about the lists of virtues and vices that typical evaluators mentally store. I shall assume that the virtues include belief formation based on sight, hearing, memory, reasoning in certain "approved" ways, and so forth. The vices include intellectual processes like forming beliefs by guesswork, wishful thinking, and ignoring contrary evidence. *Why* these items are placed in their respective categories remains to be explained. As indicated, I plan to explain them by reference to reliability. Since the account will therefore be, at bottom, a reliabilist type of account, it is instructive to see how it fares when applied to well-known problem cases for standard versions of reliabilism.

Consider first the demon-world case. In a certain possible world, a Cartesian demon gives people deceptive visual experiences, which systematically lead to false beliefs. Are these vision-based beliefs justified? Intuitively, they are. The demon's victims are presented with the same sorts of visual experiences that we are, and they use the same processes to produce corresponding beliefs. For most epistemic evaluators, this seems sufficient to induce the judgment that the victims' beliefs are justified. Does our account predict this result? Certainly it does. The account predicts that an epistemic evaluator will match the victims' vision-based processes to one (or more) of the items on his list of intellectual virtues, and therefore judge the victims' beliefs to be justified.

Turn next to Laurence BonJour's (1985) cases in which hypothetical agents are assumed to possess a perfectly reliable clairvoyant faculty. Although these agents form their beliefs by this reliable faculty, BonJour contends that the beliefs are not justified; and apparently most (philosophical) evaluators agree with that judgment. This result is not predicted by simple forms of reliabilism.[3] What does our present theory predict? Let us consider the four cases in two groups. In the first three cases (Samantha, Casper, and Maud), the agent has contrary evidence that he or she ignores. Samantha has a massive amount of apparently cogent evidence that the president is in Washington, but she nonetheless believes (through clairvoyance) that the president is in New York City. Casper and Maud each has large amounts of ostensibly cogent evidence that he/she has no reliable clairvoyant power, but they rely on such a power nonetheless. Here our theory predicts that the evaluator will match these agent's belief-forming processes to the vice of ignoring contrary evidence. Since the processes include a vice, the beliefs will be judged to be unjustified.

BonJour's fourth case involves Norman, who has a reliable clairvoyant power but no reasons for or against the thesis that he possesses it.

When he believes, through clairvoyance, that the president is in New York City, while possessing no (other) relevant evidence, how should this belief be judged? My own assessment is less clear in this case than the other three cases. I am tempted to say that Norman's belief is *non*justified, not that it is thoroughly *un*justified. (I construe unjustified as "having negative justificational status", and nonjustified as "lacking positive justificational status".) This result is also readily predicted by our theory. On the assumption that I (and other evaluators) do not have clairvoyance on my list of virtues, the theory allows the prediction that the belief would be judged neither justified nor unjustified, merely nonjustified. For those evaluators who would judge Norman's belief to be *un*justified, there is another possible explanation in terms of the theory. There is a class of putative faculties, including mental telepathy, ESP, telekinesis, and so forth that are scientifically disreputable. It is plausible that evaluators view any process of basing beliefs on the supposed deliverances of such faculties as vices. It is also plausible that these evaluators judge the process of basing one's belief on clairvoyance to be *similar* to such vices. Thus, the theory would predict that they would view a belief acquired in this way as unjustified.[4]

Finally, consider Alvin Plantinga's (1988) examples that feature disease-triggered or mind-malfunctioning processes. These include processes engendered by a brain tumor, radiation-caused processes, and the like. In each case Plantinga imagines that the process is reliable, but reports that we would not judge it to be justification conferring. My diagnosis follows the track outlined in the Norman case. At a minimum, the processes imagined by Plantinga fail to match any virtue on a typical evaluator's list. So the beliefs are at least nonjustified. Furthermore, evaluators may have a prior representation of pathological processes as examples of cognitive vices. Plantinga's cases might be judged (relevantly) similar to these vices, so that the beliefs they produce would be declared unjustified.

In some of Plantinga's cases, it is further supposed that the hypothetical agent possesses countervailing evidence against his belief, which he steadfastly ignores. As noted earlier, this added element would strengthen a judgment of unjustifiedness according to our theory, because ignoring contrary evidence is an intellectual vice. Once again, then, our theory's predictions conform with reported judgments.

Let us now turn to the question of how epistemic evaluators acquire their lists of virtues and vices. What is the basis for their classification?

As already indicated, my answer invokes the notion of reliability. Belief-forming processes based on vision, hearing, memory, and ("good") reasoning are deemed virtuous because they (are deemed to) produce a high ratio of true beliefs. Processes like guessing, wishful thinking, and ignoring contrary evidence are deemed vicious because they (are deemed to) produce a low ratio of true beliefs.

We need not assume that each epistemic evaluator chooses his/her catalogue of virtues and vices by direct application of the reliability test. Epistemic evaluators may partly inherit their lists of virtues and vices from other speakers in the linguistic community. Nonetheless, the hypothesis is that the selection of virtues and vices rests, ultimately, on assessments of reliability.

It is not assumed, of course, that all speakers have the same lists of intellectual virtues and vices. They may have different opinions about the reliability of processes, and therefore differ in their respective lists.[5] Or they may belong to different subcultures in the linguistic community, which may differentially influence their lists. Philosophers sometimes seem to assume great uniformity in epistemic judgments. This assumption may stem from the fact that it is mostly the judgments of philosophers themselves that have been reported, and they are members of a fairly homogeneous subculture. A wider pool of "subjects" might reveal a much lower degree of uniformity. That would conform to the present theory, however, which permits individual differences in catalogues of virtues and vices, and hence in judgments of justifiedness.

If virtues and vices are selected on the basis of reliability and unreliability, respectively, why doesn't a hypothetical case introducing a novel reliable process induce an evaluator to add that process to his list of virtues, and declare the resulting belief justified? Why, for example, doesn't he add clairvoyance to his list of virtues, and rule Norman's beliefs to be justified?

I venture the following explanation. First, people seem to have a trait of *categorial conservatism*. They display a preference for "entrenched" categories, in Nelson Goodman's (1955) phraseology, and do not lightly supplement or revise their categorial schemes. An isolated single case is not enough. More specifically, merely imaginary cases do not exert much influence on categorial structures. People's cognitive systems are responsive to live cases, not purely fictional ones. Philosophers encounter this when their students or nonphilosophers are unimpressed with science fiction-style counterexamples. Philosophers become impatient with this response because they presume that possible cases are on a par (for counterexample purposes) with actual ones. This phenomenon testifies, however, to a psychological propensity to take an invidious attitude toward purely imaginary cases.

To the philosopher, it seems both natural and inevitable to take hypothetical cases seriously, and if necessary to restrict one's conclusions about them to specified "possible worlds". Thus, the philosopher might be inclined to hold, "If reliability is the standard of intellectual virtue, shouldn't we say that clairvoyance is a virtue *in the possible worlds* of BonJour's examples, if not a virtue in general?" This is a natural thing for philosophers to say, given their schooling, but there is no evidence that this is how people naturally think about the matter. There is no evidence that "the folk" are inclined to relativize virtues and vices to this or that possible world.

I suspect that concerted investigation (not undertaken here) would uncover ample evidence of conservatism, specifically in the normative realm. In many traditional cultures, for example, loyalty to family and friends is treated as a cardinal virtue.[6] This view of loyalty tends to persist even through changes in social and organizational climate, which undermine the value of unqualified loyalty. Members of such cultures, I suspect, would continue to view personal loyalty as a virtue even in *hypothetical* cases where the trait has stipulated unfortunate consequences.

In a slightly different vein, it is common for both critics and advocates of reliabilism to call attention to the relativity of reliability to the domain or circumstances in which the process is used. The question is therefore raised, what is the relevant domain for judging the reliability of a process? A critic like John Pollock (1986, pp. 118–119), for example, observes that color vision is reliable on earth but unreliable in the universe at large. In determining the reliability of color vision, he asks, which domain should be invoked? Finding no satisfactory reply to this question, Pollock takes this as a serious difficulty for reliabilism. Similarly, Sosa (1988 and 1991) notes that an intellectual structure or disposition can be reliable with respect one field of propositions but unreliable with respect to another, and reliable in one environment but unreliable in another. He does not view this as a difficulty for reliabilism, but concludes that any talk of intellectual virtue must be relativized to field and environment.

Neither of these conclusions seems apt, however, for purposes of *description* of our epistemic folkways. It would be a mistake to suppose that ordinary epistemic evaluators are sensitive to these issues. It is likely—or at least plausible—that our ordinary apprehension of the intellectual virtues is rough, unsystematic, and insensitive to any theoretical desirability of relativization to domain or environment. Thus, as long as we are engaged in the description of our epistemic folkways, it is no criticism of the account that it fails to explain what domain or environment is to be used. Nor is it appropriate for the account to introduce relativization where there is no evidence of relativization on the part of the folk.

Of course, we do need an explanatory story of how the folk arrive at their selected virtues and vices. And this presumably requires some reference to the domain in which reliability is judged. However, there may not be much more to the story than the fact that people determine reliability scores from the cases they personally "observe". Alternatively, they *may* regard the observed cases as a sample from which they infer a truth ratio in some wider class of cases. It is doubtful, however, that they have any precise conception of the wider class. They probably don't address this theoretical issue, and don't do (or think) anything that commits them to any particular resolution of it. It would therefore be wrong to expect descriptive epistemology to be fully specific on this dimension.

A similar point holds for the question of process individuation. It is quite possible that the folk do not have highly principled methods for individuating cognitive processes, for "slicing up" virtues and vices. If that is right, it is a mistake to insist that descriptive epistemology uncover such methods. It is no flaw in reliabilism, considered as descriptive epistemology, that it fails to unearth them. It may well be desirable to develop sharper individuation principles for purposes of normative epistemology (a matter we shall address in section III). But the missions and requirements of descriptive and normative epistemology must be kept distinct.

This discussion has assumed throughout that the folk have lists of intellectual virtues and vices. What is the evidence for this? In the moral sphere ordinary language is rich in virtues terminology. By contrast, there are few common labels for intellectual virtues, and those that do exist—'perceptiveness', 'thoroughness', insightfulness', and so forth—are of limited value in the present context. I propose to identify the relevant intellectual virtues (at least those relevant to *justification*) with the belief-forming capacities, faculties, or processes that would be accepted as answers to the question "How does X know?". In answer to this form of question, it is common to reply, "He saw it", "He heard it", "He remembers it", "He infers it from such-and-such evidence'', and so forth. Thus, basing belief on seeing, hearing, memory, and (good) inference are in the collection of what the folk regard as intellectual virtues. Consider, for contrast, how anomalous it is to answer the question "How does X know?" with "By guesswork", "By wishful thinking", or "By ignoring contrary evidence". This indicates that *these* modes of belief formation— guessing, wishful thinking, ignoring contrary evidence—are standardly regarded as intellectual *vices*. They are not ways of obtaining knowl- edge, nor ways of obtaining knowledge, nor ways of obtaining justified belief.

Why appeal to "knowledge"-talk rather than "justification"-talk to identify the virtues? Because 'know' has a greater frequency of occurrence than 'justified', yet the two are closely related. Roughly, justified belief is belief acquired by means of the same sorts of capacities, faculties, or processes that yield knowledge in favorable circumstances (i.e., when the resulting belief is true and there are no Gettier complications, or no relevant alternatives).

To sum up the present theory, let me emphasize that it depicts justificational evaluation as involving two stages. The first stage fea- tures the acquisition by an evaluator of some set of intellectual virtues and vices. This is where reliability enters the picture. In the second stage, the evaluator applies his list of virtues and vices to decide the epistemic status of targeted beliefs. At this stage, there is no direct consideration of reliability.

There is an obvious analogy here to rule utilitarianism in the moral sphere. Another analogy worth mentioning is Saul Kripke's (1980) theory of *reference-fixing*. According to Kripke, we can use one property to fix a reference to a certain entity, or type of entity; but once this reference has been fixed, that property may cease to play a role in identifying the entity across various possible worlds. For example, we can fix a reference to heat as the phenomenon that causes certain sensations in people. Once heat has been so picked out, this property is no longer needed, or relied upon, in identifying heat. A phenomenon can count as heat in another possible world where it doesn't cause those sensations in people. Similarly, I am proposing, we initially use reliability as a test for intellectual quality (virtue or vice status). Once the quality of a faculty or process has been determined, however, it tends to retain that status in our thinking. At any rate, it isn't reassessed each time we consider a fresh case, especially a purely imaginary and bizarre case like the demon world. Nor is quality relativized to each possible world or environment.

The present version of the virtues theory appears to be a successful variant of reliabilism, capable of accounting for most, if not all, of the most prominent counterexamples to earlier variants of reliabilism.[7] The present approach also makes an innovation in naturalistic epistemology. Whereas earlier naturalistic epistemologists have focused exclusively on the psychology of the epistemic agent, the present paper also highlights the psychology of the epistemic evaluator.

III

Let us turn now to *normative* scientific epistemology. It was argued briefly in section I that normative scientific epistemology should preserve continuity with our epistemic folkways. At a minimum, it should rest on the same types of evaluative criteria as those on which our commonsense epistemic evaluations rest. Recently, however, Stephen Stich (1990) has disputed this sort of claim. Stich contends that our epistemic folkways are quite idiosyncratic and should not be much heeded in a reformed epistemology. An example he uses to underline his claim of idiosyncrasy is the notion of justification as rendered by my "normal worlds" analysis in Goldman (1986). With hindsight, I would agree that that particular analysis makes our ordinary notion of justification look pretty idiosyncratic. But that was the fault of the analysis, not the analysandum. On the present rendering, it looks as if the folk notion of justification is keyed to dispositions to produce a high ratio of true beliefs in the actual world, not in "normal worlds"; and there is nothing idiosyncratic about that. Furthermore, there seem to be straightforward reasons for thinking that true belief is worthy of positive valuation, if only from a pragmatic point of view, which Stich also challenges. The pragmatic utility of true belief is best seen by focusing on a certain subclass of beliefs, viz., beliefs about one's own *plans of action*. Clearly, true beliefs about which courses of action would accomplish one's ends will help secure these ends better than false beliefs. Let proposition $P =$ "Plan N will accomplish my ends" and proposition $P' =$ "Plan N' will accomplish my ends". If P is true and P' is false, I am best off believing the former and not believing the latter. My belief will guide my choice of a plan, and belief in the true proposition (but not the false one) will lead me to choose a plan that *will* accomplish my ends. Stich has other intriguing arguments that cannot be considered here, but it certainly appears that true belief is a perfectly sensible and stable value, not an idiosyncratic one.[8] Thus, I shall assume that normative scientific epistemology should follow in the footsteps of folk practice and use reliability (and other truth-linked standards) as a basis for epistemic evaluation.

If scientific epistemology retains the fundamental standard(s) of folk epistemic assessment, how might it diverge from our epistemic folkways? One possible divergence emerges from William Alston's (1988) account of justification. Although generally sympathetic with reliabilism, Alston urges a kind of constraint not standardly imposed by reliabilism (at least not process reliabilism.) This is the requirement that the processes from which justified beliefs issue must have as their input, or basis, a state *of which the cognizer is aware* (or can easily become aware). Suppose that Alston is right about this as an account of our folk conception of justification. It may well be urged that this ingredient needn't be retained in a scientifically sensitive epistemology. In particular, it may well be claimed that one thing to be learned from cognitive science is that only a small proportion of our cognitive processes operate on consciously accessible inputs. It could therefore be argued that a reformed conception of intellectually virtuous processes should dispense with the "accessibility" requirement.

Alston aside, the point of divergence I wish to examine concerns the psychological units that are chosen as virtues or vices. The lay epistemic evaluator uses casual, unsystematic, and largely introspective methods to carve out the mental faculties and processes responsible for belief formation and revision. Scientific epistemology, by contrast, would utilize the resources of cognitive science to devise a more subtle and sophisticated picture of the mechanisms of belief acquisition. I proceed now to illustrate how this project should be carried out.

An initial phase of the undertaking is to sharpen our conceptualization of the types of cognitive units that should be targets of epistemic evaluation. Lay people are pretty vague about the sorts of entities that qualify as intellectual virtues or vices. In my description of epistemic folkways, I have been deliberately indefinite about these entities, calling them variously "faculties", "processes", "mechanisms", and the like. How should systematic epistemology improve on this score?

A first possibility, enshrined in the practice of historical philosophers, is to take the relevant units to be cognitive *faculties.* This might be translated into modern parlance as *modules,* except that this term has assumed a rather narrow, specialized meaning under Jerry Fodor's (1983) influential treatment of modularity. A better translation might be (cognitive) *systems,* e.g., the visual system, long-term memory, and so forth. Such systems, however, are also suboptimal candidates for units of epistemic analysis. Many beliefs are the outputs of two or more systems working in tandem. For example, a belief consisting in the visual classification of an object ("That is a chair") may involve matching some information in the visual system with a category stored in long-term memory. A preferable unit of analysis, then, might be a *process*, construed as the sort of entity depicted by familiar flow charts of cognitive activity. This sort of diagram depicts a sequence of operations (or sets of parallel operations), ultimately culminating in a belief-like output. Such a sequence may span several cognitive systems. This is the sort of entity I had in mind in previous publications (especially Goldman 1986) when I spoke of "cognitive processes".

Even this sort of entity, however, is not a fully satisfactory unit of analysis. Visual classification, for example, may occur under a variety of degraded conditions. The stimulus may be viewed from an unusual orientation; it may be partly occluded, so that only certain of its parts are visible; and so forth. Obviously, these factors can make a big difference to the reliability of the classification process. Yet it is one and the same process that analyzes the stimulus data and comes to a perceptual "conclusion". So the same process can have different degrees of reliability depending on a variety of parameter values. For purposes of epistemic assessment, it would be instructive to identify the parameters and parameter values that are critically relevant to degrees of reliability. The virtues and vices might then be associated not with processes per se, but with processes operating *with specified parameter values*. Let me illustrate this idea in connection with visual perception.

Consider Irving Biederman's (1987, 1990) theory of object recognition, recognition-by-components (RBC). The core idea of Biederman's theory is that a common concrete object like a chair, a giraffe, or a mushroom is mentally represented as an arrangement of simple primitive volumes called *geons (geometrical ions)*. These geons, or primitive "components" of objects, are typically symmetrical volumes lacking sharp concavities, such as blocks, cylinders, spheres, and wedges. A set of twenty-four types of geons can be differentiated on the basis of dichotomous or trichotomous contrasts of such attributes as curvature (straight versus curved), size variation (constant versus expanding), and symmetry (symmetrical versus asymmetrical). These twenty-four types of geons can then be combined by means of six relations (e.g., top-of, side-connected, larger-than, etc.) into various possible multiple-geon objects. For example, a cup can be represented as a cylindrical geon that is side-connected to a curved, handle-like geon, whereas a pail can be represented as the same two geons bearing a different relation: the curved, handle-like geon is at the top of the cylindrical geon.

Simplifying a bit, the RBC theory of object recognition posits five stages of processing. (1) In the first stage, low-level vision extracts edge characteristics, such as Ls, Y-vertices, and arrows. (2) On the basis of these edge characteristics, viewpoint-independent attributes are detected, such as curved, straight, size-constant, size-expanding, etc. (3) In the next stage, selected geons and their relations are activated. (4) Geon activation leads to the activation of object models, that is, familiar models of simple types of objects, stored in long-term memory. (5) The perceived entity is then "matched" to one of these models, and thereby identified as an instance of that category or classification. (In this description of the five stages, all processing is assumed to proceed bottom-up, but in fact Biederman also allows for elements of top-down processing.)

Under what circumstances, or what parameter values, will such a sequence of processing stages lead to *correct,* or *accurate,* object identification? Biederman estimates that there are approximately 3,000 common basic-level, or entry-level, names in English for familiar concrete objects. However, people are probably familiar with approximately ten times that number of object models because, among other things, some entry-level terms (such as *lamp* and *chair*) have several readily distinguishable object models. Thus, an estimate of the number of familiar object models would be on the order of 30,000.

Some of these object models are simple, requiring fewer than six components to appear complete; others are complex, requiring six to nine components to appear complete. Nonetheless, Biederman gives theoretical considerations and empirical results suggesting that an arrangement of only *two* or *three* geons almost always suffices to specify a simple object and even most complex ones. Consider the number of possible two-geon and three-geon objects. With twenty-four possible geons, Biederman says, the variations in relations can produce 186,624 possible two-geon objects. A third geon with its possible relations to another geon yields over 1.4 billion possible three-geon objects. Thus, if the 30,000 familiar object models were distributed homogeneously throughout the space of possible object models, Biederman reasons, an arrangements of two or three geons would almost always be sufficient to specify any object. Indeed, Biederman puts forward a *principle of geon recovery:* If an arrangement of two or three geons can be recovered from the image, objects can be quickly recognized even when they are occluded, rotated in depth, novel, extensively degraded, or lacking in customary detail, color, and texture.

The principle of three-geon sufficiency is supported by the following empirical results. An object such as an elephant or an airplane is complex, requiring six or more geons to appear complete. Nonetheless, when only three components were displayed (the others being occluded), subjects still made correct identifications in almost 80 percent of the nine-component objects and more than 90 percent of the six-component objects. Thus, the reliability conferred by just three geons and their relations is quite high. Although Biederman doesn't give data for recovery of just one or two geons of complex objects, presumably the reliability is much lower. Here we presumably have examples of parameter values—(1) number of components in the complete object, and (2) number of recovered components—that make a significant difference to reliability. The same process, understood as an instantiation of one and the same flow diagrams, can have different levels of reliability depending on the values of the critical parameters in question. Biederman's work illustrates how research in cognitive science can identify both the relevant flow of activity and the crucial parameters. The quality (or "virtue") of a particular (token) process of belief-acquisition depends not only on the flow diagram that is instantiated, but on the parameter values instantiated in the specific tokening of the diagram.

Until now reliability has been my sole example of epistemic quality. But two other dimensions of epistemic quality—which also invoke truth or accuracy—should be added to our evaluative repertoire. These are *question-answering power* and *question-answering speed*. (These are certainly reflected in our epistemic folkways, though not well reflected in the concepts of knowledge or justification.) If a person asks himself a question, such as "What kind of object is that?" or "What is the solution to this algebra problem?", there are three possible outcomes: (A) he comes up with *no answer* (at least none that he believes), (B) he forms a belief in an answer which is *correct,* and (C) he forms a belief in an answer which is *incorrect*. Now reliability is the ratio of cases in category (B) to cases in categories (B) and (C), that is, the proportion of true beliefs to beliefs. Question-answering *power,* on the other hand, is the ratio of (B) cases to cases in categories (A), (B), and (C). Notice that it is possible for a system to be highly reliable but not very powerful. An object-recognition system that never yields outputs in category (C) is perfectly reliable, but it may not be very powerful, since most of its outputs could fall in (A) and only a few in (B). The human (visual) object-recognition system, by-contrast, is very powerful as well as quite reliable. In general, it is power and not just reliability that is an important epistemic desideratum in a cognitive system or process.

Speed introduces another epistemic desideratum beyond reliability and power. This is another dimension on which cognitive science can shed light. It might have been thought, for example, that correct identification of complex objects like an airplane or an elephant requires more time than simple objects such as a flashlight or a cup. In fact, there is no advantage for simple objects, as Biederman's empirical studies indicate. This lack of advantage for simple objects could be explained by the geon theory in terms of parallel activation: geons are activated in parallel rather than through a serial trace of the contours of the object. Whereas more geons would require more processing time under a serial trace, this is not required under parallel activation.

Let us turn now from perception to learning, especially language learning. Learnability theory (Gold 1967; Osherson, Stob, and Weinstein 1985) uses a criterion of learning something like our notion of power, viz., the ability or inability of the learning process to arrive at a correct hypothesis after some fixed period of time. This is called *identification in the limit*. In language learning, it is assumed that the child is exposed to some information in the world, e.g., a set of sentences parents utter, and the learning task is to construct a hypothesis that correctly singles out the language being spoken. The child is presumed to have a learning strategy: an algorithm that generates a succession of hypotheses in response to accumulating evidence. What learning strategy might lead to success? *That* children learn their native language is evident to common sense. But *how* they learn it—what algorithm they possess that constitutes the requisite intellectual virtue—is only being revealed through research in cognitive science.

We may distinguish two types of evidence that a child might receive about its language (restricting attention to the language's grammar): positive evidence and negative evidence. Positive evidence refers to information about which strings of words are *not* grammatical sentences. Interestingly, it appears that children do not receive (much) negative evidence. The absence of negative evidence makes the learning task much harder. What algorithm might be in use that produces success in this situation?

An intriguing proposal is advanced by Robert Berwick (1986; cf.

Pinker 1990). In the absence of negative evidence, the danger for a learning strategy is that it might hypothesize a language that is a superset of the correct language, i.e., one that includes all grammatical sentences of the target language plus some additional sentences as well. Without negative evidence, the child will be unable to learn that the "extra" sentences are incorrect, i.e., don't belong to the target language. A solution is to avoid ever hypothesizing an overly general hypothesis. Hypotheses should be *ordered* in such a way that the child always guesses the narrowest possible hypothesis or language at each step. This is called the *subset principle*. Berwick finds evidence of this principle at work in a number of domains, including concepts, sound systems, and syntax. Here, surely, is a kind of intellectual disposition that is not dreamed of by the "folk".

IV

We have been treating scientific epistemology from a purely reliabilist, or veritistic (truth-linked), vantage point. It should be stressed, however, that scientific epistemology can equally be pursued from other evaluative perspectives. You need not be a reliabilist to accept the proposed role of cognitive science in scientific epistemology. Let me illustrate this idea with the so-called *responsibilist* approach, which characterizes a justified or rational belief as one that is the product of epistemically responsible action (Kornblith 1983; Code 1987), or perhaps epistemically responsible processes (Talbott 1990). Actually, this conception of justification is approximated by my own *weak* conception of justification, as presented in chapter 7. Both depict a belief as justified as long as its acquisition is *blameless* or *nonculpable*. Given limited resources and limited information, a belief might be acquired nonculpably even though its generating processes are not virtuous according to the reliabilist criterion.

Let us start with a case of Hilary Kornblith. Kornblith argues that the justificational status of a belief does not depend exclusively on the *reasoning* process that produces that belief. Someone might reason perfectly well from the evidence he possesses, but fail to be epistemically responsible because he neglects to acquire certain further evidence. Kornblith gives the case of Jones, a headstrong young physicist eager to hear the praise of his colleagues. After Jones presents a paper, a senior colleague makes an objection. Unable to tolerate criticism, Jones pays no attention to the objection. The criticism is devastating, but it makes no impact on Jones's beliefs because he does not even hear it. Jones's conduct is epistemically irresponsible. But his reasoning process from the evidence he actually possesses—which does not include the colleague's evidence—may be quite impeccable.

The general principle suggested by Kornblith's example seems to be something like this. Suppose that an agent (1) believes P, (2) does not believe Q, and (3) would be unjustified in believing P if he did believe Q. If, finally, he is *culpable* for failing to believe Q (for being ignorant of Q), then he is unjustified in believing P. In Kornblith's case, P is the physics thesis that Jones believes. Q consists in the criticisms of this thesis presented by Jones's senior colleague. Jones does not believe Q, but if he did believe Q, he would be unjustified in believing P. However, although Jones does not believe Q, he is culpable for failing to believe it (for being ignorant of these criticisms), because he *ought* to have paid attention to his colleague and acquired belief in Q. Therefore, Jones's belief in P is unjustified.

The provision that the agent be *culpable* for failing to believe Q is obviously critical to the principle in question. If the criticisms of Jones's thesis had never been presented within his hearing, nor published in any scientific journal, then Jones's ignorance of Q would not be culpable. And he might well be justified in believing P. But in Kornblith's version of the case, it seems clear that Jones *is* culpable for failing to believe Q, and that is why he is unjustified in believing P.

Under what circumstances is an agent culpable for failing to believe something? That is a difficult question. In a general discussion of culpable ignorance, Holly Smith (1983) gives an example of a doctor who exposes an infant to unnecessarily high concentrations of oxygen and thereby causes severe eye damage. Suppose that the latest issue of the doctor's medical journal describes a study establishing this relationship, but the doctor hasn't read this journal. Presumably his ignorance of the relationship would be culpable; he *should* have read his journal. But suppose that the study had appeared in an obscure journal to which he does not subscribe, or had only appeared one day prior to this particular treatment. Is he still culpable for failing to have read the study by the time of the treatment?

Smith categorizes her example of the doctor as a case of *deficient investigation*. The question is (both for morals and for epistemology), What amounts and kinds of investigation are, in general, sufficient or deficient? We may distinguish two types of investigation: (1) investigation into the physical world (including statements that have been made by other agents), and (2) investigation into the agent's own storehouse of information, lodged in long-term memory. Investigation of the second sort is particularly relevant to questions about the role of cognitive science, so I shall concentrate here on this topic. Actually, the term 'investigation' is not wholly apt when it comes to long-term memory. But it is adequate as a provisional delineation of the territory.

To illustrate the primary problem that concerns me here, I shall consider two examples drawn from the work of Amos Tversky and Daniel Kahneman. The first example pertains to their study of the "conjunction fallacy" (Tversky and Kahneman 1983). Suppose that a subject assigns a higher probability to a conjunction like "Linda is a bank teller and is active in the feminist movement" than to one of its own conjuncts, "Linda is a bank teller". According to the standard probability calculus, no conjunction can have a higher probability than one of its conjuncts. Let us assume that the standard probability calculus is, in some sense, "right". Does it follow that a person is irrational, or unjustified, to make probability assignments that violate this calculus? This is subject to dispute. One might argue that it does not follow, in general, from the fact that M is an arbitrary mathematical truth, that anyone who believes something contrary to M is ipso facto irrational or unjustified. After all, mathematical facts are not all so transparent that it would be a mark of irrationality (or the like) to fail to believe any of them. However, let us set this issue aside. Let us imagine the case of a subject who has studied probability theory and learned the conjunction rule in particular. Let us further suppose that this subject would retract at least one of his two probability assignments if he recognized that they violate the conjunction rule in connection with the Linda example. Shall we say that the failure to recover the conjunction rule from long-term memory is a *culpable omission,* one that makes his maintenance of his probability judgments unjustified? Is this like the example of Jones who culpably fails to learn of his senior colleague's criticism? Or is it a case of nonculpable nonrecovery of a relevant fact, a fact that is, in some sense "within reach", but legitimately goes unnoticed?

This raises questions about when a failure to recover or activate something from long-term memory is culpable, and that is precisely a problem that invites detailed reflection on mechanisms of memory retrieval. This is not a matter to which epistemologists have devoted much attention, partly because little has been known about memory retrieval until fairly recently. But now that cognitive science has at least the beginnings of an understanding of this phenomenon, normative epistemology should give careful attention to that research. Of course, we cannot expect the issue of culpability to be resolved directly by empirical facts about cognitive mechanisms. Such facts are certainly relevant, however.

The main way that retrieval from memory works is by *content addressing* (cf. Potter 1990). Content addressing means starting retrieval with part of the content of the to-be-remembered material, which provides an "address" to the place in memory where identical or similar material is located. Once a match has been made, related information laid down by previously encoded associations will be retrieved, such as the name or appearance of the object. For example, if you are asked to think of a kind of bird that is yellow, a location in memory is addressed where "yellow bird" is located. "Yellow bird" has previously been associated with "canary", so the latter information is retrieved. Note, however, that there are some kinds of information that cannot be used as a retrieval address, although the information is in memory. For example, what word for a family relationship (e.g., *grandmother*) ends in *w*? Because you have probably never encoded that piece of information explicitly, you may have trouble thinking of the word (hint: not *niece*). Although it is easy to move from the word in question *(nephew)* to "word for a family relationship ending in *w*", it is not easy to move in the opposite direction.

Many subjects who are given the Linda example presumably have not established any prior association between such pairs of propositions ("Linda is a bank teller and is active in the feminist movement" and "Linda is a bank teller") and the conjunction rule. Furthermore, in some versions of the experiment, subjects are not given these propositions adjacent to one another. So it may not occur to the subject even to *compare* the two probability judgments, although an explicit comparison would be more likely to address a location in memory that contains an association with the conjunction rule. In short, it is not surprising, given the nature of memory retrieval, that the material provided in the specified task does not automatically yield retrieval of the conjunction rule for the typical subject.

Should the subject deliberately search memory for facts that might retrieve the conjunction rule? Is omission of such deliberate search a culpable omission? Perhaps, but how much deliberate attention or effort ought to be devoted to this task? (Bear in mind that agents typically have numerous intellectual tasks on their agendas, which vie for attentional resources.) Furthermore, what form of search is obligatory? Should memory be probed with the question, "Is there any rule of probability theory that my (tentative) probability judgments violate?'' This is a plausible search probe for someone who has already been struck by a thought of the conjunction rule and its possible violation, or whose prior experiences with probability experiments make him suspicious. But for someone who has not already retrieved the conjunction rule, or who has not had experiences with probability experiments that alert him to such "traps", what reason is there to be on the lookout for violations of the probability calculus? It is highly questionable, then, that the subject engaged in "deficient investigation" in failing to probe memory with the indicated question.

Obviously, principles of culpable retrieval failure are not easy to come by. Any principles meriting our endorsement would have to be sensitive to facts about memory mechanisms.

A similar point can be illustrated in connection with the so-called *availability heuristic,* which was formulated by Tversky and Kahneman (1973) and explored by Richard Nisbett and Lee Ross (1980). A cognizer uses the availability heuristic when he estimates the frequency of items in a category by the instances he can *bring to mind* through memory retrieval, imagination, or perception. The trouble with this heuristic, as the abovementioned researchers indicate, is that the instances one brings to mind are not necessarily well correlated with objective frequency. Various *biases* may produce discrepancies: biases in initial sampling, biases in attention, or biases in manner of encoding or storing the category instances.

Consider some examples provided by Nisbett and Ross: one hypothetical example and one actual experimental result. (1) (Hypothetical example) An Indiana businessman believes that a disproportionate number of Hoosiers are famous. This is partly because of a bias in initial exposure, but also because he is more likely to notice and remember when the national media identify a famous person as a Hoosier. (2) (Actual experiment) A group of subjects consistently errs in judging the relative frequency of words with *R* in first position versus words with *R* in third position. This is an artifact of how words are encoded in memory (as already illustrated in connection with *nephew*). We don't normally code words by their third letters, and hence words having *R* in the third position are less "available" (from memory) than words beginning with *R*. But comparative availability is not a reliable indicator of actual frequency.

Nisbett and Ross (p. 23) view these uses of the availability heuristic as normative errors. "An indiscriminate use of the availability heuristic," they write, "clearly can lead people into serious judgmental errors." They grant, though, that in many contexts perceptual salience, memorability, and imaginability may be relatively unbiased and well correlated with true frequency or causal significance. They conclude: "The normative status of using the availability heuristic . . . thus depend[s] on the judgmental domain and context. People are not, of course, totally unaware that simple availability criteria must sometimes be discounted. For example, few people who were asked to estimate the relative number of moles versus cats in their neighborhood would conclude 'there must be more cats because I've seen several of them but I've never seen a mole.' Nevertheless, as this book documents, people often fail to distinguish between legitimate and superficially similar, but illegitimate, uses of the availability heuristic."

We can certainly agree with Nisbett and Ross that the availability heuristic can often lead to incorrect estimates of frequency. But does it follow that uses of the heuristic are often *illegitimate* in a sense that implies the epistemic *culpability* of the users? One might retort, "These cognizers are using all the evidence that they possess, at least *consciously* possess. Why are they irresponsible if they extrapolate from this evidence?" The objection apparently lurking in Nisbett and Ross's minds is that these cognizers *should* be aware that they are using a systematically biased heuristic. This is a piece of evidence that they *ought* to recognize. And their failure to recognize it, and/or their failure to take it into account, makes their judgmental performance culpable. Nisbett and Ross's invocation of the cat/mole example makes the point particularly clear. If someone can appreciate that the relative number of cats and moles *he has seen* is not a reliable indicator of the relative number of cats and moles in the neighborhood, surely he can be expected to appreciate that the relative number of famous Hoosiers *he can think of* is not a reliable indicator of the proportion of famous people who are Hoosiers!

Is it so clear that people *ought* to be able to appreciate the biased nature of their inference pattern in the cases in question? Perhaps it seems transparent in the mole and Hoosier cases; but consider the letter *R* example. What is (implicitly) being demanded here of the cognizer? First, he must perform a feat of meta-cognitive analysis: he must recognize that he is inferring the relative proportion of the two types of English words from his own constructed samples of these types. Second, he must notice that his construction of these samples depends on the way words are encoded in memory. Finally, he must realize that this implies a bias in ease of retrieval. All these points may seem obvious in hindsight, once pointed out by researchers in the field. But how straightforward or obvious are these matters if they haven't already been pointed out to the subject? Of course, we currently have no "metric" of straightforwardness or obviousness. That is precisely the sort of thing we need, however, to render judgments of culpability in this domain. We need a systematic account of how difficult it is, starting from certain information and preoccupations, to generate and apprehend the truth of certain relevant hypotheses. Such an account clearly hinges on an account of the inferential and hypothesis-generating strategies that are natural to human beings. This is just the kind of thing that cognitive science is, in principle, capable of delivering. So epistemology must work hand in hand with the science of the mind. The issues here are not purely scientific, however. Judgments of justifiedness and unjustifiedness, on the responsibilist conception, require assessments of culpability and nonculpability. Weighing principles for judgments of culpability is a matter for philosophical attention. (One question, for example, is how much epistemic culpability depends on voluntariness.) Thus, a mix of philosophy and psychology is needed to produce acceptable principles of justifiedness.

Notes

I wish to thank Tom Senor, Holly Smith, and participants in a conference at Rice University for helpful comments on earlier versions of this paper.

1. Normative scientific epistemology corresponds to what I elsewhere call *epistemics* (see Goldman 1986). Although epistemics is not restricted to the assessment of *psychological* processes, that is the topic of the present paper. So we are here dealing with what I call *primary epistemics*.

2. Sosa's approach is spelled out most fully in Sosa 1985, 1988, and 1991.

3. My own previous formulations of reliabilism have not been so simple. Both "What Is Justified Belief?" and *Epistemology and Cognition* (1986) had provisions—e.g., the non-undermining provision of *Epistemology and Cognition*—that could help accommodate BonJour's examples. It is not entirely clear, however, how well these qualifications succeeded with the Norman case, described below.

4. Tom Senor presented the following example to his philosophy class at the University of Arkansas. Norman is working at his desk when out of the blue he is hit (via clairvoyance) with a very distinct and vivid impression of the president at the Empire State Building. The image is phenomenally distinct from a regular visual impression but is in some respects similar and of roughly equal force. The experience is so overwhelming that Norman just can't help but form the belief that the president is in New York. About half of Senor's class judged that in this case Norman justifiably believes that the president is in New York. Senor points out, in commenting on this paper, that their

judgments are readily explained by the present account, because the description of the clairvoyance process makes it sufficiently similar to vision to be easily "matched" to that virtue.

5. Since some of these opinions may be true and others false, people's lists of virtues and vices may have varying degrees of accuracy. The "real" status of a trait as a virtue or vice is independent of people's opinions about the trait. However, since the enterprise of descriptive epistemology is to describe and explain evaluators' judgments, we need to advert to the traits they *believe* to be virtues or vices, i.e., the ones on their mental lists.

6. Thanks to Holly Smith for this example. She cites Riding 1989 (chap. 6) for relevant discussion.

7. It should be noted that this theory of justification is intended to capture what I call the *strong* conception of justification. The complementary conception of *weak* justification will receive attention in section IV of this essay.

8. For further discussion of Stich, see Goldman 1991.

References

Goldman, A. I. 1986. *Epistemology and Cognition*. Cambridge, MA: Harvard University Press.

Goldman, A. I. 1991. "Review of S. Stich, *The Fragmentation of Reason*." *Philosophy and Phenomenological Research* 51, 189–193.

Riding, A. 1989. *Distant Neighbors: A Portrait of the Mexicans*. New York: Vintage Books.

Sosa, E. 1985. "Knowledge and Intellectual Virtue." *The Monist* 68, 226–263.

Sosa, E. 1988. "Beyond Scepticism, to the Best of our Knowledge." *Mind* 97, 153–188.

Sosa, E. 1991. "Reliabilism and Intellectual Virtue." In *Knowledge in Perspective*. Cambridge: Cambridge University Press.

A Bibliography on Empirical Knowledge

General Works

Achinstein, Peter, ed. *The Concept of Evidence*. Oxford: Oxford University Press, 1983.

Alcoff, Linda, and Elizabeth Potter, eds. *Feminist Epistemologies*. London: Routledge, 1993.

Almeder, Robert. *Blind Realism*. Lanham, Md.: Rowman & Littlefield, 1992.

Alston, William P. *Epistemic Justification*. Ithaca, N.Y.: Cornell University Press, 1989.

———. "Meta-Ethics and Meta-Epistemology." In *Values and Morals*, pp. 275–97. Edited by A. I. Goldman and J. Kim. Dordrecht: D. Reidel, 1978.

———. *The Reliability of Sense Perception*. Ithaca: Cornell University Press, 1993.

———. "The Role of Reason in the Regulation of Belief." In *Rationality in the Calvinian Tradition*, pp. 135–70. Edited by N. Wolterstorff et al. Lanham, Md.: University Press of America, 1983.

Armstrong, David M. *Belief, Truth, and Knowledge*. Cambridge: Cambridge University Press, 1973.

Audi, Robert. *Belief, Justification, and Knowledge: An Introduction to Epistemology*. Belmont, Cal.: Wadsworth, 1988.

———. *The Structure of Justification*. Cambridge: Cambridge University Press, 1993.

———, ed. *The Cambridge Dictionary of Philosophy*. Cambridge: Cambridge University Press, 1995.

Aune, Bruce. *Knowledge, Mind, and Nature*. New York: Random House, 1967.

———. *Knowledge of the External World*. London: Routledge, 1991.

Ayer, A. J. *Probability and Evidence*. New York: Columbia University Press, 1972.

———. *The Problem of Knowledge*. Baltimore: Penguin Books, 1956.

447

Baergen, Ralph. *Contemporary Epistemology.* New York: Harcourt Brace, 1995.

BonJour, Laurence. *The Structure of Empirical Knowledge.* Cambridge, Mass.: Harvard University Press, 1985.

Butchvarov, Panayot. *The Concept of Knowledge.* Evanston, Ill.: Northwestern University Press, 1970.

Carruthers, Peter. *Human Knowledge and Human Nature.* Oxford: Oxford University Press, 1992.

Cassam, Quassim, ed. *Self-Knowledge.* Oxford: Oxford University Press, 1994.

Chisholm, Roderick M. *The Foundations of Knowing.* Minneapolis: University of Minnesota Press, 1982.

———. *Perceiving: A Philosophical Study.* Ithaca: Cornell University Press, 1957.

———. *Theory of Knowledge, 1st ed.* Englewood Cliffs, NJ: Prentice-Hall, 1966; 2d ed., 1977; 3d ed., 1989.

Chisholm, Roderick, and Robert Swartz, eds. *Empirical Knowledge.* Englewood Cliffs, N.J.: Prentice-Hall, 1973.

Clay, Marjorie, and Keith Lehrer, eds. *Knowledge and Skepticism.* Boulder, Col.: Westview, 1989.

Coady, C. A. J. *Testimony.* Oxford: Oxford University Press, 1992.

Coffa, J. Alberto. *The Semantic Tradition from Kant to Carnap.* Cambridge: Cambridge University Press, 1991.

Cornman, James W. *Perception, Common Sense, and Science.* New Haven, Conn.: Yale University Press, 1975.

Craig, Edward. *Knowledge and the State of Nature.* Oxford: Oxford University Press, 1991.

Dancy, Jonathan. *An Introduction to Contemporary Epistemology.* Oxford: Basil Blackwell, 1985.

Dancy, Jonathan, ed. *Perceptual Knowledge.* Oxford: Oxford University Press, 1988.

Dancy, Jonathan, and Ernest Sosa, eds. *A Companion to Epistemology.* Oxford: Blackwell, 1992.

Danto, Arthur. *Analytical Philosophy of Knowledge.* Cambridge: Cambridge University Press, 1968.

Dicker, Georges. *Perceptual Knowledge.* Dordrecht: D. Reidel, 1980.

Dretske, Fred I. *Seeing and Knowing.* Chicago: University of Chicago Press, 1969.

Feldman, Richard, and Earl Conee. "Evidentialism." *Philosophical Studies* 48 (1985), 15–35. Reprinted in P. K. Moser and Arnold vander Nat, eds., *Human Knowledge,* pp. 334–45. Oxford: Oxford University Press, 1987.

Foley, Richard. *The Theory of Epistemic Rationality.* Cambridge, Mass.: Harvard University Press, 1987.

———. *Working Without a Net.* Oxford: Oxford University Press, 1993.

French, Peter, et al., eds. *Midwest Studies in Philosophy, Vol. V: Studies in Epistemology*. Minneapolis: University of Minnesota Press, 1980.

Fumerton, Richard. *Metaphysical and Epistemological Problems of Perception* Lincoln: University of Nebraska Press, 1985.

———. *Metaepistemology and Skepticism*. Lanham, Md.: Rowman & Littlefield, 1995.

Gardner, Howard. *The Mind's New Science*. New York: Basic Books, 1985; expanded ed., 1987.

Ginet, Carl. "The Justification of Belief: A Primer." In *Knowledge and Mind*. Edited by C. Ginet and S. Shoemaker. Oxford: Oxford University Press, 1983.

———. *Knowledge, Perception, and Memory*. Dordrecht: D. Reidel, 1975.

Goldman, Alvin I. *Epistemology and Cognition*. Cambridge, Mass.: Harvard University Press, 1986.

———. *Liaisons*. Cambridge, Mass.: MIT Press, 1992.

Griffiths, A. P., ed. *Knowledge and Belief*. Oxford: Oxford University Press, 1967.

Grossmann, Reinhardt. *The Fourth Way*. Bloomington, In.: Indiana University Press, 1990.

Haack, Susan. *Evidence and Inquiry*. Oxford: Blackwell, 1994.

———. "Theories of Knowledge: An Analytic Framework." *Proceedings of the Aristotelian Society* 83 (1983), 143–57.

Hamlyn, D. W. *The Theory of Knowledge*. Garden City, N.Y.: Doubleday, 1971.

Harman, Gilbert. *Change In View*. Cambridge, Mass.: MIT Press, 1986.

———. *Thought*. Princeton: Princeton University Press, 1973.

Hetherington, Stephen. *Epistemology's Paradox*. Lanham, Md.: Rowman & Littlefield, 1992.

Hill, Thomas. *Contemporary Theories of Knowledge*. New York: Macmillan, 1961.

Hintikka, Jaakko. *Knowledge and Belief*. Ithaca: Cornell University Press, 1962.

Hirsch, Eli. *Dividing Reality*. Oxford: Oxford University Press, 1993.

Hirst, R. J., ed. *Perception and the External World*. New York: Macmillan, 1965.

Jeffrey, Richard C. *The Logic of Decision, 2d ed.* Chicago: University of Chicago Press, 1983.

Kvanvig, Jonathan. *The Intellectual Virtues and the Life of the Mind*. Lanham, Md.: Rowman & Littlefield, 1992.

Lehrer, Keith. *Knowledge*. Oxford: Clarendon Press, 1974.

———. *Theory of Knowledge*. Boulder, Col.: Westview, 1990.

Levi, Isaac. *The Enterprise of Knowledge*. Cambridge, Mass.: MIT Press, 1980.

Lewis, C. I. *An Analysis of Knowledge and Valuation*. LaSalle, IL: Open Court, 1946.

Lycan, William. *Judgment and Justification*. Cambridge: Cambridge University Press, 1988.

Malcolm, Norman. *Knowledge and Certainty*. Ithaca: Cornell University Press, 1963.

McGinn, Colin. "The Concept of Knowledge." In *Midwest Studies in Philosophy, Vol. IX*, pp. 529–54. Edited by P. French et al. Minneapolis: University of Minnesota Press, 1984.

Meyers, Robert. *The Likelihood of Knowledge*. Dordrecht: Kluwer, 1988.

Montmarquet, James. *Epistemic Virtue and Doxastic Responsibility*. Lanham, Md.: Rowman & Littlefield, 1993.

Morton, Adam. *A Guide Through the Theory of Knowledge*. Encino, Cal.: Dickenson, 1977.

Moser, Paul K. *Empirical Justification*. Dordrecht: D. Reidel, 1985.

———. *Knowledge and Evidence*. Cambridge: Cambridge University Press, 1989.

———. *Philosophy after Objectivity*. Oxford: Oxford University Press, 1993.

———. "Epistemology (1900–Present)." In John Canfield, ed., *Routledge History of Philosophy, Vol. 10: Philosophy of the English Speaking World in the 20th Century*. London: Routledge, 1996.

———, ed. *A Priori Knowledge*. Oxford: Oxford University Press, 1987.

———, ed. *Rationality in Action*. Cambridge: Cambridge University Press, 1990.

Moser, Paul K., and Arnold VanderNat, eds. *Human Knowledge: Classical and Contemporary Approaches*. Oxford: Oxford University Press, 1987; 2d ed., 1995.

Nagel, Ernest, and Richard Brandt, eds. *Meaning and Knowledge*. New York: Harcourt, Brace & World, 1965.

O'Connor, D. J., and Brian Carr. *Introduction to the Theory of Knowledge*. Minneapolis: University of Minnesota Press, 1982.

Pappas, G. S., ed. *Justification and Knowledge*. Dordrecht: D. Reidel, 1979.

Pappas, G. S., and Marshall Swain, eds. *Essays on Knowledge and Justification*. Ithaca: Cornell University Press, 1978.

Plantinga, Alvin. *Warrant: The Current Debate*. Oxford: Oxford University Press, 1993.

———. *Warrant and Proper Function*. Oxford: Oxford University Press, 1993.

Pollock, John L. *Contemporary Theories of Knowledge*. Rowman & Littlefield, 1986.

———. *Knowledge and Justification*. Princeton: Princeton University Press, 1974.

Popper, Karl. *Objective Knowledge*. Oxford: Oxford University Press, 1972.

Price, H. H. *Thinking and Experience, 2d ed*. London: Hutchinson, 1969.

Quine, W. V. *Pursuit of Truth*. Cambridge, Mass.: Harvard University Press, 1990.

Quine, W. V., and J. S. Ullian, *The Web of Belief, 2d ed*. New York: Random House, 1978.

Roth, Michael, and Leon Galis, eds. *Knowing: Essays in the Analysis of Knowledge*. New York: Random House, 1970.

Russell, Bertrand. *Human Knowledge: Its Scope and Limits*. New York: Simon & Schuster, 1948.

Schlesinger, George N. *The Range of Epistemic Logic*. Atlantic Highlands, NJ: Humanities Press, 1985.

Schmitt, Frederick. *Knowledge and Belief*. London: Routledge, 1992.

————, ed. *Socializing Epistemology*. Lanham, Md.: Rowman & Littlefield, 1994.

Shope, Robert. "The Conditional Fallacy in Contemporary Philosophy." *The Journal of Philosophy* 75 (1978), 397–413.

Sosa, Ernest. *Knowledge in Perspective*. Cambridge: Cambridge University Press, 1991.

————, ed. *Knowledge and Justification*. Aldershort, Eng.: Dartmouth, 1994.

Stalnaker, Robert C. *Inquiry*. Cambridge, Mass.: MIT Press, 1984.

Stich, Stephen. *The Fragmentation of Reason*. Cambridge, Mass.: MIT Press, 1990.

Stroll, Avrum, ed. *Epistemology*. New York: Harper & Row, 1967.

Swain, Marshall. *Reasons and Knowledge*. Ithaca: Cornell University Press, 1981.

Swartz, R. J., ed. *Perceiving, Sensing, and Knowing*. Garden City, N.Y.: Doubleday, 1965.

Tomberlin, James, ed. *Philosophical Perspectives, 2: Epistemology*. Atascadero, Cal.: Ridgeview, 1988.

Epistemic Foundationalism

Almeder, Robert F. "Basic Knowledge and Justification." *The Canadian Journal of Philosophy* 13 (1983), 115–28.

Alston, William P. *Epistemic Justification*. Ithaca: Cornell University Press, 1989.

————. "Plantinga's Religious Epistemology." In *Alvin Plantinga*, pp. 287–309. Edited by J. E. Tomberlin and P. van Inwagen. Dordrecht: D. Reidel, 1985.

————. "Some Remarks on Chisholm's Epistemology." *Noûs* 14 (1980), 565–86.

Annis, David B. "Epistemic Foundationalism." *Philosophical Studies* 31 (1977), 345–52.

Armstrong, David M. *Belief, Truth, and Knowledge*. Cambridge: Cambridge University Press, 1973.

Audi, Robert. *The Structure of Justification*. Cambridge: Cambridge University Press, 1993.

Ayer, A. J. "Basic Propositions." In *Philosophical Analysis*, pp. 60–74. Edited by M. Black. Englewood Cliffs, NJ: Prentice-Hall, 1950. Reprinted in Ayer, *Philosophical Essays* (London: Macmillan, 1965).

———. *The Foundations of Empirical Knowledge.* New York: Macmillan, 1940.

BonJour, Laurence. "Externalist Theories of Empirical Knowledge." In *Midwest Studies in Philosophy, Vol. V: Studies in Epistemology*, pp. 53–74. Edited by P. French et al. Minneapolis: University of Minnesota Press, 1980.

———. *The Structure of Empirical Knowledge.* Cambridge, Mass.: Harvard University Press, 1985.

Chisholm, Roderick. "The Directly Evident." In *Justification and Knowledge*, pp. 115–27. Edited by G. S. Pappas. Dordrecht: D. Reidel, 1979.

———. "On the Nature of Empirical Evidence." In *Essays on Knowledge and Justification*, pp. 253–78. Edited by G. S. Pappas and M. Swain. Ithaca: Cornell University Press, 1978.

———. *Theory of Knowledge, 1st ed.* Englewood Cliffs, N.J.: Prentice-Hall, 1966; 2d ed., 1977; 3d ed., 1989.

———. "Theory of Knowledge in America." In Chisholm, *The Foundations of Knowing*, pp. 109–96. Minneapolis: University of Minnesota Press, 1982. See Selection 2, this book.

———. "A Version of Foundationalism." In Chisholm, *The Foundations of Knowing*, pp. 3–32. Minneapolis: University of Minnesota Press, 1982.

Churchland, Paul M. *Scientific Realism and the Plasticity of Mind*, chapter 2. Cambridge: Cambridge University Press, 1979.

Cornman, James W. "Foundational versus Nonfoundational Theories of Empirical Justification." *American Philosophical Quarterly* 14 (1977), 287–97. Reprinted in *Essays on Knowledge and Justification*, pp. 229–52. Edited by G. S. Pappas and M. Swain. Ithaca: Cornell University Press, 1978.

———. "On Acceptability Without Certainty." *Journal of Philosophy* 74 (1977), 29–47.

———. "On the Certainty of Reports about What is Given." *Noûs* 12 (1978), 93–118.

———. "On Justifying Non-Basic Statements by Basic-Reports." In *Justification and Knowledge*, pp. 129–49. Edited by G. S. Pappas. Dordrecht: D. Reidel, 1979.

———. *Skepticism, Justification, and Explanation.* Dordrecht: D. Reidel, 1980.

Dancy, Jonathan. *An Introduction to Contemporary Epistemology*, chapters 4 and 5. Oxford: Basil Blackwell, 1985.

Foley, Richard. *The Theory of Epistemic Rationality.* Cambridge, Mass.: Harvard University Press, 1987.

———. *Working Without a Net.* Oxford: Oxford University Press, 1993.

Fumerton, Richard. *Metaepistemology and Skepticism.* Lanham, Md.: Rowman & Littlefield, 1995.

————. *Metaphysical and Epistemological Problems of Perception*, chapter 2. Lincoln: University of Nebraska Press, 1985.

Goldman, Alan H. "Appearing Statements and Epistemological Foundations." *Metaphilosophy* 10 (1979), 227–46.

————. *Empirical Knowledge*. Berkeley, Calif.: University of California Press, 1988.

————. "Epistemic Foundationalism and the Replaceability of Ordinary Language." *Journal of Philosophy* 79 (1982), 136–54.

Heidelberger, Herbert. "Chisholm's Epistemic Principles." *Noûs* 3 (1969), 73–82.

Heil, John. "Foundationalism and Epistemic Rationality." *Philosophical Studies* 42 (1982), 179–88.

Kornblith, Hilary. "Beyond Foundationalism and the Coherence Theory." *The Journal of Philosophy* 72 (1980), 597–612. Reprinted in *Naturalizing Epistemology*, pp. 115–28. Edited by H. Kornblith. Cambridge, Mass.: MIT Press, 1985.

Kvanvig, Jonathan. "What is Wrong With Minimal Foundationalism?" *Erkenntnis* 21 (1984), 175–87.

Lehrer, Keith. *Theory of Knowledge*. Boulder, Colo.: Westview, 1990.

Lewis, C. I. *An Analysis of Knowledge and Valuation*, chapters 7 and 8. LaSalle, Ill.: Open Court, 1946.

————. "The Given Element in Empirical Knowledge." *The Philosophical Review* 61 (1952), 168–75.

————. *Mind and the World Order*. New York: Scribner's Sons, 1929.

McGrew, Timothy. *The Foundations of Knowledge*. Lanham, Md.: Littlefield Adams, 1995.

Moser, Paul K. "A Defense of Epistemic Intuitionism." *Metaphilosophy* 15 (1984), 196–209.

————. *Empirical Justification*, chapters 4 and 5. Dordrecht: D. Reidel, 1985.

————. *Knowledge and Evidence*. Cambridge: Cambridge University Press, 1989.

————. *Philosophy after Objectivity*. Oxford: Oxford University Press, 1993.

Pastin, Mark. "Lewis' Radical Foundationalism." *Noûs* 9 (1975), 407–20.

————. "Modest Foundationalism and Self-Warrant." In *American Philosophical Quarterly Monograph Series, No. 9: Studies in Epistemology*, pp. 141–49. Edited by N. Rescher. Oxford: Basil Blackwell, 1975. Reprinted in *Essays on Knowledge and Justification*, pp. 279–88. Edited by G. S. Pappas and M. Swain. Ithaca: Cornell University Press, 1978.

Pollock, John. *Contemporary Theories of Knowledge*. Lanham, Md.: Rowman & Littlefield, 1986.

————. *Knowledge and Justification*. Princeton: Princeton University Press, 1974.

————. "A Plethora of Epistemological Theories." in *Justification and Knowledge*, pp. 93–113. Edited by G. S. Pappas. Dordrecht: D. Reidel, 1979.

Quinton, Anthony. "The Foundations of Knowledge." In *British Analytic Philosophy*, pp. 55–86. Edited by Bernard Williams and Alan Montefiore. London: Routledge & Kegan Paul, 1966.

———. *The Nature of Things*, chapter 8. London: Routledge & Kegan Paul, 1973.

Russell, Bertrand. *Human Knowledge: Its Scope and Limits*, Part 2. New York: Simon & Schuster, 1948.

———. *An Inquiry into Meaning and Truth*, chapters 9 and 10. New York: Norton, 1940.

———. "On Verification." *Proceedings of the Aristotelian Society* 38 (1937–38), 1–15.

Scheffler, Israel. *Science and Subjectivity, 2d ed.*, chapters 2 and 5. Indianapolis: Hackett, 1982.

Sellars, Wilfrid. "Does Empirical Knowledge Have a Foundation?" in "Empiricism and the Philosophy of Mind." In *Minnesota Studies in the Philosophy of Science*, Vol. 1, pp. 293–300. Edited by H. Feigl and M. Scriven. Minneapolis: University of Minnesota Press, 1956. Reprinted in Sellars, *Science, Perception, and Reality*. London: Routledge & Kegan Paul, 1963.

Sosa, Ernest. *Knowledge in Perspective*. Cambridge: Cambridge University Press, 1991.

Strawson, Peter F. "Does Knowledge Have Foundations?" In *Teorema, Mono. 1: Conocimiento y Creencia*, pp. 99–110. Universidad de Valencia, 1974.

Swain, Marshall. "Cornman's Theory of Justification." *Philosophical Studies* 41 (1982), 129–48.

———. *Reasons and Knowledge*. Ithaca: Cornell University Press, 1981.

Van Cleve, James. "Epistemic Supervenience and the Circle of Belief." *The Monist* 68 (1985), 90–104.

———. "Foundationalism, Epistemic Principles, and the Cartesian Circle." *The Philosophical Review* 88 (1979), 55–91.

Epistemic Coherentism

Audi, Robert. *The Structure of Justification*. Cambridge: Cambridge University Press, 1993.

Bender, John, ed. *The Current State of the Coherence Theory*. Dordrecht: Kluwer, 1989.

Blanshard, Brand. *The Nature of Thought*, Vol. 2, chapters 25–27. London: Allen & Unwin, 1939.

BonJour, Laurence. "The Coherence Theory of Empirical Knowledge." *Philosophical Studies* 30 (1976), 281–312.

———. *The Structure of Empirical Knowledge*. Cambridge, Mass.: Harvard University Press, 1985.

Dancy, Jonathan. *An Introduction to Contemporary Epistemology,* chapters 8 and 9. Oxford: Blackwell, 1985.

———. "On Coherence Theories of Justification: Can an Empiricist be a Coherentist?" *American Philosophical Quarterly* 21 (1984), 359–65.

Davidson, Donald. "A Coherence Theory of Truth and Knowledge." In *Kant oder Hegel,* pp. 423–38. Edited by Dieter Henrich. Stuttgart: Klett-Cotta, 1983.

Firth, Roderick. "Coherence, Certainty, and Epistemic Priority." *The Journal of Philosophy* 61 (1964), 545–57.

Harman, Gilbert. *Change In View.* Cambridge, Mass.: MIT Press, 1986.

———. "Knowledge, Inference, and Explanation." *American Philosophical Quarterly* 5 (1968), 164–73.

———. "Knowledge, Reasons, and Causes." *The Journal of Philosophy* 67 (1970), 841–55.

———. *Thought,* chapter 8. Princeton: Princeton University Press, 1973.

Lehrer, Keith. "Justification, Explanation, and Induction." In *Induction, Acceptance, and Rational Belief,* pp. 100–33. Edited by M. Swain. Dordrecht: D. Reidel, 1970.

———. *Knowledge,* chapters 7 and 8. Oxford: Clarendon Press, 1974.

———. "The Knowledge Cycle." *Noûs* 11 (1977), 17–26.

———. "Knowledge, Truth, and Ontology." In *Language and Ontology: Proceedings of the 6th International Wittgenstein Symposium,* pp. 201–11. Edited by W. Leinfellner et al. Vienna: Holder-Pichler-Tempsky, 1982.

———. "Self-Profile." In *Keith Lehrer,* pp. 3–104. Edited by R. J. Bogdan. Dordrecht: D. Reidel, 1981.

———. *Theory of Knowledge.* Boulder, Colo.: Westview, 1990.

Lehrer, Keith, and Stewart Cohen. "Justification, Truth, and Coherence." *Synthese* 55 (1983), 191–208.

Lemos, Noah. "Coherence and Epistemic Priority." *Philosophical Studies* 41 (1982), 299–316.

Moser, Paul K. *Empirical Justification,* chapter 3. Dordrecht: D. Reidel, 1985.

———. *Knowledge and Evidence.* Cambridge: Cambridge University Press, 1989.

Pastin, Mark. "Social and Anti-Social Justification: A Study of Lehrer's Epistemology." In *Keith Lehrer,* pp. 205–22. Dordrecht: D. Reidel, 1981.

Rescher, Nicholas. "Blanshard and the Coherence Theory of Truth." In *The Philosophy of Brand Blanshard,* pp. 574–88. Edited by P. Schlipp. LaSalle, Ill.: Open Court, 1980.

———. *Cognitive Systematization.* Oxford: Blackwell, 1979.

———. *The Coherence Theory of Truth.* Oxford: Clarendon Press, 1973.

———. "Foundationalism, Coherentism, and the Idea of Cognitive Systematization." *The Journal of Philosophy* 71 (1974), 695–708.

———. *A System of Pragmatic Idealism,* 3 vols. Princeton: Princeton University Press, 1991–1994.

——. "Truth as Ideal Coherence." *The Review of Metaphysics* 38 (1985), 795–806.

Sellars, Wilfrid. "Epistemic Principles." In *Action, Knowledge, and Reality: Critical Studies in Honor of Wilfrid Sellars*, pp. 332–48. Edited by H.-N. Castañeda. Indianapolis: Boobs-Merrill, 1975.

——. "Givenness and Explanatory Coherence." *The Journal of Philosophy* 70 (1973), 612–24.

——. "More on Givenness and Explanatory Coherence." In *Justification and Knowledge*, pp. 169–81. Edited by G. S. Pappas. Dordrecht: D. Reidel, 1979.

Sosa, Ernest. "Circular Coherence and Absurd Foundations." In *A Companion to Inquiries into Truth and Interpretation*. Edited by E. Lepore. Oxford: Basil Blackwell, 1985.

——. *Knowledge in Perspective*. Cambridge: Cambridge University Press, 1991.

Epistemic Contextualism

Airaksinen, Timo. "Contextualism: A New Theory of Epistemic Justification." *Philosophia* 12 (1982), 37–50.

Annis, David. "The Social and Cultural Component of Epistemic Justification: A Reply." *Philosophia* 12 (1982), 51–55.

Morawetz, Thomas. *Wittgenstein and Knowledge*. Amherst: University of Massachusetts Press, 1978.

Moser, Paul K. *Empirical Justification*, chapter 2. Dordrecht: D. Reidel, 1985.

Rorty, Richard. *Philosophy and the Mirror of Nature*, chapter 7. Princeton: Princeton University Press, 1979.

——. "From Epistemology to Hermeneutics." In *Acta Philosophica Fennica, Vol. 30: The Logic and Epistemology of Scientific Change*. Edited by I. Niiniluoto and R. Tuomela. Amsterdam: North-Holland Publishing Co., 1978.

Schmitt, Frederick, ed. *Socializing Epistemology*. Lanham, Md.: Rowman & Littlefield, 1994.

Shiner, Roger. "Wittgenstein and the Foundations of Knowledge." *Proceedings of the Aristotelian Society* 78 (1977–78), 103–24.

Sosa, Ernest. "On Groundless Belief." *Synthese* 43 (1979), 453–60.

Williams, Michael. "Coherence, Justification, and Truth." *The Review of Metaphysics* 34 (1980), 243–72.

——. *Groundless Belief*. Oxford: Blackwell, 1977.

——. *Unnatural Doubts*. Oxford: Blackwell, 1991.

Wittgenstein, Ludwig. *On Certainty*. Edited by G. E. M. Anscombe and G. H. von Wright. Oxford: Basil Blackwell, 1969.

Epistemic Reliabilism

Armstrong, David M. *Belief, Truth, and Knowledge*. Cambridge: Cambridge University Press, 1973.

———. "Self-Profile." In *D. M. Armstrong*, pp. 30–37. Edited by R. Bogdan. Dordrecht: D. Reidel, 1984.

Audi, Robert. *The Structure of Justification*. Cambridge: Cambridge University Press, 1993.

BonJour, Laurence. "Externalist Theories of Empirical Knowledge." In *Midwest Studies in Philosophy, Vol. V: Studies in Epistemology*, pp. 53–73. Edited by P. French et al. Minneapolis: University of Minnesota Press, 1980.

Cohen, Stewart. "Justification and Truth." *Philosophical Studies* 46 (1984), 279–95.

Dretske, Fred I. "Conclusive Reasons." *The Australasian Journal of Philosophy* 49 (1971), 1–22. Reprinted in *Essays on Knowledge and Justification*, pp. 41–60. Edited by G. S. Pappas and M. Swain. Ithaca: Cornell University Press, 1978.

———. "The Pragmatic Dimension of Knowledge." *Philosophical Studies* 40 (1981), 363–78.

———. *Knowledge and the Flow of Information*, chapters 4 and 5. Cambridge, Mass.: MIT Press, 1981.

———. "Precis of *Knowledge and the Flow of Information*." *The Behavioral and Brain Sciences* 6 (1983), 55–63.

Feldman, Richard. "Reliability and Justification." *The Monist* 68 (1985), 159–74.

Feldman, Richard and Earl Conee. "Evidentialism." *Philosophical Studies* 48 (1985), 15–34.

Firth, Roderick. "Epistemic Merit, Intrinsic and Instrumental." In *Proceedings and Addresses of the American Philosophical Association* 55 (1981), 5–23.

Foley, Richard. "What's Wrong with Reliabilism?" *The Monist* 68 (1985), 175–87.

Goldman, Alvin I. "Discrimination and Perceptual Knowledge." *The Journal of Philosophy* 73 (1976), 771–91. Reprinted in *Essays on Knowledge and Justification*, pp. 120–45. Edited by G. S. Pappas and M. Swain. Ithaca: Cornell University Press, 1978.

———. *Epistemology and Cognition*. Cambridge, Mass.: Harvard University Press, 1986.

———. "The Internalist Conception of Justification." In *Midwest Studies in Philosophy, Vol. V: Studies in Epistemology*, pp. 27–52. Edited by P. French et al. Minneapolis: University of Minnesota Press, 1980.

———. *Liaisons*. Cambridge, Mass.: MIT Press, 1992.

———. "What is Justified Belief?" In *Justification and Knowledge*, pp. 1–23. Edited by G. S. Pappas. Dordrecht: D. Reidel, 1979.

Heil, John. "Reliability and Epistemic Merit." *The Australasian Journal of Philosophy* 62 (1984), 327–38.

Kornblith, Hilary. "Beyond Foundationalism and the Coherence Theory." *The Journal of Philosophy* 72 (1980), 597–612.

———. "Ever Since Descartes." *The Monist* 68 (1985), 264–76.

———. "Justified Belief and Epistemically Responsible Action." *The Philosophical Review* 92 (1983), 33–48.

———. "The Psychological Turn." *The Australasian Journal of Philosophy* 60 (1982), 238–53.

———, ed. *Naturalizing Epistemology.* Cambridge, Mass.: MIT Press, 1994.

Kvanvig, Johathan. *The Intellectual Virtues and the Life of the Mind.* Lanham, Md.: Rowman & Littlefield, 1992.

Lycan, William G. "Armstrong's Theory of Knowing." In *D. M. Armstrong,* pp. 139–60. Edited by R. Bogdan. Dordrecht: D. Reidel, 1984.

———. *Judgment and Justification.* Cambridge: Cambridge University Press, 1988.

Montmarquet, James. *Epistemic Virtue and Doxastic Responsibility.* Lanham, Md.: Rowman & Littlefield, 1993.

Moser, Paul K. *Empirical Justification,* chapter 4 and Appendix. Dordrecht: D. Reidel, 1985.

———. "Knowledge Without Evidence." *Philosophia* 15 (1985), 109–16.

Nozick, Robert. *Philosophical Explanations,* chapter 3. Cambridge: Harvard University Press, 1981.

Pappas, George S. "Non-Inferential Knowledge." *Philosophia* 12 (1982), 81–98.

Pastin, Mark. "Knowledge and Reliability: A Critical Study of D. M. Armstrong's *Belief, Truth, and Knowledge.*" *Metaphilosophy* 9 (1978), 150–62.

———. "The Multi-perspectival Theory of Knowledge." In *Midwest Studies in Philosophy, Vol. V: Studies in Epistemology,* pp. 97–111. Edited by P. French et al. Minneapolis: University of Minnesota Press, 1980.

Pollock, John. *Contemporary Theories of Knowledge.* Lanham, Md.: Rowman & Littlefield, 1986.

———. "Reliability and Justified Belief." *The Canadian Journal of Philosophy* 14 (1984), 103–14.

Schmitt, Frederick F. "Justification as Reliable Indication or Reliable Process." *Philosophical Studies* 40 (1981), 409–17.

———. *Knowledge and Belief.* London: Routledge, 1992.

———. "Knowledge as Tracking." *Topoi* 4 (1985), 73–80.

———. "Knowledge, Justification, and Reliability." *Synthese* 55 (1983), 209–29.

———. "Reliability, Objectivity, and the Background of Justification." *The Australasian Journal of Philosophy* 62 (1984), 1–15.

Shope, Robert K. "Cognitive Abilities, Conditionals, and Knowledge: A Response to Nozick." *The Journal of Philosophy* 81 (1984), 29–47.

Sosa, Ernest. *Knowledge in Perspective.* Cambridge: Cambridge University Press, 1991.

Swain, Marshall. "Justification and the Basis of Belief." In *Justification and Knowledge,* pp. 25–49. Edited by G. S. Pappas. Dordrecht: D. Reidel, 1979.

———. "Justification and Reliable Belief." *Philosophical Studies* 40 (1981), 389–407.

———. "Justification, Reasons, and Reliability." *Synthese* 64 (1985), 69–92.

———. *Reasons and Knowledge,* chapter 4. Ithaca: Cornell University Press, 1981.

Van Cleve, James. "Reliability, Justification, and the Problem of Induction." In *Midwest Studies in Philosophy, Vol. IX,* pp. 555–67. Edited by P. French et al. Minneapolis: University of Minnesota Press, 1984.

Naturalized Epistemology

Almeder, Robert. "On Naturalizing Epistemology." *American Philosophical Quarterly* 27 (1990), 263–79.

Bogen, James. "Traditional Epistemology and Naturalistic Replies to its Skeptical Critics." *Synthese* 64 (1985), 195–223.

Boyd, Richard. "Scientific Realism and Naturalistic Epistemology." In *PSA* 80 (1982), Vol. 2. East Lansing, MI: Philosophy of Science Association.

Dancy, Jonathan. *An Introduction to Contemporary Epistemology,* chapter 15. Oxford: Basil Blackwell, 1985.

Devitt, Michael. *Realism and Truth,* chapter 5. Princeton: Princeton University Press, 1984; 2d ed., 1991.

Goldman, Alvin I. *Epistemology and Cognition.* Cambridge, Mass.: Harvard University Press, 1986.

———. "Epistemology and the Psychology of Belief." *The Monist* 61 (1978), 525–35.

———. "Epistemology and the Theory of Problem Solving." *Synthese* 55 (1983), 21–48.

———. *Liaisons.* Cambridge, Mass.: MIT Press, 1992.

———. "The Relation between Epistemology and Psychology." *Synthese* 64 (1985), 29–68.

———. "Varieties of Cognitive Appraisal." *Noûs* 13 (1979), 23–38.

Haack, Susan. "The Relevance of Psychology to Epistemology." *Metaphilosophy* 6 (1975), 161–76.

Kornblith, Hilary, ed. *Naturalizing Epistemology.* Cambridge, Mass.: MIT Press, 1985; 2d ed., 1994.

Lycan, William G. "Epistemic Value." *Synthese* 64 (1985), 137–64.

Putnam, Hilary. "Why Reason Can't Be Naturalized." In Putnam, *Realism and Reason: Philosophical Papers, Vol. 3,* pp. 229–47. Cambridge: Cambridge University Press, 1983.

Quine, W. V. "The Nature of Natural Knowledge." In *Mind and Language: Wolfson College Lectures,* pp. 67–81. Edited by S. Guttenplan. Oxford: Oxford University Press, 1975.

Rorty, Richard. *Philosophy and the Mirror of Nature,* chapter 5. Princeton: Princeton University Press, 1979.

Siegel, Harvey. "Empirical Psychology, Naturalized Epistemology, and First Philosophy." *Philosophy of Science* 51 (1984), 667–676.

————. "Justification, Discovery, and the Naturalizing of Epistemology." *Philosophy of Science* 47 (1980), 297–321.

Sosa, Ernest. "Nature Unmirrored, Epistemology Naturalized." *Synthese* 55 (1983), 49–72.

Stroud, Barry. "The Significance of Naturalized Epistemology." In *Midwest Studies in Philosophy, Vol. VI: Analytic Philosophy,* pp. 455–71. Edited by P. French et al. Minneapolis: University of Minnesota Press, 1981. Reprinted in *Naturalizing Epistemology,* pp. 71–89. Edited by H. Kornblith. Cambridge, Mass.: MIT Press, 1985.

————. *The Significance of Philosophical Scepticism,* chapter 6. Oxford: Clarendon Press, 1984.

Swain, Marshall. "Epistemics and Epistemology." *The Journal of Philosophy* 75 (1978), 523–25.

The Gettier Problem

Audi, Robert. "Defeated Knowledge, Reliability, and Justification." In *Midwest Studies in Philosophy, Vol. V: Studies in Epistemology,* pp. 75–95. Edited by P. French et al. Minneapolis: University of Minnesota Press, 1980.

Chisholm, Roderick M. "Knowledge as Justified True Belief." In Chisholm, *The Foundations of Knowing,* pp. 43–49. Minneapolis: University of Minnesota Press, 1982.

Dancy, Jonathan. *An Introduction to Contemporary Epistemology,* chapter 2. Oxford: Blackwell, 1985.

Dretske, Fred I. "Conclusive Reasons." *The Australasian Journal of Philosophy* 49 (1971), 1–22.

Feldman, Richard. "An Alleged Defect in Gettier Counter-Examples." *The Australasian Journal of Philosophy* 52 (1974), 68–69. Selection 10, this book.

Gettier, Edmund. "Is Justified True Belief Knowledge?" *Analysis* 23 (1963), 121–23. Selection 9, this book.

Goldman, Alvin. "A Causal Theory of Knowing." *The Journal of Philosophy* 64 (1967), 357–72. Reprinted in *Essays on Knowledge and Justification,* pp. 67–86. Edited by G. S. Pappas and M. Swain. Ithaca: Cornell University Press, 1978.

————. "Discrimination and Perceptual Knowledge." *The Journal of Philoso-*

phy 73 (1976), 771–91. Reprinted in *Essays on Knowledge and Justification,* pp. 120–45.

Harman, Gilbert. "Knowledge, Inference, and Explanation." *American Philosophical Quarterly* 5 (1968), 164–73.

——. "Knowledge, Reasons, and Causes." *The Journal of Philosophy* 67 (1970), 841–55.

——. "Reasoning and Evidence One Does Not Possess." In *Midwest Studies in Philosophy, Vol. V: Studies in Epistemology,* pp. 163–82. Edited by P. French et al. Minneapolis: University of Minnesota Press, 1980.

——. *Thought,* chapters 7–9. Princeton: Princeton University Press, 1973.

Kaplan, Mark. "It's Not What You Know that Counts." *The Journal of Philosophy* 82 (1985), 350–63.

Klein, Peter D. "Knowledge, Causality, and Defeasibility." *The Journal of Philosophy* 73 (1976), 792–812.

——. "A Proposed Definition of Propositional Knowledge." *The Journal of Philosophy* 68 (1971), 471–82.

——. "Misleading Evidence and the Restoration of Justification." *Philosophical Studies* 37 (1980), 81–89.

——. "Real Knowledge." *Synthese* 55 (1983), 143–64.

Lehrer, Keith. "The Gettier Problem and the Analysis of Knowledge." In *Justification and Knowledge,* pp. 65–78. Edited by G. S. Pappas. Dordrecht: D. Reidel, 1979.

——. *Knowledge,* chapter 9. Oxford: Clarendon Press, 1974.

——. "Self-Profile." In *Keith Lehrer,* pp. 3–104. Edited by R. Bogdan. Dordrecht: D. Reidel, 1981.

——. *Theory of Knowledge.* Boulder, Colo.: Westview, 1990.

Moser, Paul K. *Knowledge and Evidence.* Cambridge: Cambridge University Press, 1989.

——. "Propositional Knowledge." *Philosophical Studies* 52 (1987), 91–114. Reprinted in Ernest Sosa, ed., *Knowledge and Justification,* Vol. 1, pp. 99–124. Aldershort, Eng.: Dartmouth, 1994.

Roth, Michael D., and Leon Galis, eds. *Knowing: Essays in the Analysis of Knowledge.* New York: Random House, 1970.

Shope, Robert K. *The Analysis of Knowing.* Princeton: Princeton University Press, 1983.

——. "Knowledge and Falsity." *Philosophical Studies* 36 (1979), 389–405.

——. "Knowledge as Justified Belief in a True, Justified Proposition." *Philosophy Research Archives* 5 (1979), 1–36.

Slaght, Ralph. "Is Justified True Belief Knowledge?: A Selective Critical Survey of Recent Work." *Philosophy Research Archives* 3 (1977), 1–135.

Sosa, Ernest. "How Do You Know?" *American Philosophical Quarterly* 11 (1974), 113–22. Reprinted in *Essays on Knowledge and Justification,* pp. 184–205. Edited by G. S. Pappas and M. Swain. Ithaca: Cornell University Press, 1978.

————. "Epistemic Presupposition." In *Justification and Knowledge*, pp. 79–92. Edited by G. S. Pappas. Dordrecht: D. Reidel, 1979.

Swain, Marshall. "Epistemic Defeasibility." *American Philosophical Quarterly* 11 (1974), 15–25. Reprinted in *Essays on Knowledge and Justification*, pp. 160–83. Edited by G. S. Pappas and M. Swain. Ithaca: Cornell University Press, 1978.

————. "Knowledge, Causality, and Justification." *The Journal of Philosophy* 69 (1972), 291–300. Reprinted in *Essays on Knowledge and Justification*, pp. 87–99.

————. *Reasons and Knowledge*. Ithaca: Cornell University Press, 1981.

————. "Reasons, Causes, and Knowledge." *The Journal of Philosophy* 75 (1978), 229–49.

Thalberg, Irving. "In Defense of Justified True Belief." *The Journal of Philosophy* 66 (1969), 794–803.

Williams, Michael. "Inference, Justification, and the Analysis of Knowledge." *The Journal of Philosophy* 75 (1978), 249–63.

Epistemological Skepticism

Amico, Robert. *The Problem of the Criterion*. Lanham, Md.: Rowman & Littlefield, 1993.

Brueckner, Anthony L. "Skepticism and Epistemic Closure." *Philosophical Topics* 13 (1985), 89–117.

Chisholm, Roderick M. "The Problem of the Criterion." In Chisholm, *The Foundations of Knowing*, pp. 61–75. Minneapolis: University of Minnesota Press, 1982.

Clay, Marjorie, and Keith Lehrer, eds. *Knowledge and Skepticism*. Boulder, Colo.: Westview, 1989.

Cornman, James W. *Skepticism, Justification, and Explanation*. Dordrecht: D. Reidel, 1980.

Dancy, Jonathan. *An Introduction to Contemporary Epistemology*, chapter 1. Oxford: Basil Blackwell, 1985.

Fogelin, Robert. *Pyrrhonian Reflections on Knowledge and Justification*. Oxford: Oxford University Press, 1994.

Foley, Richard. *Working Without a Net*. Oxford: Oxford University Press, 1993.

————. Review of Peter Klein's *Certainty: A Refutation of Scepticism*. *Philosophy & Phenomenological Research* 4 (1984).

Fumerton, Richard. *Metaepistemology and Skepticism*. Lanham, Md.: Rowman & Littlefield, 1995.

Hilpinen, Risto. "Skepticism and Justification." *Synthese* 55 (1983), 165–74.

Johnson, Oliver. "Ignorance and Irrationality: A Study in Contemporary Scepticism." *Philosophy Research Archives* 5 (1979), 368–417.

————. *Skepticism and Cognitivism.* Berkeley: University of California Press, 1978.

Klein, Peter D. *Certainty: A Refutation of Scepticism.* Minneapolis: University of Minnesota Press, 1981.

————. "Real Knowledge." *Synthese* 55 (1983), 143–64.

Lehrer, Keith. *Knowledge,* chapter 10. Oxford: Clarendon Press, 1974.

————. "The Problem of Knowledge and Skepticism." In James Cornman, Keith Lehrer, and George Pappas, *Philosophical Problems and Arguments: An Introduction, 3d ed.,* chapter 2. New York: Macmillan, 1982.

————. *Theory of Knowledge.* Boulder, Colo.: Westview, 1990.

————. "Why Not Scepticism?" *The Philosophical Forum* 2 (1971), 283–98. Reprinted in *Essays on Knowledge and Justification,* pp. 346–63.

Luper-Foy, Steven, ed. *The Possibility of Knowledge: Nozick and his Critics.* Lanham, Md.: Rowman & Littlefield, 1986.

Moser, Paul K. "Justified Doubt Without Certainty." *The Pacific Philosophical Quarterly* 65 (1984), 97–104.

————. *Knowledge and Evidence.* Cambridge: Cambridge University Press, 1989.

————. *Philosophy After Objectivity.* Oxford: Oxford University Press, 1993.

Naess, Arne. *Skepticism.* London: Routledge & Kegan Paul, 1969.

Nozick, Robert. *Philosophical Explanations,* chapter 3. Cambridge: Harvard University Press, 1981.

Oakley, I. T. "An Argument for Skepticism Concerning Justified Belief." *American Philosophical Quarterly* 13 (1976), 221–28.

O'Connor, D. J. and Brian Carr. *Introduction to the Theory of Knowledge,* chapter 1. Minneapolis: University of Minnesota Press, 1982.

Odegard, Douglas. "Chisholm's Approach to Scepticism." *Metaphilosophy* 12 (1981), 7–12.

————. *Knowledge and Skepticism.* Totowa, N.J.: Rowman & Littlefield, 1983.

Pappas, George. "Some Forms of Epistemological Scepticism." In *Essays on Knowledge and Justification,* pp. 309–16. Edited by G. Pappas and M. Swain. Ithaca: Cornell University Press, 1978.

Rescher, Nicholas. *Scepticism: A Critical Reappraisal.* Oxford: Basil Blackwell, 1979.

Roth, M. D., and G. Ross, eds. *Doubting.* Dordrecht: Kluwer, 1990.

Slote, Michael. *Reason and Scepticism.* New York: Humanities Press, 1970.

Strawson, Peter F. *Skepticism and Naturalism.* New York: Columbia University Press, 1985.

Stroud, Barry. *The Significance of Philosophical Scepticism.* Oxford: Clarendon, 1984.

————. "The Significance of Scepticism." In *Transcendental Arguments and Science.* Edited by P. Bieri et al. Dordrecht: D. Reidel, 1979.

————. "Skepticism and the Possibility of Knowledge." *The Journal of Philosophy* 81 (1984), 545–51.

Unger, Peter. "A Defense of Skepticism." *The Philosophical Review* 80 (1971), 198–218. Reprinted in *Essays on Knowledge and Justification*, pp. 317–36. Edited by G. Pappas and M. Swain. Ithaca: Cornell University Press, 1978.

———. *Ignorance*. Oxford: Clarendon Press, 1976.

———. "Two Types of Scepticism." *Philosophical Studies* 25 (1974), 77–96.

Vinci, Thomas. Critical Notice of Peter Klein's *Certainty: A Refutation of Scepticism*. *The Canadian Journal of Philosophy* 14 (1984), 125–45.

Watkins, John. *Science and Scepticism*. Princeton: Princeton University Press, 1984.

Williams, Michael. *Unnatural Doubts*. Oxford: Blackwell, 1991.

Wittgenstein, Ludwig. *On Certainty*. Edited by G. E. M. Anscombe and G. H. von Wright. Oxford: Basil Blackwell, 1969.

Woods, Michael. "Scepticism and Natural Knowledge." *Proceedings of the Aristotelian Society* 54 (1980), 231–48.

Index

About the Editor

Paul K. Moser is professor of philosophy at Loyola University of Chicago. He is the author of *Philosophy After Objectivity* (Oxford University Press, 1993), *Knowledge and Evidence* (Cambridge University Press, 1989), and *Empirical Justification* (Kluwer, 1985). He is the editor of *A Priori Knowledge* (Oxford University Press, 1987), *Rationality in Action* (Cambridge University Press, 1990), and *Reality in Focus* (Prentice Hall, 1990), and coeditor of *Human Knowledge, 2nd ed.* (Oxford University Press, 1995), *Contemporary Materialism* (Routledge, 1995), *Contemporary Approaches to Philosophy* (Macmillan, 1994), and *Morality and the Good Life* (Oxford University Press, 1996).

Moser is the general editor of Rowman & Littlefield's book series *Studies in Epistemology and Cognitive Theory* and Routledge's book series *Contemporary Introductions to Philosophy*. He is a contributor to *The Cambridge Dictionary of Philosophy, The Routledge Encyclopedia of Philosophy, The Blackwell Companion to Epistemology,* and various other reference works on philosophy.